Systems Security, Resilience, and Emergency Management 2018

Transportation Research Record: Journal of the Transportation Research Board

Volume 2672, Issue 1

ISBN: 978-1-5108-7821-1

Printed from e-media with permission by:

Curran Associates, Inc.
57 Morehouse Lane
Red Hook, NY 12571

Some format issues inherent in the e-media version may also appear in this print version.

Published by Sage Publications USA

ISBN (Print) 978-1-5108-7821-1
ISBN 2018 Printed Set (All Issues) 978-1-5108-7735-1

Printed by Curran Associates, Inc. (2019)

Additional copies of this publication are available from:

Curran Associates, Inc.
57 Morehouse Lane
Red Hook, NY 12571 USA
Phone: 845-758-0400
Fax: 845-758-2633
Email: curran@proceedings.com
Web: www.proceedings.com

Volume 2672 Number 1 December 2018

TRR
JOURNAL OF THE TRANSPORTATION RESEARCH BOARD

Contents

Systems Security, Resilience, and Emergency Management 2018

Articles

TRB

TRANSPORTATION RESEARCH BOARD

CORE SPONSORS

Transportation Departments of the 50 States and the District of Columbia

Federal Government

U.S. Department of Transportation
- Federal Aviation Administration (FAA)
- Federal Highway Administration
- Federal Motor Carrier Safety Administration (FMCSA)
- Federal Railroad Administration (FRA)
- Federal Transit Administration
- National Highway Traffic Safety Administration (NHTSA)
- Office of the Assistant Secretary of Transportation for Research and Technology (OST-R)

U.S. Air Force Civil Engineer Center (AFCEC)
U.S. Coast Guard (USCG)
U.S. Department of Energy
U.S. Department of the Interior (DOI)
U.S. Environmental Protection Agency (EPA)

Nongovernmental Organizations

American Public Transportation Association (APTA)
Association of American Railroads (AAR)
California Air Resources Board

The National Academies of
SCIENCES · ENGINEERING · MEDICINE

The **National Academy of Sciences** was established in 1863 by an Act of Congress, signed by President Lincoln, as a private, nongovernmental institution to advise the nation on issues related to science and technology. Members are elected by their peers for outstanding contributions to research. Dr. Marcia McNutt is president.

The **National Academy of Engineering** was established in 1964 under the charter of the National Academy of Sciences to bring the practices of engineering to advising the nation. Members are elected by their peers for extraordinary contributions to engineering. Dr. C. D. Mote, Jr., is president.

The **National Academy of Medicine** (formerly the Institute of Medicine) was established in 1970 under the charter of the National Academy of Sciences to advise the nation on medical and health issues. Members are elected by their peers for distinguished contributions to medicine and health. Dr. Victor J. Dzau is president.

The three Academies work together as the National Academies of Sciences, Engineering, and Medicine to provide independent, objective analysis and advice to the nation and conduct other activities to solve complex problems and inform public policy decisions. The Academies also encourage education and research, recognize outstanding contributions to knowledge, and increase public understanding in matters of science, engineering, and medicine.

Learn more about the National Academies of Sciences, Engineering, and Medicine at **www.national-academies.org.**

The **Transportation Research Board** is one of seven major programs of the National Academies of Sciences, Engineering, and Medicine. The mission of the Transportation Research Board is to increase the benefits that transportation contributes to society by providing leadership in transportation innovation and progress through research and information exchange, conducted within a setting that is objective, interdisciplinary, and multimodal. The Board's varied activities annually engage about 7,000 engineers, scientists, and other transportation researchers and practitioners from the public and private sectors and academia, all of whom contribute their expertise in the public interest. The program is supported by state transportation departments, federal agencies including the component administrations of the U.S. Department of Transportation, and other organizations and individuals interested in the development of transportation.

Learn more about the Transportation Research Board at **www.TRB.org**.

Article

TRR
JOURNAL OF THE TRANSPORTATION RESEARCH BOARD

Vulnerability of Traffic Control System Under Cyberattacks with Falsified Data

Transportation Research Record
2018, Vol. 2672(1) 1–11

DOI: 10.1177/0361198118756885
journals.sagepub.com/home/trr
SAGE

Yiheng Feng[1], Shihong Huang[2], Qi Alfred Chen[3], Henry X. Liu[1,2], and Z. Morley Mao[3]

Abstract

Existing traffic control systems are mostly deployed in private wired networks. With the development of wireless technology, vehicles and infrastructure devices will be connected through wireless communications, which might open a new door for cyberattackers. It is still not clear what types of cyberattacks can be performed through infrastructure-to-infrastructure and vehicle-to-infrastructure communications, whether such attacks can introduce critical failure to the system, and what the impacts are of cyberattacks on traffic operations. This paper investigates the vulnerability of traffic control systems in a connected environment. Four typical elements, including signal controllers, vehicle detectors, roadside units, and onboard units, are identified as the attack surfaces. The paper mainly focuses on attacking actuated and adaptive signal control systems by sending falsified data, which is considered as an indirect but realistic attack approach. The objective of an attacker is to maximize system delay with constraints such as budget and attack intensity. Empirical results show that different attack scenarios result in significant differences in delay, and some ineffective attacks may even improve the system performance. Simulation results from a real-world corridor show that critical intersections, which have a higher impact on network performance, can be identified by analyzing the attack locations. Identification of such intersections can be helpful in designing a more resilient transportation network.

Existing transportation infrastructure is usually isolated regarding connectivity as all vehicles are operated independently, and traffic control systems are mostly deployed in a private wired network. With the development of wireless technology, vehicles and infrastructure will be connected through wireless communications [e.g., Dedicated Short Range Communication(DSRC) or cellular network], which might open a new door for cyberattackers. Cybersecurity of transportation systems has been a growing research area in the past decade, but most efforts are focused on intervehicle communications. As a critical part of the transportation infrastructure, existing traffic control systems have a profound impact on the safety and efficiency of urban traffic flow, but are very vulnerable to cyberattacks because of the "systematic lack of security consciousness" (*1*). For example, an Argentinian security expert hacked into New York City's wireless vehicle detection system with a cheap wireless device. The vulnerabilities he found allowed anyone to take complete control of the devices and send fake data to the traffic control systems (*2*). Although traffic signals were not directly controlled, fake vehicle data could cause severe traffic congestion and increase crash risks. Another example involved hacking into a variable message sign in Austin, Texas, and displaying "Zombie Ahead" instead of correct traffic information (*3*). To systematically investigate the cybersecurity of transportation infrastructure, the NCHRP started a new project to develop guidance for transportation agencies on mitigating the risks from cyberattacks toward traffic management systems (4).

However, it is still not clear what types of cyberattacks can be performed through infrastructure-to-infrastructure (I2I) and vehicle-to-infrastructure (V2I) communications, whether such attacks can create critical failure to traffic control systems, and what are the impacts of cyberattacks on traffic operations. A systematic study of the vulnerabilities of the existing traffic control system and corresponding remedies needs to be established. The objective of this paper is to investigate potential attack surfaces and the adversarial consequences such attacks may bring to the traffic network.

[1]University of Michigan Transportation Research Institute, Ann Arbor, MI
[2]Department of Civil and Environmental Engineering, University of Michigan, Ann Arbor, MI
[3]Department of Electrical Engineering and Computer Science, University of Michigan, Ann Arbor, MI

Corresponding Author:
Address correspondence to Yiheng Feng: yhfeng@umich.edu

First, four possible attack surfaces of the traffic control system in a connected environment are identified, including signal controllers, vehicle detectors, onboard units (OBUs), and roadside units (RSUs). The focus of analysis is on attacking actuated and adaptive signal control systems by sending falsified data from either hacked vehicle detectors or compromised OBUs. The attack is modeled as an optimization problem with the objective to maximize system delay and constraints, such as the number of compromised devices and attack intensity. Analysis of a hypothetical intersection shows that some attacks are effective in increasing total delay, whereas others are not. Finally, a real-world corridor is used to evaluate the proposed attack methods.

The rest of this paper is organized as follows. The second section provides a brief review of related work. In the third section, four attack surfaces with different attack strategies are identified. The fourth section presents the traffic model to represent the vehicle dynamics and the attack model. The fifth section evaluates the effectiveness of different attack strategies at a hypothetical intersection and a real-world corridor. The final section gives the conclusions and outlines the directions of future work.

Related Work

In this section, a brief overview of cybersecurity-related transportation infrastructure studies is provided. Ghena et al. analyzed the security of a currently deployed traffic signal control system and found the controller network could be infiltrated through its wireless infrastructure (*1*). Once on the network, the controller could be accessed by the operating system's debug port or through National Transportation Communications for ITS Protocol (NTCIP) commands. Although the safety of the signal operations was protected by the Malfunction Management Unit (MMU), the attackers could generate inefficient signal timing plans that might cause traffic congestion and even denial of service. This study mainly demonstrated how to leverage existing design flaws to gain control of the signal system, but didn't provide detailed attack strategies and corresponding consequences.

Laszka et al. (*5*) and Ghafouri et al. (*6*) studied the vulnerability of fixed-time signal control to cyberattacks. The attacks were formulated as mathematical programming problems with different objectives such as worst-case network accumulation, worst-case lane accumulation, and risk-averse target accumulation. Heuristic and decomposition algorithms were implemented to solve the problem at a network level. A precondition of tampering fixed-time signal control is that the attacker can access and manipulate signal controller directly. This is not a very realistic assumption unless the attacker can access the traffic signal cabinet physically. However, it is much easier to falsify input data to influence the control decisions under actuated or adaptive control. Toward this end, Jeske investigated Google and Waze navigation systems and demonstrated how attackers could take control of the navigation system and influence the traffic flow by sending false location information (*7*). Tufnell successfully hacked into Waze maps and generated fake GPS coordinates to create virtual traffic jams (*8*). However, neither of the two studies incorporated the fake data into traffic control systems and analyzed the consequences.

Other than tampering with traffic signal operations, Reilly et al. presented a study on attacking freeway ramp metering to generate arbitrarily complex congestion patterns (*9*). Finite-horizon optimal control and multiobjective optimization techniques were used to launch attacks on coordinated ramp metering controllers. Different attack scenarios were designed and conducted. Results showed that arbitrary congestion-on-demand patterns could be created with enough controlled ramps.

Although how to detect and protect the system from cyberattacks is beyond the scope of this paper, some approaches have been proposed to detect anomaly from traffic-flow patterns. For example, Canepa and Claudel tried to detect falsified probe-based vehicle data using the Lighthill-Whitham-Richards (LWR) traffic-flow model (*10*). The detection was posed as a mixed integer linear feasibility problem. Zhang et al. quantified anomaly by proposing an anomaly index in both spatial and temporal perspectives, founded on dictionary-based compression theory (*11*). The original intention was to identify non-recurrent traffic-flow pattern caused by incidents, but this method could be potentially used to detect cyberattacks and design defense strategies.

Threat Model

Before presenting the attack model, the threat model of a "connected" intersection is introduced and shown in Figure 1. Here, "connected" refers to that the intersection and vehicles are equipped with wireless communication devices (such as RSUs and OBUs) and can communicate with each other. Whereas exact deployments are different from location to location, we consider four typical elements in the traffic control system as possible attack surfaces:

- Traffic signal controller: used to generate signal timing plans based on different control strategies, including fixed-time, actuated and adaptive strategies. If traffic signals are under actuated or adaptive control, the controller utilizes data from vehicle detectors and RSUs to make control decisions.
- Vehicle detectors: used to detect vehicles and generate service calls to signal controller. If vehicle detectors are configured as system detectors, they can also be used to provide volume and speed information.
- OBUs in vehicles: used to generate vehicle-related information [e.g., basic safety messages (BSMs)] and broadcast the information to other vehicles and infrastructure through over the air messages.
- RSUs at intersections: used to broadcast infrastructure-related messages (e.g., Signal Phasing and

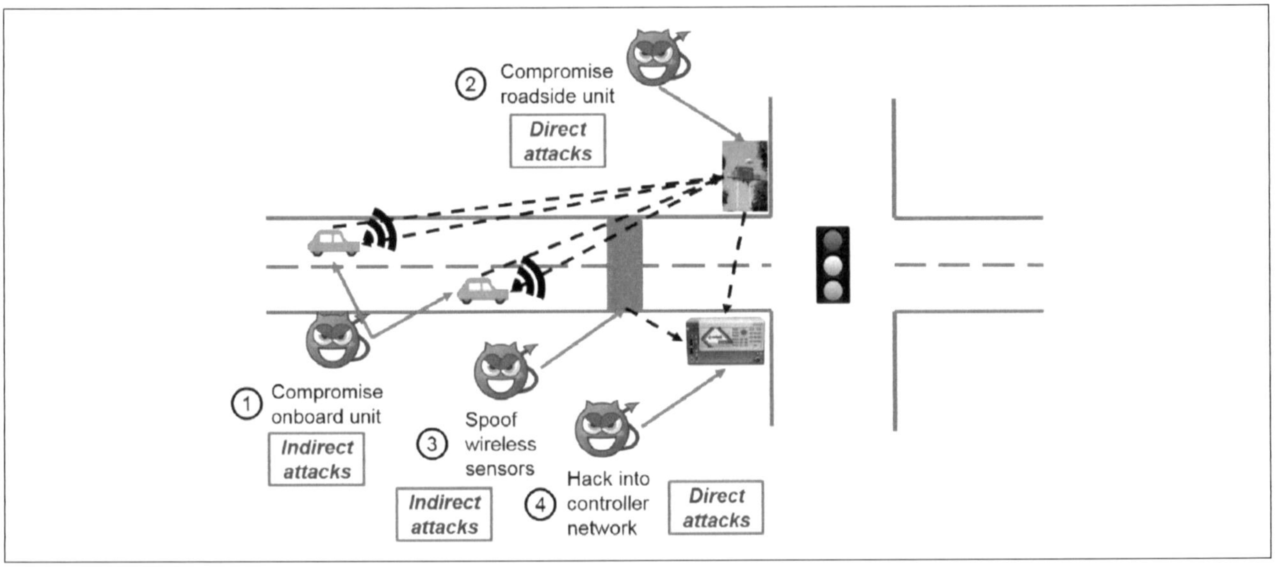

Figure 1. Intersection threat model.

Timing (SPaT), and MAP) and receive vehicle information. It provides input data (e.g., trajectory) to signal controller.

Based on the attack surfaces, two types of attacks are identified: direct attack and indirect attack. Direct attack refers to hacking into the signal controller and RSU and changing the signal timing plans directly. To launch direct attacks, an attacker needs physical access to the devices, which requires the attacker to open the signal controller cabinet and connect to the signal controller or the RSU using an Ethernet cable. Indirect attack refers to tampering with data from vehicle detectors and OBUs. Usually, only part of the input data can be falsified. Indirect attacks are more realistic to conduct. For example, spoofing into wireless vehicle detectors (*2*), or compromising OBUs in private vehicles. This paper focuses on indirect attacks under actuated and adaptive signal control. For actuated signal control, it is assumed that the signal controller utilizes vehicle detector data to perform actuation logic. Compromised vehicle detectors may generate fake vehicle calls or cancel real vehicle calls. For adaptive signal control, it is assumed that the signal controller generates optimal signal plans based on BSM data from connected vehicles (CVs). Compromised OBUs may insert virtual vehicles on the roadway that don't exist.

Traffic And Attack Models

To model the transportation network and quantify the consequences of cyberattacks, a traffic-flow model is needed. The cell transmission model (CTM) (*12, 13*) is applied for two reasons. The CTM is a macroscopic traffic model that can be used to simulate network traffic with thousands of vehicles and attack scenarios in an efficient way. However, compared with other link-based flow and density models, CTM divides roadway into homogeneous segments, so that attacks can be launched at different locations (cells). This section first introduces how to model actuated and adaptive signal control with CTM, and then presents the attack model.

Cell Transmission Model

CTM is a first-order approximation to LWR partial differential equation. The model assumes a triangular fundamental diagram and discretizes space into homogeneous cells and time into intervals. The cell length is equal to one-time interval multiplied by free-flow speed defined in the fundamental diagram. CTM was originally developed to model highway traffic with a single entrance and exit (*12*). Later, the model was extended to represent network traffic, which allows it to model traffic flows at signalized intersections (*13*). A typical intersection in CTM is shown in Figure 2.

There are six types of cells: ordinary cell, merging cell, diverging cell, intersection cell, source cell, and sink cell. An ordinary cell has one preceding cell and one following cell, and has limited jam density and capacity. A merging cell has multiple preceding cells and one following cell, whereas a diverging cell has one preceding cell and multiple following cells. An intersection cell is similar to an ordinary cell, except that the flow is controlled by signal timing. Source cells and sink cells are responsible for generating and exiting vehicles. Due to space limitation, for the detailed formulation of CTM, refer to (*12, 13*).

The parameters of the CTM model in this paper are configured as follows. Free-flow speed v is set to 54 km/h (15 m/s). Backward shockwave speed w is set to 18 km/h (5 m/s). The maximum flow rate Q_m is set to 1800 vph and the corresponding critical density k_m is 33.33 vpkm and jam density

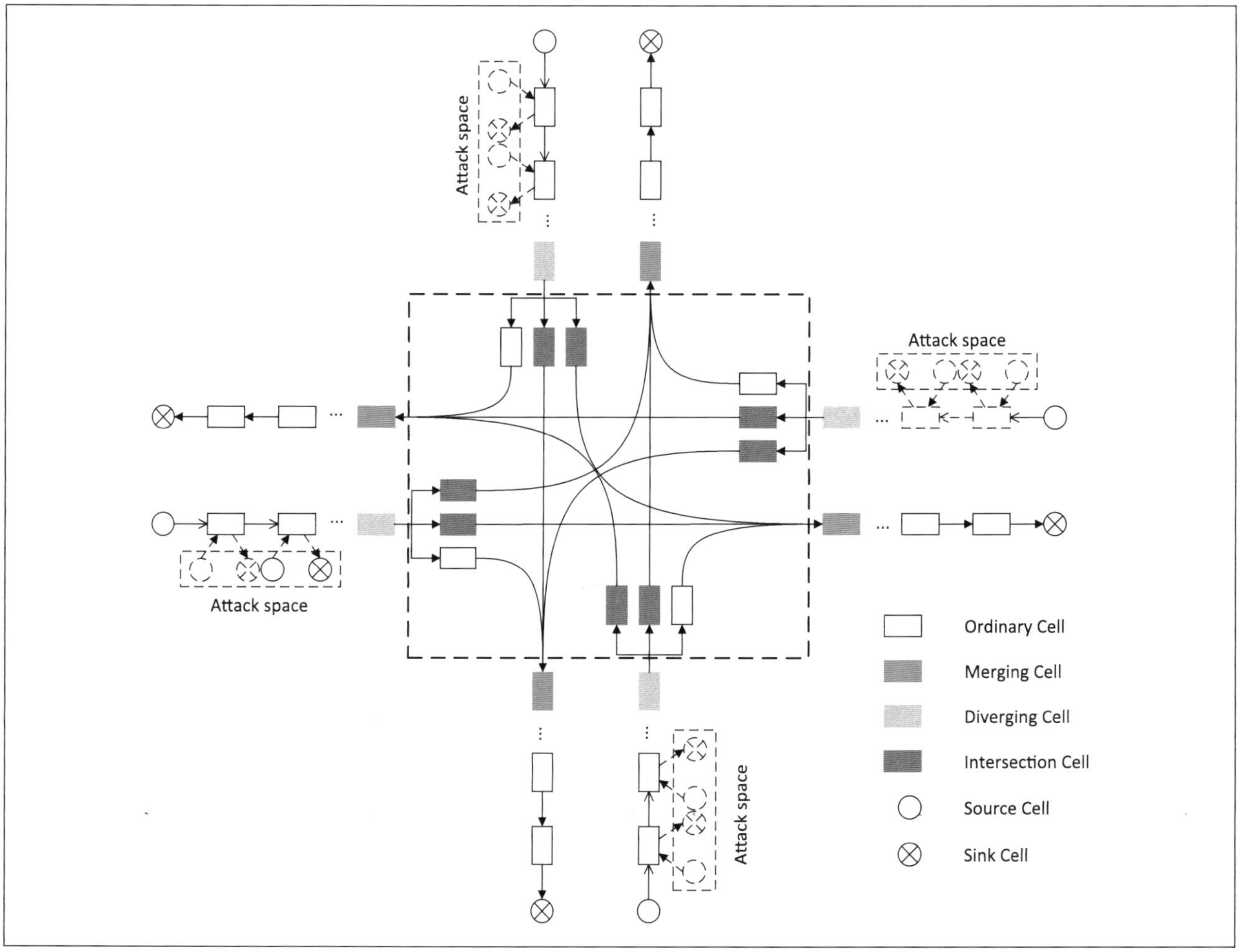

Figure 2. Intersection representation of CTM with attack spaces.

k_j is 133.33 vpkm. Time step is set to 2 s, which is similar to the unit extension time in actuated control. As a result, the cell length is 30 m.

Model Actuated Signal Control with CTM. To model actuated signal control with CTM, it is assumed that stop-bar detectors are installed in intersection cells. The density ratio of cell i at time t can be calculated as $d(i,t) = n(i,t)/N(i)$, where, $n(i,t)$ is the number of vehicles in cell i at time t, and $N(i)$ is the maximum number of vehicles in cell i. A critical density ratio d_c is defined for each intersection cell. The actuation logic is modeled based on d_c. If the density ratio of intersection cell i at time t is less than the critical density ratio, then the current phase is terminated. Otherwise, extend green to the next time step.

To determine the best d_c, a series of simulations with different values of d_c were run. Vehicle delay in cell i at time t is defined as the difference between $n(i,t)$ and the number of vehicles that can be discharged from the cell $y(i,t)$, because in CTM vehicles are either in free-flow speed (discharged to the following cell) or in queuing state (remained in current cell):

$$D = \sum_{t \in T} \sum_{i \in I} [n(i,t) - y(i,t)]. \quad (1)$$

Results show that both lower and higher critical density ratios generate higher vehicle delay. Lower critical density ratios correspond to longer unit extension times whereas higher critical density ratios correspond to shorter unit extension times. The lowest delay occurs when $d_c = 0.25$, which is the critical density ratio ($k_m/k_j = 0.25$) that separates free-flow and congestion regime. For a vehicle actuation logic, it is appropriate to terminate green when the traffic state changes from congested to free flow. Therefore, $d_c = 0.25$ is used in numerical experiments.

Model Adaptive Signal Control with CTM. The adaptive control algorithm is adapted from Sen and Head (*14*) and Feng et al. (*15*). Signal optimization is formulated as a dynamic programming (DP) problem, in which each phase is considered as one stage in DP. A forward recursion is used to calculate performance measures and record optimal value function.

The objective of the forward recursion is to choose an optimal signal plan with minimal total vehicle delay. A backward recursion is used to retrieve the optimal solution.

The major difference of the algorithm applied in this study from the original algorithm is the performance function calculation. In previous DP formulations, the performance measures were calculated from an arrival table that included estimated time of arrival and requested phase of each vehicle. However, CTM is a macroscopic model in which individual vehicle information is not available. To accurately calculate vehicle delay, a snapshot of the current network condition is taken at the beginning of each signal optimization. A parallel CTM simulation is executed based on the snapshot as an initial network condition to generate vehicle delays in each DP iteration.

The signal optimization algorithm plans as many stages (phases) as necessary until all vehicles in the snapshot pass the intersection cells. A rolling horizon scheme is adopted in which the optimization is performed at the beginning of each phase to include recent vehicle arrivals.

Attack Model

It is assumed that an attacker has limited resources. For example, the number of vehicle detectors that can be tampered with, or the number of OBUs that can be compromised are limited. As a result, an attacker needs to choose a subset of locations or devices to maximize the profit, which is defined as maximization of network congestion. Formally, an attack A is defined as

$$A = \left(S, \left\{n'(i,t) \mid \forall i \in S\right\}\right)$$

where S is the set of cells can potentially be under attack and $n'(i,t)$ is the number of vehicles in the cell under attack. Attacks are conducted through increasing or decreasing number of vehicles in a cell to mimic the change of stop-bar detector data and BSM distribution.

The attack model can be expressed as

$$\max_{A} D(A) \tag{2}$$

s.t.

$$|S| \leq B \tag{3}$$

$$n'(i,t) \leq \min\left(n(i,t) + \varepsilon, N(i)\right), \forall i \in S \tag{4}$$

$$n'(i,t) \geq \max\left(n(i,t) - \varepsilon, 0\right), \forall i \in S \tag{5}$$

The objective function means an attacker intends to maximize total vehicle delay. The first constraint indicates that the attacker is limited by budget B. The next two constraints represent the cautiousness of the attacker. The number of vehicles can be changed, is limited by a threshold ε and the physical limits of the road. If the data deviate a lot from the normal range, the attacker can be easily detected by the defender.

Numerical Examples

In this section, numerical results based on a hypothetical intersection, and insights on the effectiveness of the attacks are presented first. Then a real-world six-intersection corridor in Ann Arbor, MI, is built to evaluate the attack models. All models are coded in Matlab.

A Hypothetical Intersection

The layout of a hypothetical intersection is shown in Figure 2. This is a typical four-leg intersection with all vehicle movements. There are four signal phases: eastbound and westbound left turn (Phase 1), eastbound and westbound through (Phase 2), northbound and southbound left turn (Phase 3), and northbound and southbound through (Phase 4). Right-turn vehicles are not restricted by traffic signals. The minimum green time is set to five time steps (10 s) and the maximum green time is set to 20 steps (40 s) for each phase. The length of each approach is 10 cells (including the intersection cells), which is similar to the DSRC communication range. Traffic demand is set to 1,000 vph eastbound/westbound and 800 vph northbound/southbound. Vehicle arrivals follow the Poisson distribution. Turning ratios of each approach are the same and set to 0.2/0/7/0.1 for left turn, through, and right turn respectively. Simulation runs for 1,000 time steps with 100 steps as a warm-up period.

Figure 3 shows the total delay and congestion pattern of eastbound approach under actuated and adaptive control without attacks. This serves as the baseline for our comparison. Different colors represent different congestion levels, with green being no congestion and red being the severest congestion.

Attack Under Actuated Control. Under actuated control, it is assumed that stop-bar detector data can be manipulated by the attacker so that the number of vehicles at intersection cells can be added (generate fake vehicle calls) or subtracted (cancel real vehicle calls). This results in two attack modes $M = 2$. To cause maximum damage, the attacker changes the detector data as much as possible, but within the threshold $\varepsilon = 0.5$. Then $n'(i,t)$ is equal to either $\min\left(n(i,t) + \varepsilon, N(i)\right)$ or $\max\left(n(i,t) - \varepsilon, 0\right)$. The budget B is set to 4 so that all phases can be attacked. To thoroughly analyze the effectiveness of all attack scenarios, all the possibilities are enumerated, which results in 80 different cases:

$$\sum_{p=1,2,3,4} C_P^p * M^p = 80 \tag{6}$$

where p is the signal phase index, and
P is the total number of phases.

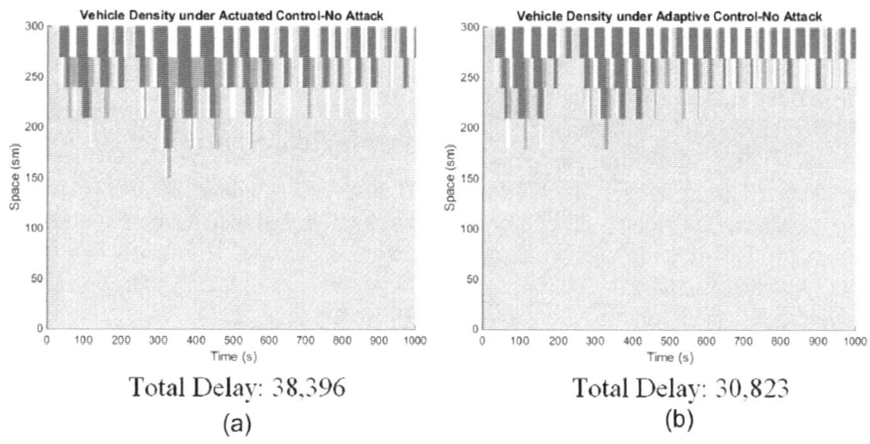

Figure 3. Vehicle delay comparison without attack: (*a*) actuated control (*left*) and (*b*) adaptive control (*right*).

Figure 4 shows total vehicle delay and average total delay by the number of attacking phases. It can be seen from Figure 4*a* that the effectiveness of different attacks varies a lot. Figure 4*b* shows the trend that attacks cause more vehicle delay when the number of attacking phases increases.

Figure 5 shows the comparison between the most effective attack and least effective attack at the eastbound approach. The most effective attack occurs when Phases 2, 3 and 4 are under attack, with subtracting vehicles at intersection cells corresponding to Phase 2, and adding vehicles at intersection cells corresponding to Phases 3 and 4. The resultant total delay can be six times higher than the baseline scenario. The least effective attack occurs when attacking intersection cells related to Phases 1 and 3, with subtracting vehicles on both phases. The resultant total delay (36,441) is even smaller than the baseline scenario (38,396), which indicates that the attack improves the system performance. The reason is that actuated control is not the optimal control strategy. In certain cases, when a phase is green with lower demand while other phases are red with higher demand, it is more efficient to terminate the lower demand phase earlier to serve other phases. In this case, Phases 1 and 3 are left-turn phases with lower demand. Subtracting vehicles shortens both phases, which gives more time to higher demand Phases 2 and 4.

Attack Under Adaptive Control. Under adaptive control, the control algorithm utilizes data from CVs (e.g., BSMs) to generate optimal signal plans, so that every cell within the communication range can be potential targets. It is assumed that an attacker is only interested in manipulating the number of vehicles in ingress cells because vehicles in egress cells don't affect the signal optimization. The attacker can add or subtract vehicles at a different number of approaches with the maximum number of attacking approaches $A = 4$. If the attacker decides to attack one approach, then all ingress cells on that approach are affected. The threshold ε is also set to 0.5. In total, 30 attack scenarios are generated.

Figure 6 shows the total vehicle delay of all attack scenarios and the average total delay by the number of attack approaches. Figure 6*a* compares the total delay of adding vehicles or subtracting vehicles when attacking the same approach(es). In general, adding fake vehicles is more effective than removing real vehicles. Under current demand level (medium), green time wasted under longer cycle lengths can cause more delay than increased percentage of lost time with shorter cycle lengths. Figure 6*b* shows a similar pattern: the average total delay increases with the number of attack approaches.

Another finding is that attacks under adaptive control are far less effective than under actuated control with the same attack intensity ε. The most effective attack under adaptive control causes a 41,199s of total vehicle delay, which is only 33.66% more than the baseline scenario. However, the most effective attack under actuated control generates a delay of six times more than the baseline scenario. Under adaptive control, all ingress cells (including intersection cells) are affected, whereas under actuated control, only intersection cells are affected. However, results suggest that adaptive control is more robust than actuated control. Becuase adaptive control tries to minimize total delay under the impact of attacks, based on inputs from all ingress cells, whereas actuated control logic can only accommodate instantaneous arriving flow at intersection cells, but doesn't have an overall picture of the current traffic condition.

Plymouth Road Corridor

Six consecutive intersections along Plymouth Road at Ann Arbor, MI, are modeled in CTM to evaluate the attack model under actuated control. To calibrate the model, video data were collected on May 16, 2017, from 4:00 to 5:00 p.m. Traffic volume of each approach, turning ratio of each

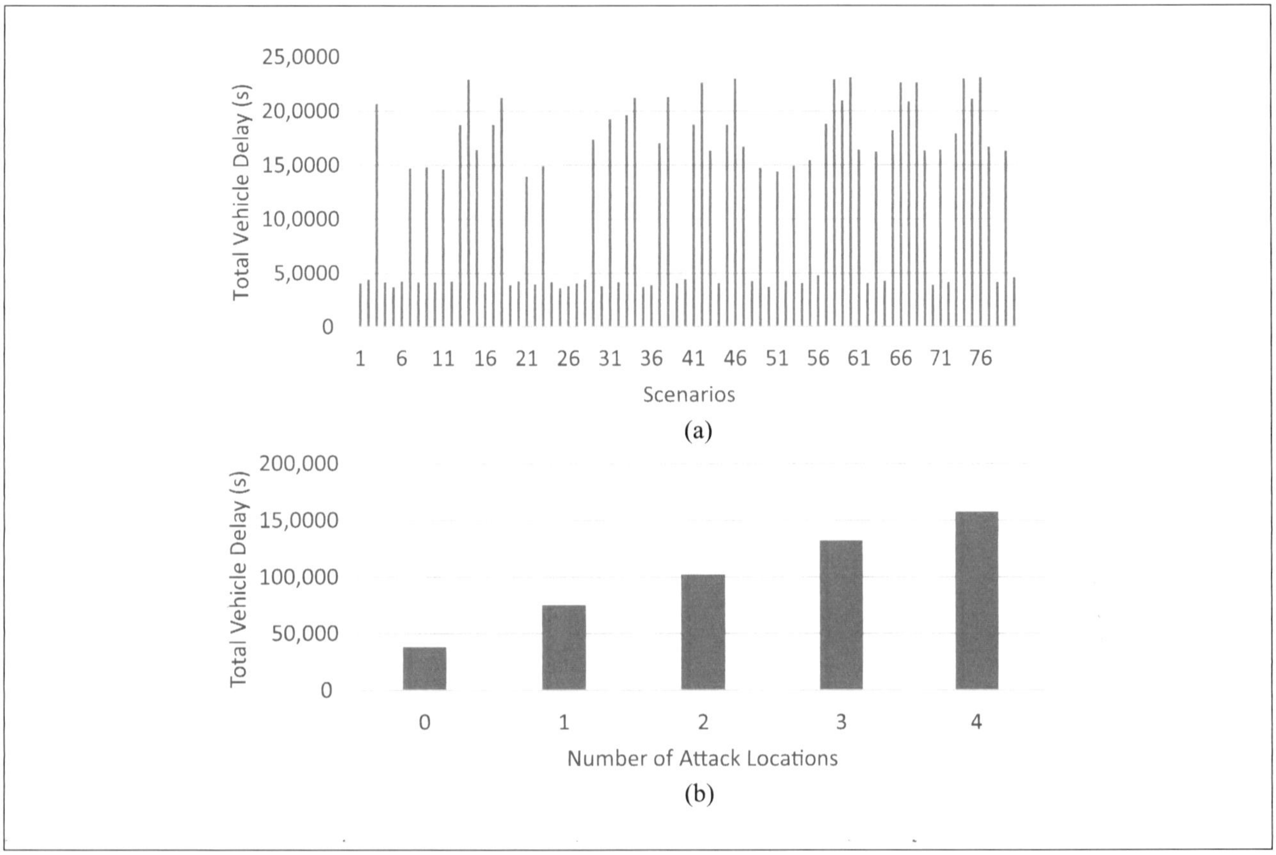

Figure 4. Vehicle delay by attack scenarios and number of attacking phases (actuated control): (*a*) vehicle delay under all attack scenarios (*top*) and (*b*) average total vehicle delay by number of attacking phases (*bottom*).

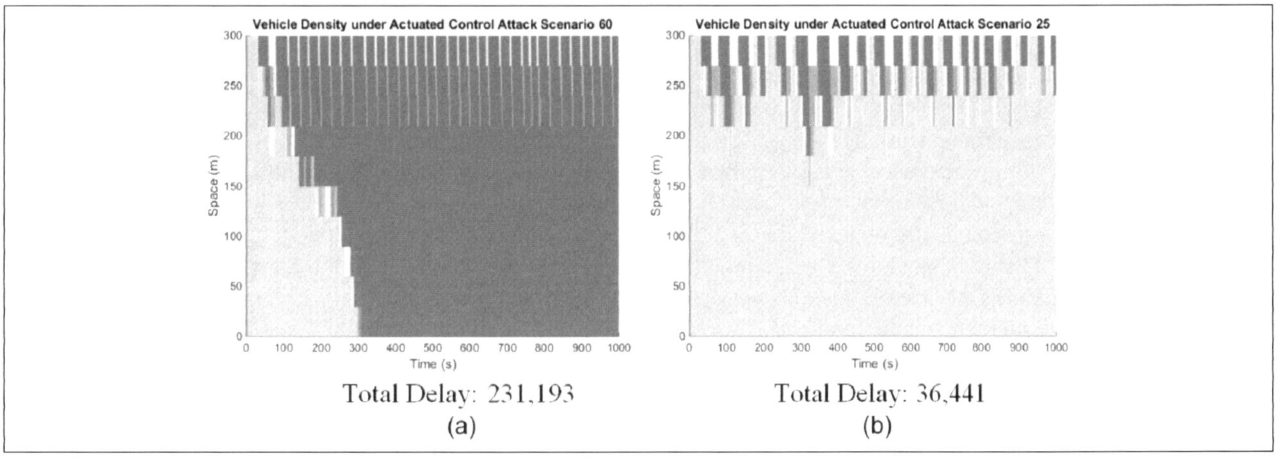

Figure 5. Comparison between the most effective attack and the least effective attack: (*a*) most effective attack (*left*) and (*b*) least effective attack (*right*).

movement, and signal timing of each intersection were extracted from the video and used as input to the CTM model.

The corridor contains two T-shape intersections and four standard intersections. The standard intersections have four phases as defined in the previous section, whereas the T

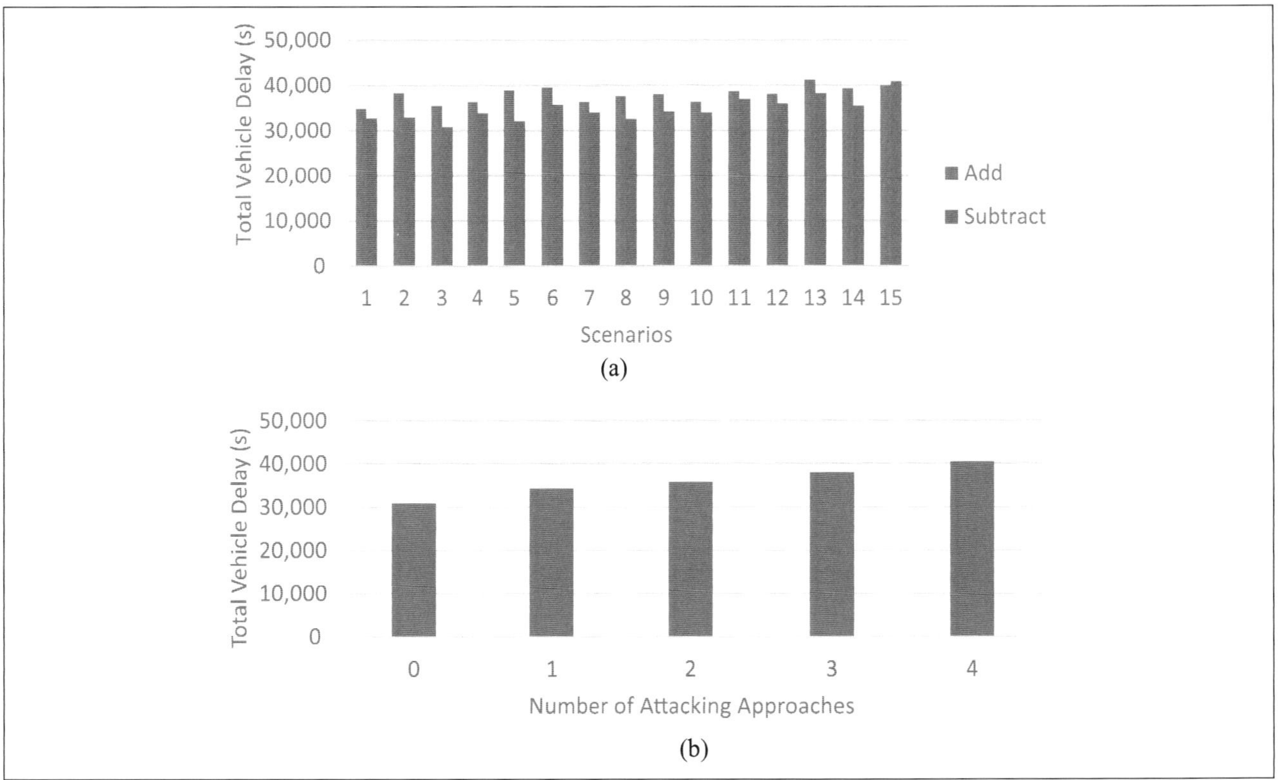

Figure 6. Vehicle delay by attack scenarios and number of attacking approaches (adaptive control): (*a*) vehicle delay under all attack scenarios (*top*) and (*b*) average total vehicle delay by number of attacking approaches (*bottom*).

intersections have only two phases. Therefore, there are 20 signal phases along the corridor. Each is identified as a potential attack location. The minimum and maximum green time for the through movement along Plymouth Road is set to 10 and 30 time steps, whereas the rest of the phases have 5 and 15 time steps for minimum and maximum green time respectively. Each attack scenario lasts 2,100 time steps, which contains 300 time steps of warm-up (no attack) period and 1,800 time steps for performance evaluation (under attack). It is assumed that the attacker has a budget limit so that a maximum of four phases can be attacked. This results in a total number of 87,440 attack scenarios. Due to this large number, the attack scenarios are carried out by Flux, a Linux-based high-performance computing cluster at the University of Michigan. A total of 20 central processing units cores are used to run the attack scenarios in parallel, and the total computation time is about 14 h.

When only one phase is under attack, Scenario 7, which attacks Phase 2 (the through phase on the main arterial) at Intersection 2 by subtracting vehicles, has the highest delay. A snapshot of the corridor at the final simulation time step is shown in Figure 7. The intersection on the left is numbered as 1, and the intersection on the right is numbered as 6. When under attack, the signal controller always terminates Phase 2 on the minimum green time, which causes oversaturation for westbound through traffic and the vehicle queue starts to accumulate. The queue eventually propagates to Intersection 3 and causes spillover. The spillover prevents westbound through traffic at Intersection 3 from entering the downstream link during green. The through traffic constantly calls for green extensions, which generates more delay for the cross-street traffic because of the long waiting time. The same situation happens when the queue propagates to Intersections 4 and 5. Notice that there is a long queue in the northbound approach of Intersection 5 because this approach has heavy left-turn traffic. The spillover on the main arterial prevents vehicles turning left from the cross street. The result indicates that Phase 2 at Intersection 2 is the critical phase along the corridor.

Figure 8 shows vehicle delay under all attack scenarios with a different number of attacking phases. The average total vehicle delay increases with the number of attacking phases, which is consistent with previous results. If all four phases are under attack, the most effective way is to subtract vehicles from Phase 2 (through movement on Plymouth Road) at Intersection 2, and Phase 3 (left-turn phase on cross street) at Intersection 6, at the same time adding vehicles to Phase 3 at Intersection 2, and Phase 2 at Intersection 6.

Although trying all attack scenarios guarantees the optimal solution, it is unrealistic for an attacker to enumerate all the possibilities and find the best strategy in real time. Thus, a simple greedy attack policy is implemented to find an

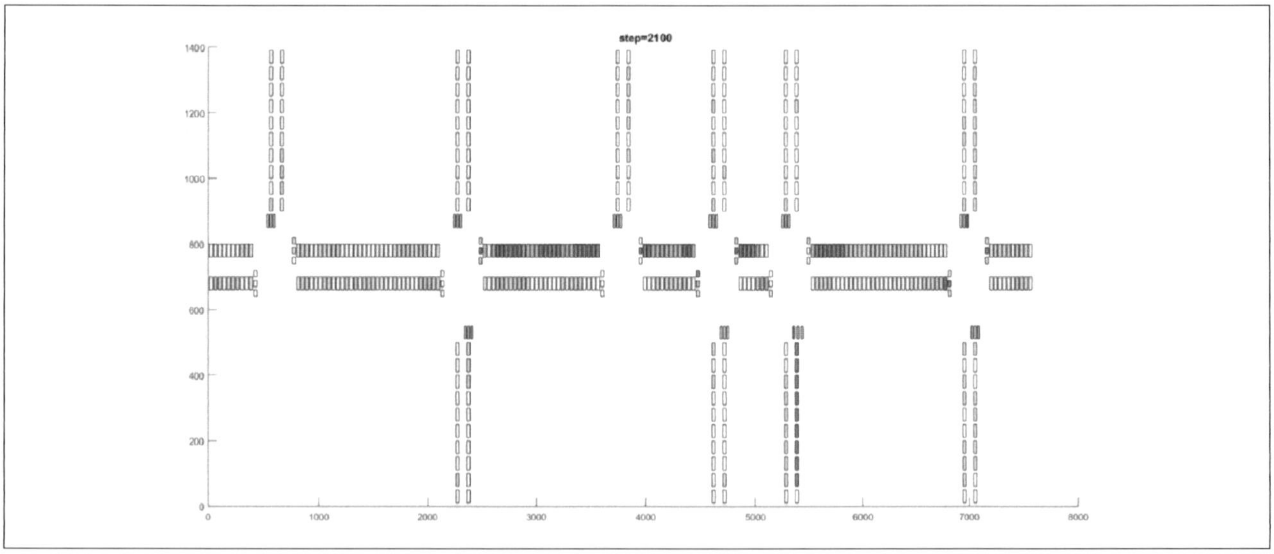

Figure 7. Snapshot of the corridor at the final simulation step.

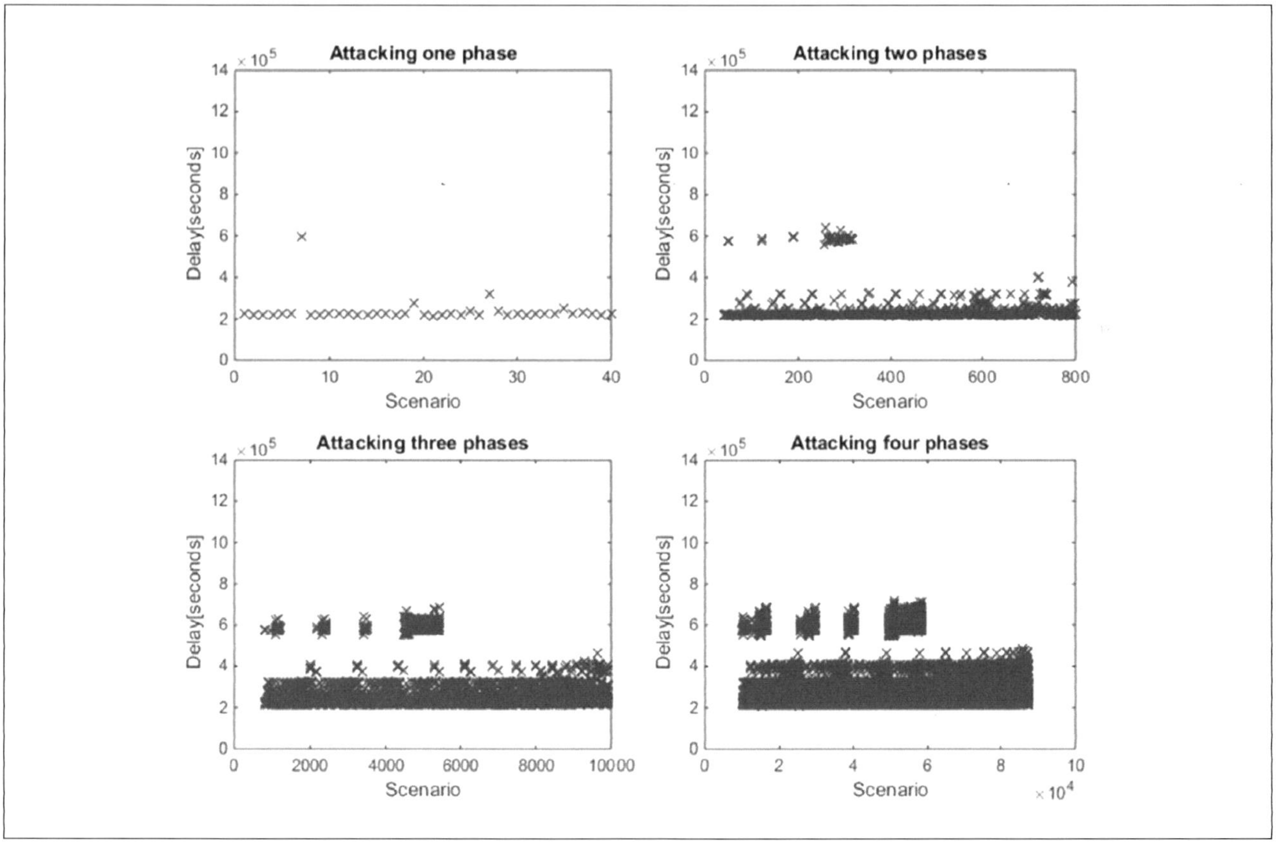

Figure 8. Vehicle delay under all attack scenarios.

effective attack strategy. The attacker starts with attacking one phase and enumerates all the scenarios to find the critical phase that causes the highest delay. Given the previous attacking phase, the attacker adds another phase and again enumerates all possibilities to find the second critical phase. This process is repeated until the budget (maximum number of phases) limit is reached. Take Plymouth corridor as an example: the total number of scenarios needed to be

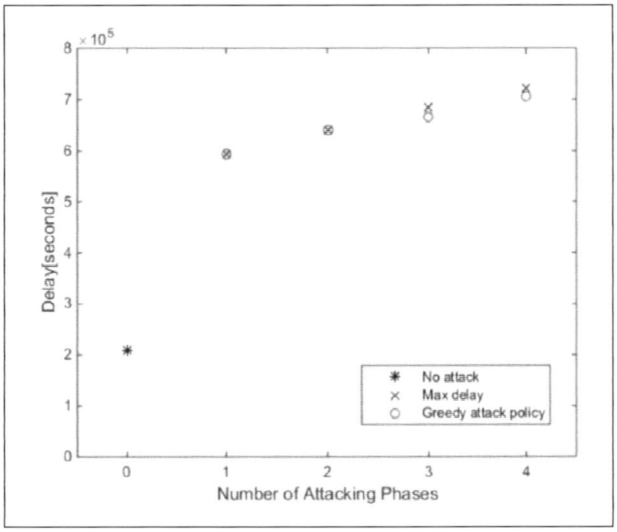

Figure 9. Comparison between enumeration and greedy attack policy.

simulated by the greedy attack policy is 40 + 38 + 36 + 34 = 148, assuming the budget is four phases. This number is significantly smaller than the total number of scenarios by enumeration. The maximum delay generated by the greedy attack policy is shown in Figure 9, as well as the delay from optimal attack strategy by enumeration. When only one or two phases are under attack, the attack strategies found by the greedy attack policy are the same as the optimal attack strategy. When more phases are under attack, the greedy attack policy can still find an attack strategy that is very close to the optimal solution, within much less time.

Conclusion and Further Research

This paper investigated the vulnerabilities of traffic control systems under cyberattacks. It focused on attacking actuated and adaptive signal control systems by sending falsified data to influence signal timing plan generation. The primary goal of an attacker was to maximize networkwide vehicle delay, with constraints such as budget and attack intensity. Results from a hypothetical intersection showed that some attacks could be very effective and cause severe congestion, whereas others may even reduce the total delay. Results from a real-world corridor showed that critical intersections, which had a higher impact on congestion, can be identified by analyzing the attack locations. Identification of critical intersections would be helpful in designing a more resilient transportation network.

To launch attacks, an attacker needs to collect necessary information about the signal control system, such as phase sequence, minimum green time and maximum green time, and so forth. However, this paper focuses mainly on the consequences of the traffic systems under cyberattacks. The authors will explore other steps in the end-to-end exploitation, for example, reconnaissance, in future work. Moreover, the authors will extend the current results toward two directions. First, besides maximizing total delay, attackers may have other objectives, such as obtaining personal gain (e.g., minimizing personal delay) or creating safety risks (e.g., causing more vehicles in dilemma zone). With different objectives, the attack strategies can be different. Second, it is necessary to consider the cybersecurity problem from a defender's point of view. Defense models need to be developed to detect attacks and protect the transportation infrastructure.

Acknowledgment

The authors would like to thank Mcity at the University of Michigan for financial support.

References

1. Ghena, B., W. Beyer, A. Hillaker, J. Pevarnek, and J. A. Halderman. Green Lights Forever: Analyzing the Security of Traffic Infrastructure. Presented at 8th USENIX Conference on Offensive Technologies, Berkeley, Calif., 2014.
2. Prigg, M. Has New York's Traffic Light System Been Hacked? *Daily Mail*, May, 2014. http://www.dailymail.co.uk/sciencetech/article-2617228/New-Yorks-traffic-lights-HACKED-technique-work-world.html. Accessed Oct. 30, 2017.
3. Miller, J. R. Hackers Crack into Texas Road Sign, Warn of Zombies Ahead. *Fox News*, Jan, 2009. http://www.foxnews.com/story/2009/01/29/hackers-crack-into-texas-road-sign-warn-zombies-ahead.html. Accessed Oct. 7, 2017.
4. Ramon. M. C,. and D. A. Zajac. *Cybersecurity Literature Review and Efforts Report*. Prepared for NCHRP Project 03-127 Cybersecurity of Traffic Management Systems. Southwest Research Institute, 2018.
5. Laszka, A., B. Potteiger, Y. Vorobeychik, S. Amin, and X. Koutsoukos. Vulnerability of Transportation Networks to Traffic-Signal Tampering. In Proceedings of the Presented at 2016 ACM/IEEE 7th International Conference on Cyber-Physical Systems (ICCPS), Vienna, Austria, 2016.
6. Ghafouri, A., W. Abbas, Y. Vorobeychik, and X. Koutsoukos. Vulnerability of Fixed-Time Control of Signalized Intersections to Cyber-Tampering. Presented at 2016 Resilience Week (RWS), Chicago, USA, 2016.
7. Jeske, T. Floating Car Data from Smartphones: What Google and Waze Know about You and How Hackers Can Control Traffic. Presented at BlackHat Europe, Amsterdam, Netherlands 2013.
8. Tufnell, N. Students Hack Waze, Send in Army of Traffic Bots. *Wired*, 2014. http://www.wired.co.uk/article/waze-hacked-fake-traffic-jam. Accessed Feb. 11, 2018.
9. Reilly, J., S. Martin, M. Payer, and A. M. Bayen. Creating Complex Congestion Patterns via Multi-Objective Optimal Freeway Traffic Control with Application to Cyber-Security. *Transportation Research Part B: Methodological*, Vol. 91, 2016, pp. 366–382. https://doi.org/10.1016/j.trb.2016.05.017.
10. Canepa, E. S., and C. G. Claudel. Spoofing Cyber Attack Detection in Probe-Based Traffic Monitoring Systems Using

Mixed Integer Linear Programming. Presented at 2013 International Conference on Computing, Networking and Communications (ICNC), San Diego, USA, 2013.
11. Zhang, Z., Q. He, H. Tong, J. Gou, and X. Li. Spatial-Temporal Traffic Flow Pattern Identification and Anomaly Detection with Dictionary-Based Compression Theory in a Large-Scale Urban Network. *Transportation Research Part C: Emerging Technologies*, Vol. 71, 2016, pp. 284–302. https://doi.org/10.1016/j.trc.2016.08.006.
12. Daganzo, C. F. The Cell Transmission Model: A Dynamic Representation of Highway Traffic Consistent with the Hydrodynamic Theory. *Transportation Research Part B: Methodological*, Vol. 28, No. 4, 1994, pp. 269–287. https://doi.org/10.1016/0191-2615(94)90002-7.
13. Daganzo, C. F. The Cell Transmission Model, Part II: Network Traffic. *Transportation Research Part B: Methodological*, Vol. 29, No. 2, 1995, pp. 79–93. https://doi.org/10.1016/0191-2615(94)00022-R.
14. Sen, S., and K. L. Head. Controlled Optimization of Phases at an Intersection. *Transportation Science*, Vol. 31, No. 1, 1997, pp. 5–17.
15. Feng, Y., K. L. Head, S. Khoshmagham, and M. Zamanipour. A Real-Time Adaptive Signal Control in a Connected Vehicle Environment. *Transportation Research Part C: Emerging Technologies*, Vol. 55, 2015, pp. 460–473. https://doi.org/10.1016/j.trc.2015.01.007.

The Standing Committee on Critical Transportation Infrastructure Protection (ABR10) peer-reviewed this paper (18-04831).

The views presented in this paper are those of the authors alone.

Article

TRR
JOURNAL OF THE TRANSPORTATION RESEARCH BOARD

Transportation Research Record
2018, Vol. 2672(1) 12–21

DOI: 10.1177/0361198118793297
journals.sagepub.com/home/trr
SAGE

Assessing Transportation Assets for Vulnerability to Extreme Weather and Other Natural Hazards

Benjamin Blandford[1], Scott Schurman[2], and Candice Wallace[1]

Abstract

The Kentucky Transportation Cabinet (KYTC) and Kentucky Transportation Center (KTC) at the University of Kentucky developed and performed an assessment of the vulnerability of the state's national highway system (NHS) to extreme weather events and natural hazards. The assessment was designed as a hybrid process that 1) gathered and analyzed existing and available quantitative data on natural hazards, including meteorological hazards and geological hazards; 2) executed a series of workshops at KYTC districts throughout the state to elicit local expert knowledge on transportation system vulnerabilities; and 3) incorporated all data into a single vulnerability assessment. The workshops involved several facilitated exercises, including map-based discussions and keypad exercises to identify vulnerabilities along highway segments. Data from the mapping and keypad exercises were gathered in a geographic information system (GIS) and combined with the other hazard data to identify vulnerabilities for each highway segment. FHWA's Vulnerability Assessment Scoring Tool (VAST) was used to process all the data into a single and ultimate assessment. Results from this analysis include a district-level prioritized ranking of NHS segments based on their vulnerability to extreme weather and natural hazard events. Additionally, a district vulnerability assessment report for each district highlighted the most vulnerable assets in the district and discussed their vulnerabilities in detail. This research is intended to aid KYTC in identifying at-risk infrastructure and direct the implementation of preparedness and resilience measures, thus helping to protect the integrity of the transportation system and the sound investment of taxpayer dollars.

In December of 2014, FHWA issued Order 5520, which holds it is the FHWA's policy to "integrate consideration of climate and extreme weather risks into its planning, operations, policies and programs" (*1*). The directive instructs state transportation agencies to evaluate and then implement risk-based, cost-effective strategies to minimize risks associated with climate change and extreme weather and protect critical infrastructure using the best available science, technology, and information. This Order supplements MAP-21 (continued through the FAST Act) requirement that state agencies develop a risk-based asset management plan for the national highway system (NHS) (*2*).

This research's objective was to assist the Kentucky Transportation Cabinet (KYTC) in its efforts to develop strategies to address natural hazard vulnerabilities and improve the resiliency of Kentucky's transportation infrastructure. KYTC owns and maintains billions of dollars of transportation-related assets across the commonwealth of Kentucky. While this transportation infrastructure has been designed to handle a broad range of climatic impacts based on historic observations and trends, less is known about how the system will respond to potential impacts from extreme weather events and other natural hazards. These issues can pose a significant threat to the safety, reliability, effectiveness, and sustainability of transportation infrastructure and operations. Examples of extreme weather events in Kentucky include sustained higher temperatures; intense, prolonged downpours and subsequent flooding; and high wind events associated with thunderstorms and tornadoes. Climate change projections show there is a higher likelihood of each of these weather phenomena over the coming decades. Geologic hazards in Kentucky include earthquakes, sinkholes, and landslides. Each of these hazards

[1]Kentucky Transportation Center, University of Kentucky, Lexington, KY
[2]Kentucky Transportation Cabinet, Frankfort, KY

Corresponding Author:
Address correspondence to Benjamin Blandford:
Benjamin.Blandford@uky.edu

has the potential to affect the lifecycle of transportation systems, resulting in higher maintenance costs and shorter replacement cycles.

This natural hazard vulnerability assessment process solicits participation from KYTC Divisions to:

a) Develop a method for assessing vulnerability of these identified assets to extreme weather events and geological hazards;
b) Perform a vulnerability assessment that identifies KYTC's assets that are at risk to extreme weather events;
c) Identify the assets that are at greatest risk to extreme weather and other natural events; and
d) Incorporate the findings and results formulated from the vulnerability assessment into the Cabinet's ongoing decision-making for planning, design, operations and maintenance processes.

The output of the assessment process includes a data system based on a geographic information system (GIS) compatible with existing Cabinet systems. The extent of the assessment is the NHS as defined by MAP-21. The choice to focus on NHS assets was deliberate, and there were two main reasons for doing so. First, MAP-21 provisions required state transportation agencies to develop a risk-based asset management plan for the NHS. Focusing vulnerability assessments on the NHS will let KYTC improve its understanding of the risks that natural hazards present to NHS assets. In turn, KYTC can develop a more targeted and accurate risk-based asset management plan. Secondly, by focusing the assessment solely on the NHS, it allowed a small, manageable set of transportation assets to be evaluated.

Literature Review

FHWA Order 5520 establishes the agency's policy on climate change and extreme weather event preparedness, and it encourages state transportation agencies to develop, prioritize, implement, and evaluate risk-based and cost-effective strategies to minimize the adverse effects of climate change and extreme weather events on critical infrastructure (*3–5*). For this project, FHWA guidance was used to identify the key components for Kentucky's vulnerability assessment. Other state vulnerability assessments, specifically, those conducted by Washington State Department of Transportation (WSDOT) and Tennessee State Department of Transportation (TDOT), also informed the project.

FHWA's *Climate Change and Extreme Weather Vulnerability Assessment Framework* provides guidance that state transportation agencies can use to perform vulnerability assessments (*6*). FHWA identifies a set of tasks which should be part of this assessment. These include: 1) gathering and integrating data and information on asset location, characteristics, and climate sensitivities; 2) collecting data on historical weather events and climate change projections; 3) combining asset and climate information to identify vulnerabilities; and potentially 4) assigning the level of risk climate change and natural hazards pose to the assets.

Vulnerability assessments begin with a research team clearly identifying project objectives. This clarifies the level of detail that will be needed for subsequent analysis and products. Identifying objectives may require identification of the end products' target audience and how it will use them. Another issue that informs the process of selecting transportation assets is a consideration of their criticality. *Criticality* is the relative importance of an asset within the overall transportation system. FHWA defines criticality as "a filter for screening the universe of assets in a particular geographic area so that the resulting list can be evaluated for exposure, sensitivity, and adaptive capacity" (*7*). Criticality relates to an asset's physical characteristics, such as its replacement value, and its function in the transportation system (e.g., an emergency response route, evacuation route, average annual daily traffic, and/or key commercial route).

FHWA identifies three approaches for assessing criticality: desk review, stakeholder solicitation, and a hybrid approach (*7*):

- A *desk review* emphasizes objective and easily obtainable empirical data to use in asset ranking. Where possible, already existing prioritization schemes are used, such as average daily traffic, functional classification, and expert judgment. Advantages of this approach are that it is transparent and easily replicable.
- *Stakeholder solicitation* utilizes expert knowledge of transportation assets provided by local officials or authorities. This can be achieved through a series of focus groups to elicit feedback on the criticality of assets. Advantages of this approach are that it encourages buy-in from relevant stakeholders and promotes collaboration and communication among stakeholders and those likely to implement adaptation strategies.
- These can also be combined into a *hybrid approach* that incorporates both a desk review and stakeholder solicitation. This strategy generally begins with a desk review to identify a list of possible critical assets and the relevant data pertaining to them. It then incorporates this information within the stakeholder solicitation process to inform and structure feedback from stakeholders and local experts.

Collecting data on assets and hazards is central to a vulnerability assessment. The specific types of data gathered should relate directly to the assessment's objectives. Hazards data serve as inputs to the analysis, whereby different scenarios of varying magnitude and probability can be evaluated. This requires an examination of historical data and projected climate change. Other factors to consider are the issues of probability and risk. Probability refers to the likelihood that a particular scenario or extreme event both occurs and impacts transportation assets included in the study. Risk refers to the magnitude of damage or destruction that would likely occur as a result of the hazard.

Transportation agencies and the scientific community have heeded FHWA's call and initiated efforts to assess transportation system vulnerabilities to extreme weather and climate change. Research has identified strategies for incorporating climate science data into transportation system engineering practices (*8*). Review of these efforts has yielded insights on which types of adaptation measures are most likely to enhance transportation system resilience within budgetary constraints (*9*). Researchers have identified data-driven methods for assessing the impacts of extreme weather events (*10*). GIS has been identified as a useful tool for identifying hot spots where transportation system vulnerabilities are elevated, and further analysis or resources are needed (*11*).

The Washington State Department of Transportation (WSDOT) was one of five state transportation agencies to receive federal grants as part of FHWA's Climate Change Vulnerability Assessment Pilot Program to test a conceptual climate risk assessment model that was developed specifically for transportation infrastructure (*12*). WSDOT applied the model using scenario planning, which involved conducting a series of statewide workshops. Workshop participants, which included local subject matter experts in a variety of fields (such as design engineers, planning, environmental, and maintenance), qualitatively assessed the vulnerability of WSDOT assets to climate extremes. WSDOT adopted an asset management approach to its vulnerability assessment. A qualitative analysis was chosen for numerous reasons—it provided a preliminary understanding of an issue, enabled analysis when the only available information is experience and subjective opinions, proved useful when quantitative analysis is beyond the scope of an investigation, and it provided a quick assessment.

The strategy initially adopted by WSDOT focused heavily on details of specific assets and mile points. WSDOT eventually hit on a more meaningful approach to data collection—roads were divided into segments and subsequently analyzed. After analyzing roadway segments, a question was posed to workshop participants that yielded significant dividends. This question was "What keeps you up at night?" This question got to the heart of what workshop participants thought about hazard vulnerabilities, and what the most pressing district concerns were.

As part of FHWA's Phase II of Climate Change Resiliency Projects, Tennessee State Department of Transportation (TDOT) was selected to conduct a transportation infrastructure vulnerability assessment for climate change and extreme weather events (*13*). Because of Tennessee's proximity to Kentucky, as well as its similar array of geologic features and climate conditions, the TDOT assessment was particularly relevant when designing an assessment project for Kentucky. TDOT's assessment looked at a range of transportation assets, including roads (interstate, state, and U.S. highways), railroads, rail yards, navigable waterways, ports, locks, bridges over navigable waterways, airport runways, maintenance and salt facilities, TDOT buildings (administration, operations), transit facilities (transfer hubs, terminals, fleet storage), and pipelines (oil and natural gas). TDOT developed criteria for determining the criticality of each asset. Criticality was based on factors such as volume of activity, strategic importance, use as an emergency response route, redundant capability, network connectivity, local knowledge and experience.

Transportation assets were evaluated for their vulnerability to a variety of extreme weather events, including extreme heat and cold, wind, tornadoes, hydrologic events (heavy rain, flash flooding, flooding), lightning, hail, drought, and winter storms (*14*). Historical data on these hazards (dating back to 1950) were drawn from the National Weather Service's Storm Events Database. Information about historical climate trends and projected climate change were combined with stakeholder feedback to rank transportation assets' vulnerability to extreme weather events. Stakeholders included representatives from a range of governing agencies, economic sectors, transportation modes, geographic regions, and political jurisdictions. Stakeholder focus group meetings were held in each of TDOT's four regions to inform the assessment and develop vulnerability rankings for transportation assets.

Assessment Framework

FHWA guidance documentation, as well as insights from the other state department of transportation vulnerability assessments, was used to develop this project design. Ultimately the research team settled on a hybrid approach to conduct the vulnerability assessment. This hybrid approach was divided into three major components: 1) gathering and analysis of available data on assets and hazards; 2) development and implementation of district-level workshops to elicit local expert knowledge on asset vulnerabilities; and 3) incorporating all the data gathered into a single and ultimate assessment. The results from these efforts were incorporated into a GIS to develop a central repository of data, facilitate easy data analysis of hazards and transportation assets, and enable KYTC users to produce maps and geovisualizations to communicate vulnerabilities.

One major departure in this project's design from the aforementioned studies was the inclusion of natural hazards beyond those associated with extreme weather events and climate change. KYTC has identified geologic hazards related to extreme weather that directly affect asset condition and life cycle. For example, extreme weather events can negatively affect surface and subsurface geology, triggering hazards such as landslides or sinkhole subsidence. Other geologic hazards, such as earthquakes, are also a concern. The western half of Kentucky is at increased risk to a large seismic event because of its proximity to the New Madrid seismic zone. Since these geological hazards pose such a large threat to KYTC assets, the research team believed it was vital to factor vulnerability and risk from these events into risk-based asset management strategies.

Data Gathering

Meteorological data used in this assessment were obtained from a number of sources. The research team collaborated with the Midwest Regional Climate Center (MRCC) to obtain county-level historical climate data for the years 1981–2015. These data pertained to extreme heat events (days with temperature exceeding 95°F), extreme precipitation events (days with 3" or more of rainfall), extreme wind events (days on which wind speed exceeds 75 mph), and freeze/thaw cycles (days on which temperature fluctuates above and below 32°F). Collected data were mapped and analyzed using GIS. An example is provided in Figure 1.

State-level historical climate information dating back to 1895 were provided by the Kentucky State Climate Center. Historical data on severe storms, including tornados, were obtained from the National Climatic Data Center's Storm Events Database (*15*). More general data on drought (*16*), and wildfire (*17*) were gathered where available. Climate change projection data were acquired from the North American Climate Assessment Program (*18*) and the Climate Change Institute at Oak Ridge National Laboratory (ORNL) (*19*). Geological data were similarly gathered on sinkholes (*20*), landslides (*21*), and seismic vulnerability (*22*).

KYTC Survey

As the data-gathering phase proceeded, it quickly became apparent that there was a potentially overwhelming amount of data that could be incorporated in the assessment. The project team decided it would be beneficial to hone in on the most significant hazards for analysis. To determine which hazards to include, the project team administered a survey to assess perceptions of risk and vulnerability of the transportation system. The survey was conducted online and was sent to 50 KYTC Central Office and District personnel. 25 surveys were completed, resulting in a survey completion percentage of 50 percent. Survey participants were asked to evaluate a list of natural hazards for how significant their impacts are to Kentucky's transportation system. For each hazard, participants indicated the short-term impacts, long-term impacts, overall impacts, and probability of impacts on a scale of 1–5, with 1 being the lowest and 5 the highest.

The results revealed how KYTC personnel view certain types of hazards as having a greater impact than others on the transportation system. Flooding, flash flooding, earthquakes, and landslides ranked in the top five for all criteria. Sinkholes were ranked highly in terms of long-term impacts and overall impacts, while snow/ice was ranked highly for its short-term impacts and probability of impacts. In light of the survey results, the project team decided to focus primarily on the following hazards: flood (both regional flood and flash flood), landslide, sinkhole, and earthquake.

Hazards Data

KYTC NHS assets, including highway segments, bridges, culverts, and other structures (overpasses) were assessed for vulnerabilities in a GIS using gathered data. For each of the four major hazards, indicators were developed to identify potential risk:

- Earthquake: United States Geological Survey (USGS) data on peak ground acceleration (PGA) zones (*22*). PGA measures the intensity of the shaking of the earth associated with seismic activity, and is expressed as a percent of gravity, with a two percent probability of exceedance in the next 50 years. When compared against the Modified Mercalli Scale, a PGA zone of greater than 60 indicates "Ruinous" impacts, a PGA zone of 30 to 60 indicates "Destructive" impacts, and a PGA zone of 18 to 30 indicates "Very Strong" impacts (*23*). Western areas of Kentucky are most vulnerable to earthquake due to proximity to the New Madrid seismic zone.
- Flood: The Federal Emergency Management Agency's (FEMA) flood hazard mapping program data for 100-year digital floodplain. A flood with a 100-year recurrence interval has an approximately 1% chance of occurring in a given year.

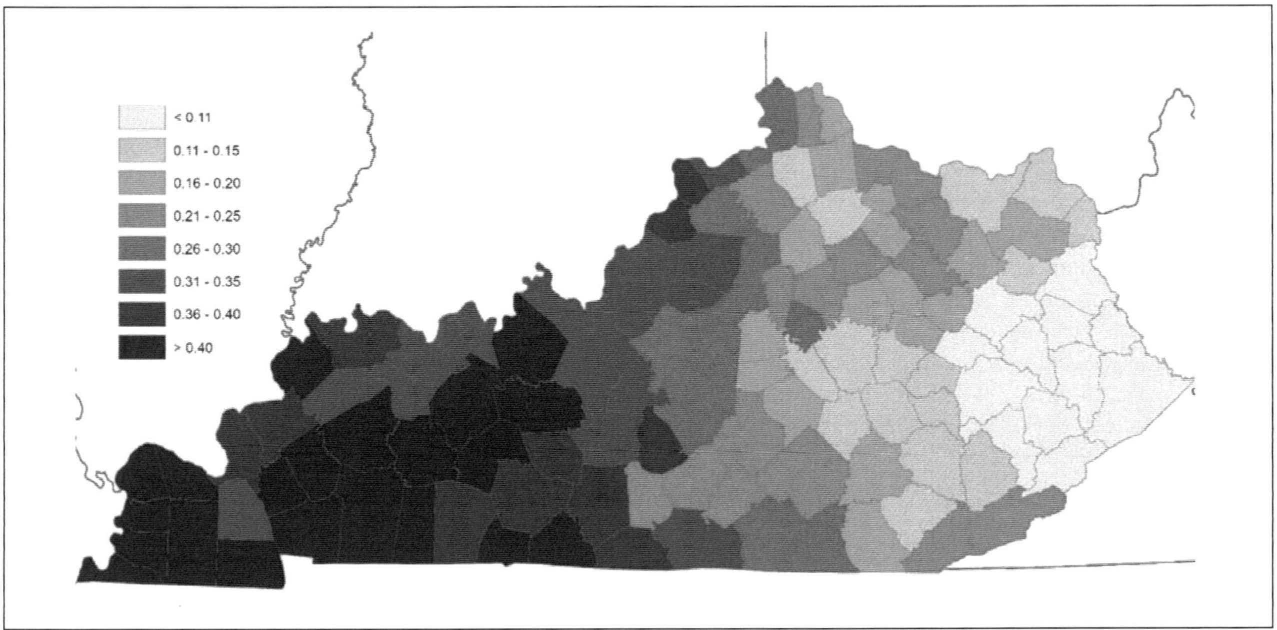

Figure 1. Average annual number of days with three or more inches of rainfall; 1981–2015.
*Source Data Provided by MRCC.

Kentucky has over 90,000 miles of streams in the state, and the transportation system is vulnerable to both regional river flooding and localized flash flooding.

- Sinkhole: Karst topography is terrain characterized by high rock solubility and porosity. Sinkholes, caves, sinking streams, and springs are commonly found in karst landscapes. Kentucky Geological Survey (KGS) data on karst potential estimates indicate that 55% of the land in the state has the potential for karst development (*20*). Additionally, 38% of the state has enough karst development to be recognized topographically, and 25% has well-developed karst features.
- Landslide: KGS landslide inventory database (*21*). In Kentucky, landslides are most common in the mountainous eastern areas of the state as well as in northern Kentucky in the Ohio River Valley.

Table 1 summarizes the relationship between NHS assets statewide and selected natural hazard indicators.

Individual transportation assets, including bridges and highway segments, were also identified and assessed. Highways were segmented based upon their county location, route designation, and other factors. Bridges, culverts and other structures were identified according to their National Bridge Inventory database identification. For each asset, fields pertaining to the indicators and vulnerabilities were produced in a GIS. Table 2 provides a snapshot from District 1 of individual assets and vulnerabilities.

The vulnerability associated with each of these four hazards varies considerably throughout the state. For example, western Kentucky is located in a significant seismic hazard zone, while central and northern areas of Kentucky have very little seismic risk. Landslide, sinkhole, and flood vulnerabilities are also geographically uneven. Understanding the extent of these vulnerabilities across the 12 KYTC Districts in Kentucky became the goal of the next phase of the project.

District Workshops

For the second phase of this project, KYTC personnel and Kentucky Transportation Center (KTC) researchers developed a workshop-based methodology to elicit local expert knowledge from transportation personnel on highway system vulnerabilities. KYTC district workshops were completed for each of the 12 districts. Each completed workshop consisted of 12 to 20 KYTC personnel representing the transportation areas of maintenance, planning, design, and environmental analysis. KYTC personnel were identified and invited by KYTC leadership to the workshops based on their knowledge, experience, and expertise of their respective district transportation systems. At the workshops, data was collected from participants through a series of facilitated exercises, including; 1) qualitative mapping exercises, and 2) quantitative keypad exercises.

Table 1. Summary of KYTC Statewide NHS Assets and Selected Natural Hazard Indicators

		Earthquake			Flood	Sinkhole		Landslide	
Asset type	KYTC total	PGA ≥ 60	PGA ≥ 30 and < 60	PGA ≥ 18 and < 30	100-year floodplain	Major karst potential	Moderate karst potential	High landslide potential	Moderate landslide potential
Total NHS centerline miles[a]	6,151	194	425	771	229	1,921	848	1,372	401
Interstate	2,064	31	161	241	49	762	398	290	56
Parkway	1,034	60	61	218	33	209	174	59	77
US-	1,932	101	128	295	97	694	136	613	181
KY-	1,076	2	76	17	49	227	141	409	86
Local road	44	0	0	0	0	29	0	0	0
Bridges	772	51	93	143	431	149	46	231	67
Culverts	292	12	21	21	166	86	40	75	27
Structures	1,409	56	109	178	26	462	188	205	86

[a]Centerline miles are doubled for dual-carriage highway segments.

Table 2. Sample of KYTC District 1 Assets and Vulnerabilities

District	County	Route	Start MP	End MP	Miles	PGA zone	Karst potential	Landslide susceptibility	Floodplain
1	Ballard	US 60	0	16.94	16.86	120 (MP 0–8.43); 80 (MP 8.43–16.94)	Low	High (MP 0–8.16); Low (MP 8.16–16.94)	MP range 1.49–12.60
1	Caldwell	I 24	54.84	57.39	2.56	30	Major (MP 54.84–57.39)	Low	MP range 56.53–56.60
1	Hickman	JC 9003	3.43	8.35	4.88	60	Low	Low	None
1	Livingston	I 24	29.35	33.88	4.5	40	Low	Low	MP range 29.35–33.88

Qualitative Assessments

The first component of the workshops involved facilitated mapping exercises during which KYTC district personnel identified, provided details on, and discussed with workshop participants highway locations that have flooded, experienced landslide activity (slips, slides, rock falls), or been impacted by sinkholes or karst-related flooding. This local knowledge yielded such insights as which areas not located in the 100-year floodplain are vulnerable to flooding due to issues such as poor drainage, blocked drains, sinkholes, land use changes, or stream channel maintenance problems, among others. Local personnel also indicated the frequency of such hazards (e.g., "This road floods every time we experience a heavy rainfall"), the duration of the events, and the severity of impacts to the transportation system. Following the mapping exercises, the groups identified and discussed what they believe the worst-case scenario could plausibly be for the district's transportation system. This discussion required personnel to consider a number of factors affecting vulnerability, including the importance of the asset, its replacement cost, network redundancy, likelihood of hazard occurring, and severity and extent of potential impacts.

Keypad Assessments

The second component of the workshops involved keypad exercises during which workshop participants rated each NHS highway segment for both its criticality to the overall system and the potential impact of hazard events. The NHS was divided into 287 segments for evaluation. Segmentation of the NHS incorporated several factors, including route number, county, length, functional class, and Average Annual Daily Traffic (AADT). Keypad rankings for criticality and impacts are combined to produce an overall vulnerability score for each NHS segment. Workshop participants viewed the rankings in real-time and discussed why certain segments were considered more vulnerable than others. The scales used to define criticality and potential impacts were adapted from the WSDOT vulnerability assessment (*12*) and are described below.

Criticality scale

- 1–3: Non-NHS; Low AADT; Alternate routes available (because all segments in this analysis are NHS, no segments were scored in this category)
- 4–6: Some NHS; Low to medium AADT; Alternative for other state routes

- 7–9: Interstate; Lifelines; Some NHS; Sole access; No alternate routes

Impacts scale

- 1–3: Results in little or negligible impact to asset. Asset would be available with full use within ten days and has immediate limited use still available.
- 4–6: Results in minor damage and/or disruption to asset. Asset would be available with either full or limited use within sixty days.
- 7–9: Results in total loss or ruin of asset. Asset may be available for limited use after at least sixty days and would require major repair or rebuild over an extended period of time.

Keypad rankings were conducted in real time. Participants were encouraged to discuss the highway segments as the exercise proceeded through the highway segments, and they were able to view the results immediately. If the rankings varied significantly across the group, a discussion followed to explore why diverging opinions existed, and if necessary, another ranking was conducted to reflect the conversations. Data from the mapping exercises and keypad ranking exercises were gathered and digitized in a GIS.

Assessment

The third phase of the project entailed incorporating all the data gathered into a single comprehensive assessment. In order to incorporate all the varying data types, FHWA's Vulnerability Assessment Scoring Tool (VAST) was used to refine the results. This allowed the project team to include quantitative data gathered from available sources as well as the data generated in the workshops.

FHWA has supported usage of VAST to help transportation organizations implement an indicator-based assessment of transportation assets. VAST provides users with a high-level screening tool to identify asset vulnerability. Data inputs for VAST are organized into three types:

1. Exposure—potential exposure of transportation assets to climate stressors and other hazards. Exposure data may include average number of extreme heat days or extreme precipitation events.
2. Sensitivity—how sensitive the transportation asset is to the particular climate stressors and other hazards. Sensitivity data may include characteristics of the asset or historical occurrences.
3. Adaptive Capacity—how resilient the system is to disruption and/or damage. Adaptive Capacity data may include indicators of asset criticality, such as AADT, functional class, asset replacement value, or detour route viability.

Data are furthermore organized according to what class of asset and what type of hazard are being assessed. VAST is set up to allow users to assess multiple transportation asset classes, including, highways, bridges, railways, ports and waterways, airports, and pipelines. For hazard types, VAST is set up to incorporate data on temperature, flooding, sea level rise, storm surge, wind, drought, dust storms, wildfires, winter storms, freeze/thaw cycles, and permafrost thaw. Users can also input a new type of asset class or hazard if desired.

For each indicator, transportation assets are given a score from one to four. For exposure and sensitivity indicators, a low score means less exposure or sensitivity while a high score means more exposure or sensitivity. For adaptive capacity indicators, a low score indicates better adaptive capacity while a high score indicates less adaptive capacity. Within each indicator category, individual datasets can be weighted. For example, if one adaptive capacity indicator is believed to be more impactful than the others, it can be assigned a higher percentage of the adaptive capacity indicator score. In addition, each indicator type—exposure, sensitivity, and adaptive capacity—is assigned an overall contribution to the final vulnerability score as a percentage. The output from VAST is a score from one to four indicating vulnerability of each transportation asset to each hazard type, where a higher score indicates higher vulnerability.

For this assessment, additional data were gathered and analyzed to provide a more robust set of indicators of asset vulnerability. In all, 22 indicator datasets were included to assess NHS vulnerability to earthquake, flooding, landslide, and sinkholes. NHS assets were defined as the same highway segments used in the district workshop keypad exercises. Table 3 lists all the datasets used in the VAST analysis. Further detail on each of these indicators is available in the full project report (*24*).

Results and Discussion

The VAST analysis resulted in a scoring from one to four of each of the 287 Kentucky NHS segments for vulnerability to earthquakes, flooding, landslides, and sinkholes. Assets with a score of 3.5 or greater to any of the hazards were flagged for further analysis and discussion. Table 4 shows a sample of the VAST analysis results for KYTC District 8.

Data generated from the VAST assessment and from the district workshops resulted in a GIS database of transportation assets that includes the prioritized ranking of vulnerabilities derived from the keypad exercises, local expert knowledge on historical vulnerabilities elicited

Table 3. Indicators Used to Assess NHS Asset Vulnerability to Earthquake, Flood, Landslide, and Sinkhole

Indicator type	Hazard type	Description	Measure
Exposure	Earthquake	Peak ground acceleration	Max PGA of asset
Exposure	Flood	Historical precipitation	Average annual # of days precipitation >3"
Exposure	Flood	Projected precipitation	Average annual # of days precipitation >1"
Exposure	Flood	KGS flood hazard layer	Segments intersecting this layer
Exposure	Flood	Location via 100-year floodplain	Percent of asset located in flood zone
Exposure	Landslide	USGS landslide susceptibility	Percent of asset located in High landslide zone
Exposure	Landslide	Max slope	Maximum slope along NHS segment
Exposure	Landslide	Mean slope	Mean slope along NHS segment
Exposure	Landslide	KGS slope stability/landslide layer	Segments intersecting this layer
Exposure	Landslide	KGS landslide inventory	Landslides within 500 feet of NHS per centerline mile
Exposure	Sinkhole	KGS karst potential	Percent of asset located in high karst zone
Exposure	Sinkhole	KGS karst hazard layer	Segments intersecting this layer
Exposure	Sinkhole	KGS sinkhole database	Acres sinkholes/centerline miles
Sensitivity	Flood	Workshop mapping results	NHS assets identified in workshops as having flooded
Sensitivity	Flood	FEMA flood event	NHS assets that experienced FEMA response flood events
Sensitivity	Landslide	Workshop landslide results	NHS assets identified in workshops as having landslides
Sensitivity	Landslide	FEMA slide event	NHS assets that experienced FEMA response landslide events
Sensitivity	Sinkhole	Workshop sinkhole results	NHS assets identified in workshops as having sinkholes
Adaptive capacity	All	AADT	Average annual daily traffic per NHS segment
Adaptive capacity	All	Criticality rating	Workshop keypad criticality rating of NHS asset
Adaptive capacity	All	Functional class	Functional classification of NHS asset
Adaptive capacity	All	Impacts rating	Workshop keypad impacts rating of NHS asset

Table 4. Sample of VAST Analysis Results for KYTC District 8

District	County	Route	Vulnerability score	Hazard
8	Pulaski	US-27	3.54	Flood
8	Casey	US-127	3.63	Flood
8	Rockcastle	I-75	3.71	Flood
8	Pulaski	US-27	3.57	Landslide
8	Rockcastle	I-75	3.87	Landslide
8	Pulaski	KY-914	3.58	Sinkhole
8	Pulaski	US-27	3.72	Sinkhole

from the workshops, and other relevant vulnerability data gathered from available data sources. This resulting geodatabase is designed to serve as an exhaustive and easily accessible inventory of natural hazard risk and vulnerabilities for Kentucky's NHS.

In addition to the geodatabase, a vulnerability assessment report was produced for each district to summarize the findings from the project. In the district reports, the most vulnerable assets for the district are identified and discussed in detail. The reports also summarize the workshop discussions on the worst case plausible scenario for the district in terms of natural hazards. Finally, a full report on the project was produced to comprehensively detail the development of the project (*25*). This report details the federal guidance, acquisition of data, analysis of meteorological and geological patterns and trends, and development of the assessment process.

Discussion of Assessment

The vulnerability assessment has gathered and documented extensive data regarding asset vulnerability to extreme weather and natural hazards. Information gathered to establish risk and vulnerability during the assessment is now being utilized to develop a risk-based asset management plan on the NHS as required by MAP-21. As a part of this asset management plan, strategies are being developed to help mitigate the effects from extreme weather and natural hazards on NHS transportation assets. These strategies are being integrated into KYTC's planning, design, and maintenance efforts. Design mitigation strategies may include implementing enhanced design standards or utilizing construction techniques or materials that make an asset more resilient to the effects from extreme weather and natural hazards. Maintenance mitigation activities may include removing debris flows near structures on a regular basis or placing protective fencing adjacent to roadway where rock falls occur regularly. These mitigation strategies can be specific to the hazard or hazards that pose the greatest threat to NHS assets for specific areas of the state identified during the assessment district workshops. For instance, mitigation in the western part of the state will focus more on seismic and flooding hazards while the northern part of the state will focus on landslides and freeze–thaw cycles. In addition, as part of the asset management plan, cost–benefit analyses will be developed and performed to determine whether the proposed mitigation strategies are cost effective for implementation.

Vulnerability assessments identify facets of exposure in the transportation system that could be addressed through policy or infrastructure enhancements. They also provide decision makers with information on where the transportation system's resiliency could be improved. With this knowledge, decision makers can be more proactive in addressing issues related to transportation vulnerabilities. These assessments illuminate potential issues before they result in a major incidents or closures. Drawing on information from these assessments will create a more up-to-date and well-maintained transportation system by helping officials identify and resolve issues before they grow and become unmanageable. Additionally, conducting and acting upon vulnerability assessments reduces financial losses that would occur if the system were to fail completely. Maintaining a secure and safe transportation system mitigates negative publicity that may arise were the system to fail. Identifying and resolving vulnerabilities was the primary goal of this vulnerability assessment.

Author Contributions

The authors confirm contribution to the paper as follows: study conception and design: Blandford, Schurman; data collection: Blandford, Schurman, Wallace; analysis and interpretation of results: Blandford, Schurman; draft manuscript preparation: Blandford, Schurman. All authors reviewed the results and approved the final version of the manuscript.

References

1. Federal Highway Administration. *Transportation System Preparedness and Resilience to Climate Change and Extreme Weather Events*. FHWA Order 5520. Federal Highway Administration, U.S. Department of Transportation, 2014.
2. *Transportation Asset Management Plans. 23 U.S.C. 119(e)(1), MAP-21 § 1106*, n.d.
3. ICF International. *Integrating Climate Change into the Transportation Planning Process*. Federal Highway Administration, U.S. Department of Transportation, 2008. https://s3.amazonaws.com/nyclimatescience.org/climatechange.pdf
4. ICF International. *Climate Change Vulnerability Assessment, Risk Assessment, and Adaptation Approaches*. Federal Highway Administration, U.S. Department of Transportation, 2009. www.fhwa.dot.gov/environment/climate_change/adaptation/publications_and_tools/vulnerability_assessment/
5. ICF International. *Assessment of the Body of Knowledge on Incorporating Climate Change Adaptation Measures into Transportation Projects*. Federal Highway Administration, U.S. Department of Transportation, 2013. https://ntl.bts.gov/lib/51000/51500/51556/transportationprojects.pdf
6. Federal Highway Administration. *Climate Change & Extreme Weather Vulnerability Assessment Framework*. Federal Highway Administration, U.S. Department of Transportation, 2012. https://www.fhwa.dot.gov/environment/sustainability/resilience/publications/vulnerability_assessment_framework/fhwahep13005.pdf
7. ICF International. *Assessing Criticality in Transportation Adaptation Planning*. Federal Highway Administration, U.S. Department of Transportation, 2011. https://s3.amazonaws.com/nyclimatescience.org/cta092111.pdf
8. Douglas, E., J. Jacobs, K. Hayhoe, and L. Silka. Progress and Challenges in Incorporating Climate Change Information into Transportation Research and Design. *Journal of Infrastructure Systems*, Vol. 23, No. 4, 2017.
9. Stamos, I., E. Mitsakis, and J. Maria Salanova Grau. Roadmaps for Adaptation Measures of Transportation to Climate Change. *Transportation Research Record: Journal of the Transportation Research Board*, 2015. 2532: 1–12.
10. Stamos, I., E. Mitsakis, J. Maria Salanova Grau, and G. Aifadopoulou. Impact Assessment of Extreme Weather Events on Transport Networks: A Data-driven Approach. *Transportation Research Part D: Transport and Environment*, Vol. 34, 2015, pp. 168–178
11. Camp, J. V., and M. D. Abkowitz. *Estimating the Effects of Extreme Weather on Transportation Infrastructure*. CFIRE RI-05. National Center for Freight & Infrastructure Research & Education, 2015. https://ntl.bts.gov/lib/61000/61500/61514/CFIRE_RI05_FinalReport.pdf
12. Washington State Department of Transportation. *Climate Impacts Vulnerability Assessment*. Washington State Department of Transportation, 2015. http://www.wsdot.wa.gov/NR/rdonlyres/B290651B-24FD-40EC-BEC3-EE5097ED0618/0/WSDOTClimateImpactsVulnerabilityAssessmentforFHWAFinal.pdf
13. Abkowitz, M. D., J. V. Camp, and L. Dundon. *Assessing the Vulnerability of Tennessee Transportation Assets to Extreme Weather*. Tennessee Department of Transportation, 2015.
14. Abkowitz, M. D., A. Jones, L. Dundon, and J. V. Camp. Performing a Regional Transportation Asset Extreme Weather Vulnerability Assessment. *Transportation Research Procedia*, Vol. 25, 2017, pp. 4422–4437.
15. National Oceanic and Atmospheric Administration. *Storm Prediction Center Warning Coordination Meteorologist's Page*. National Oceanic and Atmospheric Administration, 2016. http://www.spc.noaa.gov/wcm/
16. *United States Drought Monitor*. The National Drought Mitigation Center, University of Nebraska-Lincoln. http://droughtmonitor.unl.edu/
17. United States Geological Survey. *GeoMAC Wildland Fire Support*. United States Geological Survey, 2016. http://geomac.gov
18. Kunkel, K. E., L. E. Stevens, S. E. Stevens, L. Sun, E. Janssen, D. Wuebbles, and C. E. Konrad, II. *Regional Climate Trends and Scenarios for the U.S. National Climate Assessment: Part 2. Climate of the Southeast U.S.* National Oceanic and Atmospheric Administration Technical Report, 2013.

19. Ashfaq, M., D. Rastogi, R. Mei, S. Kao, S. Gangrade, B. S. Naz, and D. Touma. High-Resolution Ensemble Projections of Near-Term Regional Climate over the Continental U.S. *Journal for Geophysical Research-Atmospheres*, Vol.121, No. 17, 2016, pp. 9943–9963.
20. Paylor, R. L., and J. C. Currens. *Karst Occurrence in Kentucky*. Kentucky Geological Survey, 2001. http://kgs.uky.edu/kgsweb/olops/pub/kgs/mc33_12.pdf
21. Kentucky Geological Survey. *Landslide Information Map*. Kentucky Geological Survey, 2016. http://kgs.uky.edu/kgsmap/kgsgeoserver/viewer.asp?layoutid = 25
22. United States Geological Survey. *PGA 2% in 50 Yrs*. United States Geological Survey, Earthquake Hazards Program, 2014. 2014. http://earthquake.usgs.gov/hazards/products/conterminous/
23. Kentucky Emergency Management. *Commonwealth of Kentucky Enhanced Hazard Mitigation Plan*. Kentucky Emergency Management, 2013. http://kyem.ky.gov/recovery/Documents/CK-EHMP%202013,%20Standard%20and%20Enhanced.pdf
24. Blandford, B. L., S. Schurman, and C. Wallace. *Kentucky NHS Vulnerability Assessment for Extreme Weather and Other Natural Hazards*. Forthcoming.
25. Blandford, B. L., S. Schurman, C. Wallace, and S. McCormack. *Transportation System Vulnerability and Resilience to Extreme Weather Events and Other Natural Hazards*. KTC-16-20/SPR16-524-1F, 2016.

The Standing Committee on Critical Transportation Infrastructure Protection (ABR10) peer-reviewed this paper (18-05906).

Article

TRR
JOURNAL OF THE TRANSPORTATION RESEARCH BOARD

Transportation Research Record
2018, Vol. 2672(1) 22–29

DOI: 10.1177/0361198118756875
journals.sagepub.com/home/trr
SAGE

Exit Choice Behavior of Pedestrians Involving Individuals with Disabilities During Building Evacuations

Nirdosh Gaire[1], Ziqi Song[1], Keith M Christensen[2], Mohammad Sadra Sharifi[1], and Anthony Chen[3]

Abstract

Pedestrian evacuation studies are critical in obtaining information about evacuation scenarios and in preparing to face the challenges of actual evacuations. Studies have examined evacuation policies, exit choice modeling, and evacuation curve analysis. Some studies have addressed the evacuation behavior of individuals with disabilities (IWDs), although this important aspect of evacuation seems to be missing from modeling of the exit choice in many studies. In modeling of the exit choice for evacuation, many studies have been found to be based on the stated preference survey method, where evacuees are asked to choose an exit based on descriptions, without an actual experiment taking place. This study focuses on the discrete choice model of the exit choice in the room for both IWDs and individuals without disabilities (IWODs). The results demonstrate that the presence of IWDs in the group plays a crucial role in the exit choice for all evacuees. The results demonstrate that there are significant differences in exit choice between IWDs and IWODs. Current evacuation policies have been found to be more focused on visual signs, while this study demonstrates that these visual signs are of little importance to individuals with visual disabilities.

Emergency evacuation models have been developed to estimate the evacuation time of evacuees. The danger of accidents caused as a result of panic during a real evacuation makes it difficult for researchers to observe pedestrian evacuation behavior (*1*). Evacuation models are used to study the various effects on both evacuation behavior and evacuation time. The exit behavior of people, both individually and in a group, has been studied using the evacuation simulation model (*2*). An evacuation model was built to understand the effects of both fast flows of evacuees and interactions among individuals on evacuation time (*3*). Varas et al. (*4*) built a cellular automaton model to study evacuation behavior in a closed room with obstacles. Their simulation considered two different situations by varying the type of exit doors from the room, with single and double doors, which demonstrated that evacuation time is minimized using multiple doors rather than a single door.

Pedestrian evacuation behavior in a single room with single or multiple exits has previously been studied, and results have been drawn from both experimental studies and from modeling of the exit choice. Helbing et al. (*5*) performed experiments in a classroom simulating the evacuation process, which revealed that escape time distribution is affected by jamming of students at the exit. Isobe et al. (*6*) performed an experimental study to evaluate the evacuation process using students, which also revealed that jamming of the exit doors affects the exit time, concluding that studies are necessary to plan safe evacuations from buildings. Shi et al. (*7*) performed controlled laboratory experiments to examine the safety of pedestrians at merging angles, suggesting these have a significant influence during emergency evacuation of pedestrians. Pedestrians were found to use rational knowledge in making exit choices in cases of fire (*8*). If pedestrians make exit choices rationally rather than panicking, it is likely that many factors influence the exit choice of pedestrians.

The selection of exits during evacuation is a stochastic process that is defined by the behavioral uncertainty of pedestrians (*1*). Exit choice is also influenced by the familiarity of the exits (*9*). Pedestrians tend to choose the exit that is nearest them in most scenarios (*10–13*). Density of

[1]Department of Civil and Environmental Engineering, Utah State University, Logan, UT
[2]Department of Landscape Architecture and Environmental Planning, Utah State University, Logan, UT
[3]Department of Civil and Environmental Engineering, The Hong Kong Polytechnic University, Hung Hom, Kowloon, Hong Kong, China

Corresponding Author:
Address correspondence to Ziqi Song: ziqi.song@usu.edu

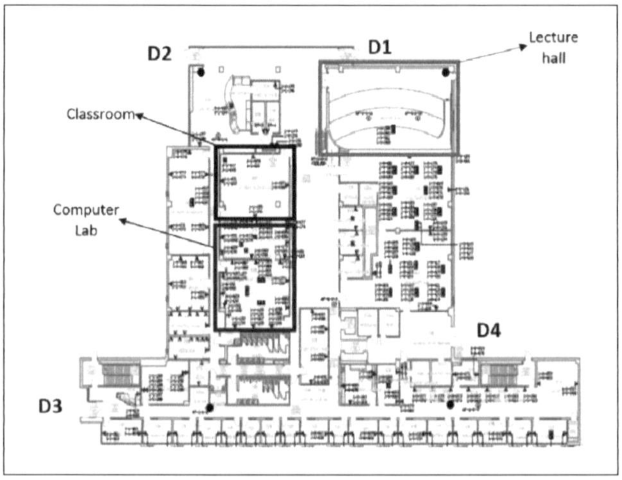

Figure 1. Layout of the building with different components.

pedestrians around the exit also plays a significant role in exit choice selection (*14*). If the exits are of different widths, the width of an exit also influences exit choice (*15*). Additional information, such as green flashing lights (*15*), lights above the exits (*10*) and, the availability of staff to assist evacuees (*16*), is also considered in the choice of exit. These factors act as a positive influence on the evacuation, resulting in less evacuation time. It is necessary to study group behavior of individuals during an evacuation. A pedestrian evacuation study that employed a simulation model from a single room revealed that arching, clogging, and irregular outflows were seen at an exit for a group of pedestrians (*17*). The tendency to follow other individuals while making an exit choice (*12*) also demonstrates the group behavior of individuals during emergency evacuation. A study of exit choice with a focus on human factors, such as social influence and proximity behavior during emergency evacuation, revealed that herding behavior was found to be present in individuals (*13*). The authors of the study found that group dynamics influence exit choice and should be considered during modeling of exit choice. Herding behavior of individuals has been explained as an individual's trust of others during selection of an exit or preferring not to be embarrassed by being the only one to select an exit (*14*). It was revealed that group behavior has a negative effect on evacuation time, as evacuation time was found to be faster if individuals egress independently rather than cooperating with others (*18*). The questionnaire method of an experimental study was employed by Chen et al. (*19*). In their study, children were asked different questions about exit choice in a classroom. The study revealed that position, congestion, group behavior and backtracking behavior play significant roles in the determination of evacuation route choice in a classroom.

Although these studies focused on exit choice during an emergency evacuation, they have not focused on the significant percentage of the individuals with disabilities (IWDs) population in the U.S. For example, the percentage of people with disabilities increased from 11.9% in 2010 to 12.6% in 2014 (*20*). The flow of pedestrians during evacuation has been found to be affected by mobility-impaired participants due to their slower walking speed, which affects exit choice (*21*). An experimental study on the evacuation behavior of the visually impaired revealed mixed behavior (*22*). The effects of a mobility stick, a guide dog, hand rails and walls on navigation toward an exit were studied. Even though many of the evacuation models reviewed were based on consideration by the models, no study founded on the models was based on IWDs (*23*). To address this limitation, an empirical study was performed at Utah State University (USU) to observe the evacuation behavior of IWDs in heterogeneous population contexts. The purpose of this study is to understand which parameters have effects on individuals' exit door choices in evacuation situations.

The experiments for this study were performed in the Agricultural Sciences (Ag Science) building at USU. This building possesses the necessary environmental conditions: various walkway configurations (resulting in directional changes and cross-flows), stairways, and queuing areas (exits), which meet standards promulgated by the International Building Code (IBC) (*24*) and the Americans with Disabilities Act (ADA) Standards for Accessible Design Accessible Means of Egress (*25*).

The building contains a large lecture hall, wide and narrow hallways, classrooms, offices, and study rooms in which participants were distributed for the experiments. There are four exits on the ground floor, which are accessible to all individuals. Three are main exits, and one is for emergencies. Figure 1 reveals the layout and various components of the floor. Doors D1, D2, and D3 have similar dimensions, and they are wider than the emergency door (D4). A total of 47 individuals participated in the experiments, including 12 individuals with various mobility-related disabilities, including physical (requiring use of a wheelchair) and sensory (visually impaired) disabilities. Participants were positioned throughout the building and when prompted by an alarm were asked to evacuate the building through an exit of their choice at their maximum comfortable speed. To minimize the learning bias, we conduct a tour for all participants before running various scenarios and showed them all parts and exit doors of the building, so that all participants were familiar with the building layout. A total of 16 evacuation experiments were conducted, with participant distribution, exit door availability, and evacuation strategy modified in each scenario. Participants' evacuation behavior was recorded using radiofrequency identification (RFID) tracking technology, supplemented by video tracking where desirable to verify the accuracy of the collected data (exits and congested areas). RFID is an automatic

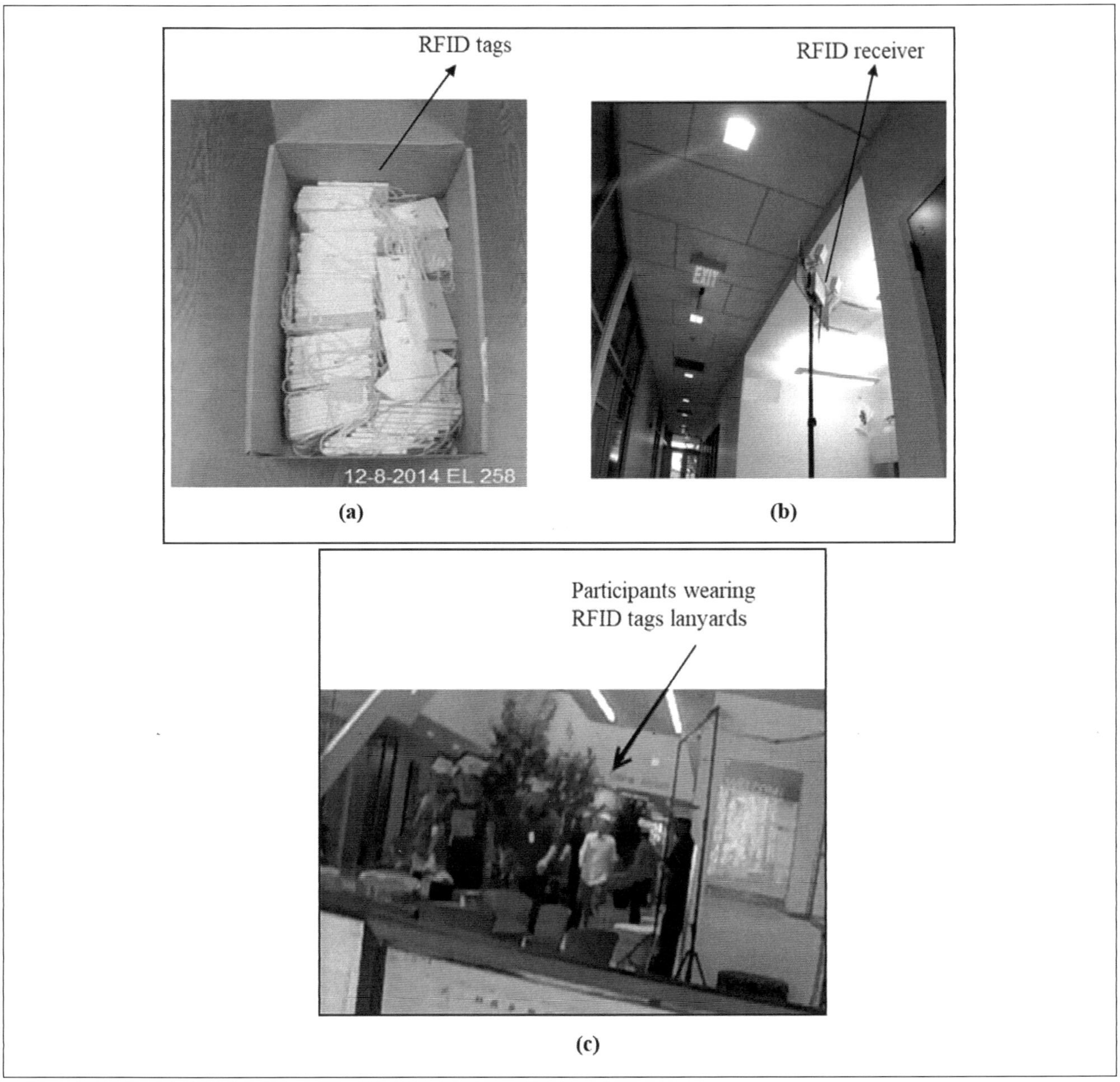

Figure 2. (*a*) RFID tags used for the experiment. (*b*) RFID receiver mounted in a stand to receive signals from the RFID tags. (*c*) Participants wear markers on their heads and RFID tags with lanyards.

identification system that consists of two components: a reader and tags (Figure 2). An RFID reader can recognize tags at high speed and send data within various distances. It is cost effective, small, and capable of storing more than enough information.

Methods

The RFID data were studied, and an analysis was then conducted on the exit choice model. The exit choice pattern was examined by developing a binary logit model for the exit choices of the scenarios in the classroom and the computer lab. For modeling purposes, from all the scenarios, there were nine similar scenarios, five in a classroom and four in a computer lab, in which two doors were available for evacuees to exit from the room. Exit choice analysis was conducted to observe how different parameters affect methods of exit door selection. Prior to model calibration, it is necessary to extract potential variables that may affect individuals' exit choices. These variables include distance to exit doors, pedestrian density at exit doors, and number of IWDs at exit doors.

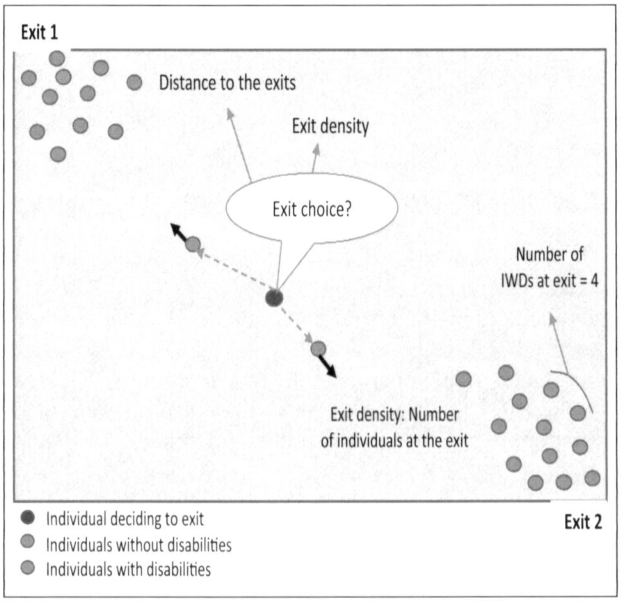

Figure 3. Exit choices in a room with two exits.

Distance to Exits

The distance of individuals from the exit doors can be calculated based on the start position and the exit door position. Based on the RFID data trajectory of the individuals, it was possible to determine the start position of individuals in the room. Distance was calculated based on the initial position of individuals and the exit coordinates of the doors. In general, distance calculation assumes the shortest distance between two points, which is given by a distance formula:

$$\text{Distance } (D) = \sqrt[2]{(x_2 - x_1)^2 + (y_2 - y_1)^2} \qquad (1)$$

where:

(x_1, y_1) = initial position (start coordinates) of the individual; and

(x_2, y_2) = position (coordinates) of the doors.

Total Density at Exits

The population density at each exit during exiting of the rooms appears to play a greater role than reflected in existing exit choice models. The density of individuals at the exit determines the number of individuals at the exit at a certain period of time. The higher the number of individuals at an exit, the lower the probability of choosing that exit. The density was calculated at the exits by counting the number of individuals leaving the exit within a certain time period. Traffic flow properties were taken as analogous to define the density of individuals leaving the exits. Density is defined as the number of people present at a specified section for a given time interval. Different time intervals were analyzed to observe the density pattern in conjunction with elapsed time, and various time intervals, such as 5 and 10 s, were also analyzed to generate the density graph. The density pattern was then analyzed to determine whether the pattern was smooth. Analysis revealed that the pattern was found to be smooth when the time interval was taken as 10 seconds.

Number of IWDs at Exits

Because some IWDs may walk slowly, which results in congestion at the exits, this may affect the exit choice behavior of other individuals with or without disabilities. The counting was done at 10-s intervals, as there were 12 IWDs out of 47 total individuals in the room. This variable was considered to determine whether the presence of IWDs at the exit doors had any effect on the exit choice analysis. Our experiment included 12 IWDs, out of which 11 had visual disabilities. The walking speed of IWDs has been found to be lower than that of individuals without disabilities (IWODs) in previous studies. Hence, this factor may lead to heightened congestion at the exit doors, which in turn may affect the exit choice of IWODs. This variable was analyzed by counting the number of IWDs at the exit doors for 10-s intervals, as we did in the calculation of the exit density to determine whether the presence of IWDs has any effect on the selection of a particular exit.

Exit Choice

Exit choice is a crucial factor that affects the quality of evacuation (Figure 3). In the experiment, different exit choices were available for evacuation under various scenarios. Discrete choice modeling was used to analyze and predict a decision maker's choice of one alternative from a finite number of alternatives. The variables to be considered for the exit choice are believed to have certain connections to IWDs. The primary focus of this study was IWDs and the heterogeneity of the population. Based on the data available regarding exit time information and movement of individuals at every 2-s interval, we identified the exit coordinates of every individual. A binary logit model was constructed for exit choice evaluation. The variables used for the logit model were distance to the exits, exit density, and number of IWDs at the exit (Figure 3).

Discrete choice models are the choice made by an individual based on different statistical parameters of the attributes that could relate to the alternatives and the decision maker. The approach used for the determination of the choice is based entirely on probability (*26*). The problem is defined in the modeling process, in which a choice must be made. The situation is then provided with a finite set of alternatives, which may be two or more. Every alternative is examined, and a set of attributes are defined, which will influence the selection of the alternative. Individuals will select based on utility function: the higher the utility for an alternative, the higher the probability of selecting that alternative. Let the utility that determines the outcome k be represented by U_i. The general representation of the utility function is then given by

$$U_i = V_i + e_i \tag{2}$$

where:

U_i = total utility of alternative i;

V_i = deterministic component of alternative i; and

e_i = stochastic component (nonmeasurable component) of alternative i.

The deterministic component of the utility function is the sum of different attributes that affect choice among the alternatives, multiplied by parameters that define the weight of the attributes based on the importance of the attributes.

$$V_i = \sum_{i=1}^{k} \beta_i X_i \tag{3}$$

where:

k = the number of attributes used for the utility function;

β_i = the parameter that defines the weight of the attribute; and

X_i = the attribute for selection.

Let A define the set of all of the alternatives for the discrete choice model. Another assumption in discrete choice modeling is the assumption of error distribution. The stochastic component (nonmeasurable component), which is often referred to as the error term, is assumed to be distributed as a Gumbel distribution. The probability that an individual will choose discrete alternative i among l different alternatives is given by

$$P_i = \frac{\exp(V_i)}{\sum_{l \in A} \exp(V_l)} \tag{4}$$

The binary choice model is a discrete choice model in which an individual must choose between only two alternatives. In our case, two doors were available as exit alternatives for any individual, and it was possible to use the binary logit model to model exit choice. A total of nine scenarios were noted to have two doors as alternatives for individuals to make exit choices. Other scenarios consisted of the lecture hall as the experiment area, which had only one exit door. As a result, these scenarios were excluded in the analysis, and the nine scenarios with two exit doors were chosen. Utility function was defined for the two alternatives with different attributes that may have an impact on selection of a particular exit.

Two different utility functions were created for IWDs and IWODs. The deterministic components of the exit doors were constructed using three different variables as described above. The deterministic components (V_i) for the two doors were as follows:

$$\mathrm{V_{door1}} = \mathrm{CONS1} + \mathrm{BETA1} * D_{d1} + \mathrm{BETA2} * K_{e1} + \mathrm{BETA3} * N_{d1} \tag{5}$$

$$\mathrm{V_{door2}} = \mathrm{BETA1} * D_{d2} + \mathrm{BETA2} * K_{e2} + \mathrm{BETA3} * N_{d2} \tag{6}$$

where:

D_{d1} and D_{d2} = distance of the individual's initial position from the doors (meters);

K_{e1} and K_{e2} = exit density at the two doors;

N_{d1} and N_{d2} = number of IWDs at doors at different time intervals;

BETA1, BETA2, and BETA3 = coefficients of the variable distances, exit density and number of IWDs, respectively, at exit doors; and

CONS1 = constant.

The probability of choosing an exit among two alternatives is given by (assuming error distribution is modeled as a Gumbel distribution):

$$\mathrm{P_{door1}} = \frac{\exp(\mathrm{V_{door1}})}{\exp(\mathrm{V_{door1}}) + \exp(\mathrm{V_{door2}})} \tag{7}$$

Model Calibration

IWODs

The data sets for nine different scenarios were combined, and the model was calibrated using "BIOGEME" (*27*) with 90% of the data sets and was also validated with the remaining 10% of the data sets. This BIOGEME software is designed to estimate discrete choice models, allowing for the estimation of parameters of the various discrete choice models, including logit, binary probit, nested logit, cross-nested logit, and so forth. The results from the BIOGEME were as follows.

Table 1 presents results of calibration, including coefficient value, coefficient standard error, t-statistic, and p-value for different attributes. In addition, model performance indices, including final log-likelihood, likelihood ratio test, and rho-squared, can be found in the table. The results indicate that exit density has less relevance in exit choice than other attributes. Presence of IWDs has the highest negative impact on individuals' exit choices. This finding suggests that IWODs believe that slow walking speed of IWDs may slow down their exit considerably.

IWDs

The data sets for the nine different scenarios were combined, the model calibrated with 90% of the data sets and validated

Table 1. Statistical Parameters from the Logit Model for IWODs

Attributes	Value	Standard Error	*t*-statistic	*p*-value
Distance	−0.23	0.0321	−7.14	0.00
Exit density	−0.08	0.0554	−1.33	0.18
Number of IWDs at door	−0.33	0.08	−4.1	0.00
Constant1 (CONS1)	−0.38	0.151	−2.55	0.01
Constant 2 (CONS2)	0	0	0	1
Summary of statistics				
Final log-likelihood			−157.321	
Likelihood ratio test			90.156	
Rho-squared			0.223	
Adjusted rho-squared			0.198	
Number of observations			292	
Number of individuals			292	

Table 2. Statistical Parameters from the Logit Model for IWDs

Attributes	Value	Standard Error	*t*-statistic	*p*-value
Distance	−0.21	0.0585	−3.65	0.00
Exit density	−0.16	0.131	−1.21	0.2
Number of IWDs at door	0.39	0.181	2.14	0.03
Constant1 (CONS1)	−0.26	0.322	−0.8	0.42
Constant 2 (CONS2)	0	0	0	1
Summary of statistics				
Final log-likelihood			−34.98	
Likelihood ratio test			28.457	
Rho-squared			0.29	
Adjusted rho-squared			0.19	
Number of observations			71	
Number of individuals			71	

with the remaining 10% of the data sets. The dependence of the exit choice was found to be similar to the exit choice model of IWODs, although a variance was found in the number of IWDs at the door (Table 2). The logit model for IWODs had a negative impact on exit choice, due to the number of IWDs at the door, compared with the logit model for IWDs had a positive impact on exit choice. This indicates that the presence of IWDs at the door renders it less likely that IWODs will choose that exit, although IWODs appear to express an opposite opinion in the evacuation surveys conducted. Conversely, IWDs appear to choose the same exits chosen by other IWDs.

The model was then validated with the remaining 10% of the data sets. The models were found to be valid with the validation data sets. The validation process is displayed in Table 3.

This experiment was undertaken to derive a binary logit model and predict the exit choice of study participants. The calibrated and validated model demonstrates that different factors are considered by pedestrians when selecting an exit. The results of the logit model of the exit choice for both IWDs and IWODs are presented below:

IWODs

$$\text{Door 1:} -0.38 - 0.23 * D_{d1} - 0.08 * K_{e1} - 0.33 * N_{d1} \quad (8)$$

$$\text{Door 2:} - 0.23 * D_{d2} - 0.08 * K_{e2} - 0.33 * N_{d2} \quad (9)$$

IWDs

$$\text{Door 1:} - 0.26 - 0.21 * D_{d1} - 0.16 * K_{e1} + 0.39 * N_{d1} \quad (10)$$

$$\text{Door 2:} - 0.21 * D_{d2} - 0.16 * K_{e2} + 0.39 * N_{d2} \quad (11)$$

The role of distance to the exit in the selection of an exit door is found to be inversely related suggesting that individuals tend to choose the exit that is nearest to their initial position. The negative sign on the coefficient of the distance variable in the logit model indicates that individuals are less likely to choose an exit if the distance to the exit from their initial position is greater.

Table 3. Validation of the Binary Logit Model for IWDs

ID	U_{D1}	U_{D2}	EXP (U_{D1})	EXP (U_{D2})	Probability		Actual Choice	Validation
					Door 1	Door 2		
41	−0.58	−1.68	0.56	0.19	0.75	0.25	1	Yes
36	−0.25	−2.47	0.78	0.08	0.90	0.10	1	Yes
18	−0.65	−1.54	0.52	0.22	0.71	0.29	1	Yes
22	−0.34	−1.74	0.71	0.17	0.80	0.20	1	Yes
33	0.23	−2.57	1.25	0.08	0.94	0.06	1	Yes
52	−0.61	−1.56	0.54	0.21	0.72	0.28	1	Yes
52	−1.87	−0.17	0.15	0.85	0.15	0.85	2	Yes
23	−1.69	−0.38	0.18	0.68	0.21	0.79	2	Yes
72	−1.87	−0.17	0.15	0.85	0.15	0.85	2	Yes
18	−1.06	−0.05	0.35	0.95	0.27	0.73	2	Yes

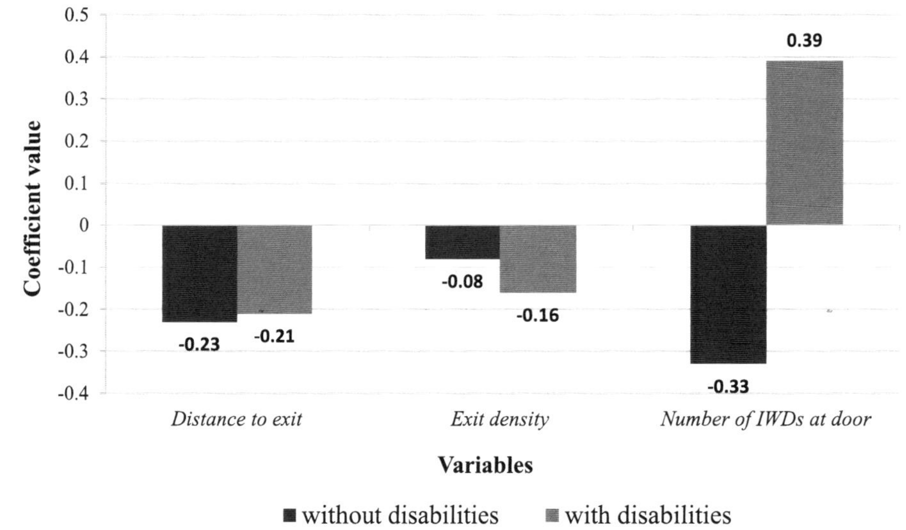

Figure 4. Comparison of the coefficients of the variables in the logit model for IWDs and IWODs.

Exit density plays an inverse role in the selection of an exit door, and also had a negative relationship with the selection of an exit door. The number of IWDs at the exit assumes a totally different role in the selection of an exit door in the case of IWDs and IWODs. The presence of IWDs at an exit door has an inverse relationship with the selection of that exit for IWODs. The presence of IWDs at an exit door has a positive relationship with the selection of that particular exit for IWDs. Figure 4 shows the comparison of the coefficients of the variables in the logit models for IWDs and IWODs. The figure demonstrates that there is a marked difference in dependence on the exit choice based on the number of IWDs at the exit door between IWDs and IWODs. In terms of variable distance and exit density, the dependence of the model on the variables is seen to have a similar trend, that is, it is inversely related (Figure 4).

The results from the model suggest that IWDs and IWODs differ in their exit choice based on the presence of IWDs at the exit doors. IWODs may choose an exit where there are no or fewer (compared to other exits) IWDs at the exit door, and likely they make this decision because they believe the slow walking speed of IWDs may impede their exit from the room (*28*). Based upon the results, it was determined that IWDs trust other IWDs, which compels them to choose the same exit as other IWDs.

Although the ADA accessibility guidelines (*29*), IBC (*24*), and ADA Standards for Accessible Design (*25*) identify signage requirements at the exit doors for egress, these signs may not be as useful to individuals with visual disabilities as they could be, because individuals with visual disabilities follow other individuals with visual disabilities during an emergency exit. The signage requirements, as provided by the applicable codes (*24, 25, 29*), require the exit doors to feature the visual signs. IBC (*24*) mandates that the signage requirements be illuminated and raised chartered and braille signage must also be provided. The use of braille signage is generally found to be on the decline, and this is attributable to the introduction of many user-friendly devices for individuals with visual disabilities. These signs offer greater and easier access to exits to individuals with mobility disabilities, not individuals with visual disabilities, despite braille signage being provided. From this study, we found that individuals with visual disabilities do not choose an emergency exit based exclusively on the distance to the exit and exit density, instead also considering their trust of other individuals with visual disabilities. Audible indicators provided in an emergency egress in conjunction with visual signs may prove more helpful to individuals with visual disabilities. Assistance from trained personnel during an evacuation may also serve to provide additional aid to individuals with visual disabilities. Note that this study focused on IWODs and individuals with visual impairments. Individuals with other types of disabilities may have different exit choice behaviors due to their characteristics. Therefore, further empirical studies are needed to examine their behaviors.

References

1. Huang, H. J., and R. Y. Guo. Static Floor Field and Exit Choice for Pedestrian Evacuation in Rooms with Internal Obstacles and Multiple Exits. *Physical Review E*, Vol. 78, No. 2, 2008, 021131.
2. Helbing, D., I. Farkas, and T. Vicsek. Simulating Dynamical Features of Escape Panic. *Nature*, Vol. 407, No. 6803, 2000, pp. 487–490.
3. Song, W., X. Xu, B. Wang, and S. Ni. Simulation of Evacuation Processes using a Multi-Grid Model for Pedestrian Dynamics. *Physica A: Statistical Mechanics and its Applications*, Vol. 363, No. 2, 2006, pp. 492–500.
4. Varas, A., M. D. Cornejo, D. Mainemer, B. Toledo, J. Rogan, V. Munoz, and J. A. Valdivia. Cellular Automaton Model for Evacuation Process with Obstacles. *Physica A: Statistical Mechanics and its Applications*, Vol. 382, No. 2, 2007, pp. 631–642.
5. Helbing, D., M. Isobe, T. Nagatani, and K. Takimoto. Lattice Gas Simulation of Experimentally Studied Evacuation Dynamics. *Physical Review E*, Vol. 67, No. 6, 2003, 067101.
6. Isobe, M., D. Helbing, and T. Nagatani. Experiment, Theory, and Simulation of the Evacuation of a Room without Visibility. *Physical Review E*, Vol. 69, No. 6, 2004, 066132.
7. Shi, X., Z. Ye, N. Shiwakoti, and Z. Li. A Review of Experimental Studies on Complex Pedestrian Movement Behaviors. *CICTP*, 2015, pp. 1081–1096.
8. Proulx, G. Occupant Behaviour and Evacuation. *Proceedings of the 9th International Fire Protection Symposium*, 2001, pp. 219–232.
9. Ren-Yong, G., and H. Hai-Jun. Logit-Based Exit Choice Model of Evacuation in Rooms with Internal Obstacles and Multiple Exits. *Chinese Physics B*, Vol. 19, No. 3, 2010, 030501.
10. Lovreglio, R., A. Fonzone, and L. dell'Olio. A Mixed Logit Model for Predicting Exit Choice during Building Evacuations. *Transportation Research Part A: Policy and Practice*, Vol. 92, 2016, pp. 59–75.
11. Liu, S., L. Yang, T. Fang, and J. Li. Evacuation from a Classroom Considering the Occupant Density Around Exits. *Physica A: Statistical Mechanics and its Applications*, Vol. 388, No. 9, 2009, pp. 1921–1928.
12. Haghani, M., O. Ejtemai, M. Sarvi, A. Sobhani, M. Burd, and K. Aghabayk. Random Utility Models of Pedestrian Crowd Exit Selection based on SP-off-RP Experiments. *Transportation Research Procedia*, Vol. 2, 2014, pp. 524–532.
13. Lovreglio, R., D. Borri, L. dell'Olio, and A. Ibeas. A Discrete Choice Model based on Random Utilities for Exit Choice in Emergency Evacuations. *Safety Science*, Vol. 62, 2014, pp. 418-426.
14. Lovreglio, R., A. Fonzone, L. dell'Olio, and D. Borri. A Study of Herding Behaviour in Exit Choice during Emergencies based on Random Utility Theory. *Safety Science*, Vol. 82, 2016, pp. 421–431.
15. Nilsson, D., H. Frantzich, and W. L. Saunders. Influencing Exit Choice in the Event of a Fire Evacuation. *Fire Safety Science*, Vol. 9, 2008, pp. 341–352.
16. Shields, T. J., and K. E. Boyce. A Study of Evacuation from Large Retail Stores. *Fire Safety Journal*, Vol. 35, No. 1, 2000, pp. 25–49.
17. Fu, L., W. Song, and S. Lo. A Fuzzy-Theory-based Behavioral Model for Studying Pedestrian Evacuation from a Single-exit Room. *Physics Letters A*, Vol. 380, No. 34, 2016, pp. 2619–2627.
18. Heliövaara, S., J. M. Kuusinen, T. Rinne, T. Korhonen, and H. Ehtamo. Pedestrian Behavior and Exit Selection in Evacuation of a Corridor—An Experimental Study. *Safety Science*, Vol. 50, No. 2, 2012, pp. 221–227.
19. Chen, L., T. Tang, H. Huang, and Z. Song. Elementary Students' Evacuation Route Choice in a Classroom: A Questionnaire-based Method. *Physica A*, Vol. 492, 2018, pp. 1066-1074.
20. Kraus, L. *Disability Statistics Annual Report*. University of New Hampshire, 2015.
21. Sørensen, J. G., and A. S. Dederichs. Evacuation from a Complex Structure—The Effect of Neglecting Heterogeneous Populations. *Transportation Research Procedia*, Vol. 2, 2014, pp. 792–800.
22. Sørensen, J. G., and A. S. Dederichs. Evacuation Characteristics of Visually Impaired People—A Qualitative and Quantitative Study. *Fire and Materials*, Vol. 39, No. 4, 2015, pp. 385–395.
23. Kuligowski, E. D., R. D. Peacock, and B. L. Hoskins. *A Review of Building Evacuation Models*. U.S. Department of Commerce, National Institute of Standards and Technology, 2005.
24. International Code Council. *International Building Code*. Washington, D.C.: ICC, 2012.
25. *ADA Standards for Accessible Design*. Guidance on the 2010 ADA Standards for Accessible Design. U.S. Department of Justice, 2010.
26. McFadden, D, and K. Train. Mixed MNL Models for Discrete Response. *Journal of Applied Econometrics*, 2000, pp. 447-470.
27. Bierlaire, M. BIOGEME: A Free Package for the Estimation of Discrete Choice Models. Swiss Transport Research Conference, 2003.
28. California Employment Law. Emergency Evacuation Must Comply with ADA. *Safety for All: Ensure Evacuation Plans Comply with the ADA*. California Employment Law, 2011, p. 4.
29. *ADAAG (Americans with Disabilities Act Accessibility Guidelines)*. ADA Accessibility Guidelines for Buildings and Facilities. U.S. Department of Justice, Washington, D.C., 2006.

The Standing Committee on Critical Transportation Infrastructure Protection (ABR30) peer-reviewed this paper (18-05439).

Article

TRR
JOURNAL OF THE TRANSPORTATION RESEARCH BOARD

Transportation Research Record
2018, Vol. 2672(1) 30–42

DOI: 10.1177/0361198118786622
journals.sagepub.com/home/trr
SAGE

Learning to Build Resilience into Transportation Systems

Karl Kim[1], Oceana Francis[2,3], and Eric Yamashita[1]

Abstract

There is increased recognition of the value of resilience thinking yet few studies have examined key experiences and lessons for the transportation industry. This paper presents eight case studies on natural and man-made disasters focusing on transportation impacts, lessons learned, and efforts to increase resilience. The cases are the Atlanta snowstorm, I-35W bridge collapse, Boston Marathon bombing, Puna lava highway closing, superstorm Sandy, Hampton Roads sea level rise study, Fukushima disaster, and the Oregon Resilience Plan. The cases are used to inform transportation agencies to better respond to and recover from disasters, as well as improve the resilience of transportation systems. In addition to integrating resilience into the management of transportation assets, new technologies and tools relevant to transportation resilience are described.

Resilience is a *national imperative* (*1*). In this study, lessons for transportation officials interested in building resilience are identified. Transportation leaders face daunting challenges with natural, man-made and cascading disasters. Without functioning transportation systems, emergency response, search and rescue, relief and recovery, health, safety, and security can be imperiled. Roads and transportation networks are needed to restore critical infrastructure as well as key societal functions. Transportation is so important that it has been designated by the federal government as Emergency Support Function (ESF) #1. ESFs identify capabilities to assist communities to return to normalcy after an incident (*2*). The research was motivated by requirements of the National Disaster Preparedness Training Center (ndptc.hawaii.edu), funded by the U.S. Department of Homeland Security and the Federal Emergency Management Agency to develop training courses to increase the preparedness of state and local governments. Research conducted by the authors as was part of an NCHRP project to develop a primer on resilience for transportation officials. AASHTO, moreover, is implementing a "roadmap for resilience" and working to the increase resilience of transportation systems to extreme weather and climate change (*3*, *4*).

Like other complex systems, transportation requires communications and coordination across different time and spatial scales for operations, management, repair and maintenance, and construction (*5*, *6*). With diverse stakeholders, concerns include safety and security, congestion management, efficient operations, as well as economic, environmental, and social considerations.

Disasters test the capabilities of transportation agencies. Disasters expose underlying weaknesses and deficiencies. How robust is the system of detection and alert for disruptions? How quickly is information communicated so that resources can be deployed to minimize system interruptions? What is the time to full or partial recovery? What outside resources, including mutual aid and federal government assistance, can help to hasten response and recovery? Will rebuilding lead to mitigation and adaptation so that re-built systems are stronger and safer than before? Were risks, hazards, and threats anticipated, planned for, and managed prior to the event (*7*, *8*) Foresight is essential to effective leadership and management and involves understanding the threat landscape. Earthquakes and volcanoes arise from geological conditions. Tornadoes, hurricanes, snowstorms, rains, and flooding result from atmospheric, weather, and climatic conditions. Other disasters occur because of human actions or inactions. Drought, sea level rise, or climate change occurs more gradually than earthquakes or terroristic acts which can happen with little or no warning. Others such as the Fukushima disaster can involve multiple, cascading hazards: Fukushima is particularly challenging because it is an ongoing disaster involving long-term cleanup and remediation of contaminated communities.

Table 1 provides a summary of the eight case studies. The cases were categorized by: type of disaster, disruption type, transportation impacts, key lessons, and resilience takeaways. Of the eight, four (Atlanta snowstorm, Puna lava highway closing, superstorm Sandy, and Hampton Roads sea

[1]Department of Urban and Regional Planning, University of Hawaii, Honolulu, HI
[2]Department of Civil and Environmental Engineering, University of Hawaii, Honolulu, HI
[3]Hawaii Sea Grant College Program, University of Hawaii, Honolulu, HI

Corresponding Author:
Address correspondence to Karl Kim: karlk@hawaii.edu

Table 1. Attributes of Disaster Cases

Name of case	Type of disaster	Date	Location	Duration	Fatalities	Disruption	Losses (million)	Transportation impacts	Time to recovery (days)	Key lessons	Resilience takeaways
Atlanta snow storm	N	1/28/2014	Atlanta, GA	1 day	13	TD, BD, GD, SD, F, ID, TI	2.8	GL, TA, SP, C	3	S-P, EM, CC	SE
I-35W bridge collapse	I	8/1/2007	Minneapolis, MN	1 year	13	TD, F, ID	17 (2007), 43 (2008)	TS	21+	NI, PM	MA, PTC
Boston Marathon bombing	T	4/15/2013	Boston, MA	5 days	3	TD, F, ID	333	TS, E, SP, SH	Weeks	NI, CC, EM, PM	UC, EP
Puna lava highway closing	N	2014-2015	Puna, Hawaii	1 year	0	TD, BD, SD, DS	15	RC, LC, AR, E	Months	S-P, EM, CC	JIC, CP, PE, CAC, RM
Superstorm Sandy	N	8/29/2012	NE U.S.	8 days	72	TD, BD, TI, GD, SD, DS, F, ID	50,000	RD, RC, GL, E, SH	1095+	EM, CC, NI	SL, V
Hampton Roads sea level rise study	N	Reoccurring	Virginia	Long duration	Indeterminate	TD, DS	12,000–87,000	RC, RD	Decades	S-P	LD, TMS
Fukushima disaster	NT	3/11/2011	Japan	Long duration	15,894	TD, B, TI, GD, SD, DS, F, ID	360,000	RC, RD, TS, TD, LC, E	1825+	PM	PTC
Oregon Resilience Plan	NT	N/A	Oregon	N/A	N/A	TD, BD, TI, GD, SD, DS, F, ID	Indeterminate	RC, RD, TS, SP, C, LC, E, SH	N/A	S-P, EM, CC, NI, PM	MPR

Note: N = natural disaster; NT = NATECH; T = terrorism; I = infrastructure failure; TD = transportation disruption; BD = business disruption; TI = technology critical infrastructure disruption; GD = government disruption; SD = school closure; DS = destruction of homes, buildings and infrastructure; F = fatalities; ID = injuries; RC = road closures; RD = road damage; TS = transportation stopped/suspended; TA = traffic accidents; GR = gridlock; SP = stranded people; C = disruption of transportation for children; LC = lifeline cutoff; E = evacuation; AR = alternative routes; TD = transportation modes damaged; SH = sheltering; S-P = science/policy; EM = emergency management; CC = communication/coordination; NI = national incident management system/incident command systems; PM = change in policy or mitigation; SE = snowball effect; MA = mutual aid; PTC = preparedness, training, coordination, and exercise; UC = establishing unified command; EP = need for evacuation planning; JIC = establish joint information center; CP = contingency planning; PE = public engagement; CAC = cross-agency collaboration; RM = risk management; SL = need for strong leadership; V = needs of vulnerable populations; LD = local data importance; TMS = importance of tools, methods, and engagement strategies; MPR = model for resilience planning.
Source: See References (*9–32*).

level rise) involved natural hazards. The I-35W bridge collapse was an infrastructure failure. Boston Marathon bombing was terrorism, and two (Fukushima disaster and the Oregon Resilience Plan) were disasters combining natural hazards and technological failures.

Categories of disruption are transportation, business, technology/critical infrastructure, government, school closure, destruction of structures, fatalities, and injuries. The Atlanta snow storm case involved six types of disruptions: transportation, business, government, schools, fatality, injury, and technology/critical infrastructure. Both the I-35W bridge collapse and the Boston Marathon bombing resulted in transportation disruptions as well as fatalities and injuries. The Puna lava highway case had potential disruptions to transportation, business, school, and destruction of structures. Superstorm Sandy, the Fukushima disaster, and the Oregon Resilience Plan have all eight types of disruptions. The Hampton Roads sea level rise case potentially involves disruptions to transportation and destruction of homes/buildings/infrastructure.

There are 12 types of transportation impacts: road closures; road damage, stoppage or suspension; accidents; gridlock; stranding of people; disruption in the transport of children; lifeline cutoff; evacuation; need for alternative routes; damage to different modalities; and need for sheltering.

Lessons from these cases include the need to better integrate science and policy, strengthen emergency management, enhance communications/coordination, implement national incident management system (NIMS)/incident command systems (ICS), and foster changes in policy to strengthen mitigation efforts.

Takeaways include the need for strong leadership and the importance of a unified command structure especially during the response and relief phases of a disaster. There also could be snowball effects and the need for not just updated situational awareness, but strong contingency planning. Mutual aid from other areas can provide immediate assistance and critical support when demands for service overwhelm existing capacity. This necessitates cross-agency collaboration and strong communications and coordination across agencies and with the public. This is especially needed with evacuation and sheltering planning. The establishment of a joint information center (JIC) is useful for complex disasters involving multiple agencies. The JIC can help to engage the public. Special effort is needed to identify the requirements of at-risk vulnerable populations as well as improve risk assessment and management using local data. Efforts to develop new tools, methods and engagement strategies can support models of resilience planning which can be applied across different sectors as well as across communities.

Following this introduction, the cases are presented. The first part describes what happened while the second focuses on lessons for the transportation sector. Some cases are large, complicated events with many lessons for transportation officials and implications for emergency management, ICS and NIMS, communications and coordination, integration of science and policy, and mitigation, adaptation, and resilience. Conclusions highlight lessons across hazards and jurisdictions including the potential for new technologies to support transportation resilience.

Case Studies

Atlanta Snow Storm

What started as a typical southeastern winter storm became a major disaster for the Atlanta region in January 2014. It began with a strong cold front moving from the central part of the nation which mixed with warmer, moist air from the coasts producing a mid-latitude cyclone over eastern Texas which pushed east increasing rainfall and snow over the Atlanta region. On Sunday, January 24, the National Weather Service issued a winter storm watch for Georgia. By Monday, the National Weather Service warned of significant snow on Tuesday. On Tuesday morning, the watches were upgraded to warnings. At 10 a.m. the Governor issued an executive order allowing state agencies to release employees early. The accumulated snow was only one to three inches, although there was freezing rain to the south (Figure 1).

By noon, major roads were congested with people leaving early and parents traveling to pick up children at schools. Efforts to plow and salt the roadways were hampered by heavy congestion. Snow and ice led to numerous accidents, stranded motorists, and massive delay in a region which normally has heavy congestion during peak times. School buses were also stuck, forcing students and teachers to shelter-in-place at schools and other locations. At 5:15 p.m. the Governor declared a state of emergency (*9–11*).

This case demonstrates how a seemingly insignificant event can snowball into a larger emergency. The disruption of the transportation of school children can quickly turn into a crisis. It heightens the extent to which users are central to the perception of threats and hazards. Transportation officials also need to respond to rapidly changing weather conditions. It highlights the importance of communications and coordination not just between the National Weather Service, local forecasters, and state and local governments, but with schools, businesses, and transportation service providers. Varying conditions across the region led to confusion and conflict over decision-making regarding school closings and transportation assets management. There was limited purchase and stockpiling of winter equipment because Atlanta rarely experiences snow and ice. Following this event, a weather alert system was created so that the Georgia Emergency Management Agency can send messages to authorities and the public similar to the Amber Alert System. (*12*, *13*). Such improvements are facilitated by improvements in technology and communications systems including smartphones, social media, and other information technologies. A weather advisory group was also created to improve

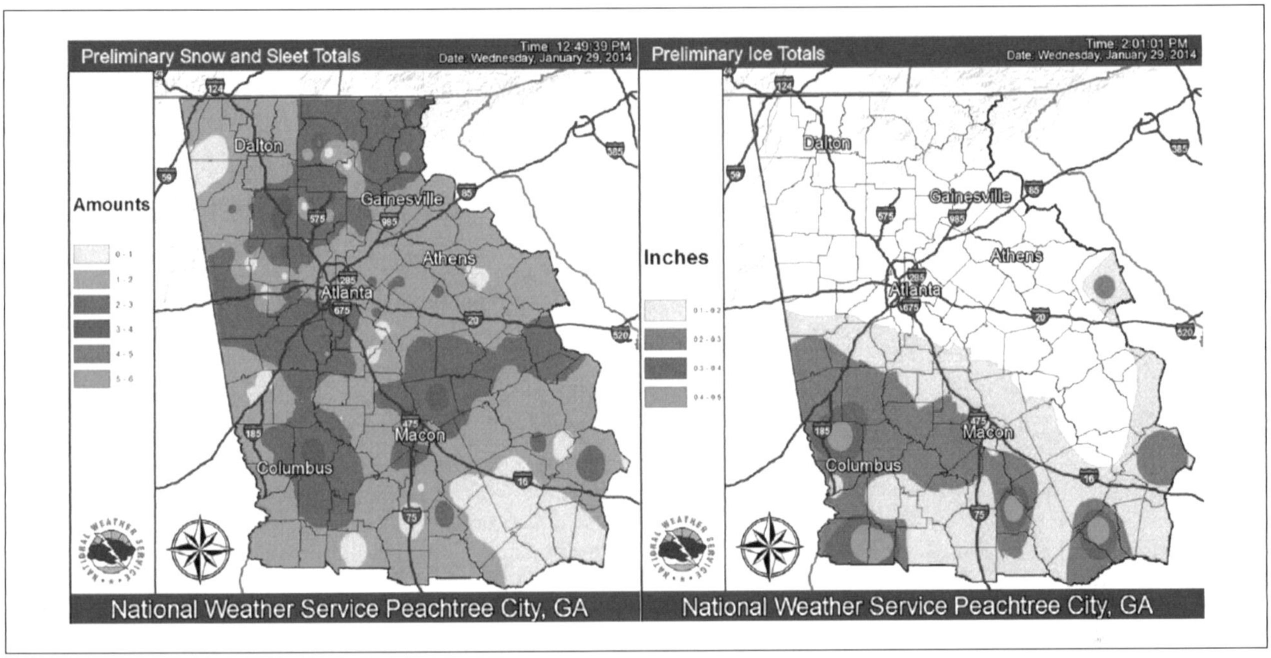

Figure 1. 2014 Atlanta Snow Storm Map.
Note: Units are in inches.
Source: National Oceanic and Atmospheric Administration, National Weather Service, Peachtree City, Georgia.

communications between meteorologists and state and local officials, enhancing coordination between scientists, policy-makers, and decision-makers (*14*).

I-35W Bridge Collapse

The Interstate 35 West Bridge over the Mississippi River connects Minneapolis to St. Paul, Minnesota. In 2007, the bridge experienced a structural failure, causing 1,000 feet of the eight-lane bridge to plummet to the river below (Figure 2). The collapse occurred on Wednesday, August 1, 2007, during the evening rush hour at 6:07 p.m. There were 13 fatalities and over 140 injuries. The National Transportation Safety Board (NTSB) determined the cause was a design defect of the load capacity of the bridge's gusset plates (*15*). The Minnesota Department of Transportation did not have an adequate bridge inspection program and suffered from poor communications (*15*, *16*, *33*). Maintenance crews had also stockpiled building materials and supplies on the bridge deck, further increasing the load. This was a complicated disaster because of the heavy traffic over the bridge and the difficulties associated with the coordination of search and rescue. The bridge connected two local jurisdictions was managed by the state department of transportation and funded by the federal government. As an interstate, it contained a mix of different vehicles which complicated search and rescue efforts. The collapse affected pedestrians, bicyclists, and rail and river traffic as well. Traffic needed to be re-routed after the collapse and during the reconstruction period (*17*, *18*, *34*, *35*).

Mutual aid in law enforcement, fire, and emergency medical services from surrounding localities poured in. Because many had received ICS training, the first responders and emergency managers from different agencies knew their roles and had practiced for this type of disaster. The disaster points to the importance of preparation and training and coordination during complex emergencies. In addition to the search and rescue operations, there were also problems with the routing of emergency vehicles and the transport of victims and patients needing hospitalization. Following NIMS,

Figure 2. I-35W bridge collapse.
Source: CBS Minnesota. http://minnesota.cbslocal.com/tag/i-35w-bridge-collapse/

a command center was created in proximity to the site and an emergency operations center was established at 6:20 p.m. in the Minneapolis City Hall. While the disaster led to improvements in the oversight of bridge design as well as in the inspection and maintenance of bridges in Minnesota, this case shows how technological failures, maintenance practices, design flaws, and poor communications between bridge operators, construction companies, and agencies providing oversight can lead to disasters. Improvements in scanning technologies and in the use of devices such as drones can also be used to support more robust inspection programs (*22*).

Boston Marathon Bombing

On April 15, 2013, during the Boston Marathon, near the finish line, two homemade bombs exploded, killing three people and injuring several hundred others (Figure 3). The suspects were two Chechen-American brothers who later went on to kill a policeman and kidnap another man. One of the brothers was killed during a shootout with police when the other brother ran over him while escaping in a stolen vehicle. A massive manhunt was launched on April 19 involving thousands of law enforcement officers. The public transit system, as well as Amtrak service, was shut down. Many businesses, schools, and other facilities closed as a city-wide shelter-in-place order was implemented. Social media played a critical role in pushing information from agencies out to the public, and in collecting information about the suspects as well as understanding the impacts of the disaster on the transportation system. While the suspects were eventually captured, there were also significant impacts associated with the shutdown of transportation services which in turn affected businesses and residents throughout the metropolitan region (*9*, *19*).

Lessons regarding response and recovery during the marathon also emerged. Almost immediately after the explosions, a unified command (UC) was established at the scene with senior law enforcement and emergency managers coming together to coordinate the response. The UC shifted into a UC Center based in the Westin Hotel to provide periodic updates and briefings to the media and public. There were significant coordination issues with many first responders and medical personnel on site and from surrounding jurisdictions providing aid and assistance. Many volunteers and personnel on staff from the marathon's Alpha Medical Tent, as well as bystanders, rendered aid and care for the injured (*20*, *37–39*).

Among the significant problems was the evacuation of runners and spectators after the marathon was shut down. Approximately 5,000 runners still were on the course when the race was stopped. Temporary shelters for stranded

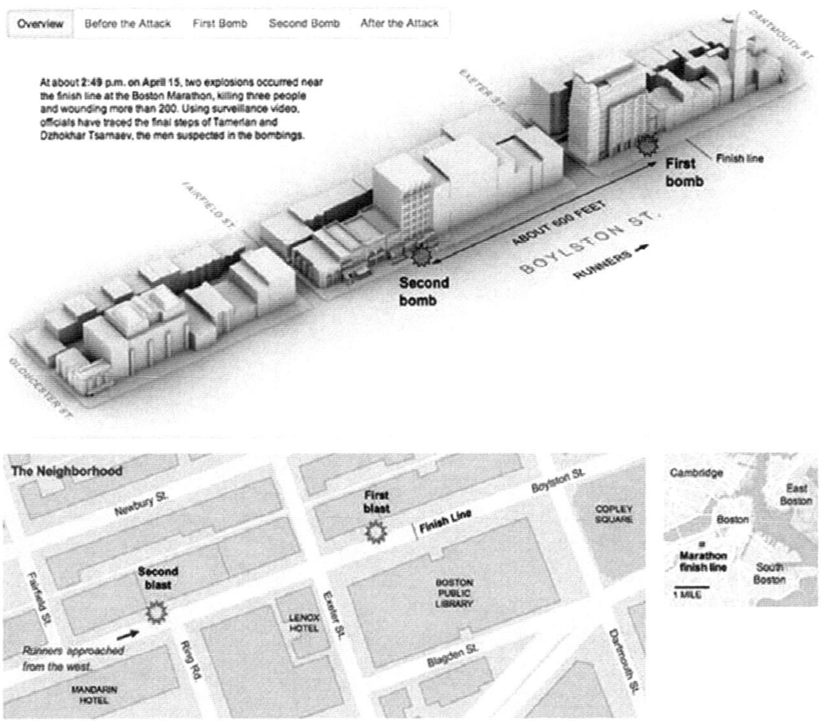

Figure 3. Boston Marathon finish line and bomb locations.
Source: http://elearningexamples.com/caught-in-the-blasts-at-the-boston-marathon/.

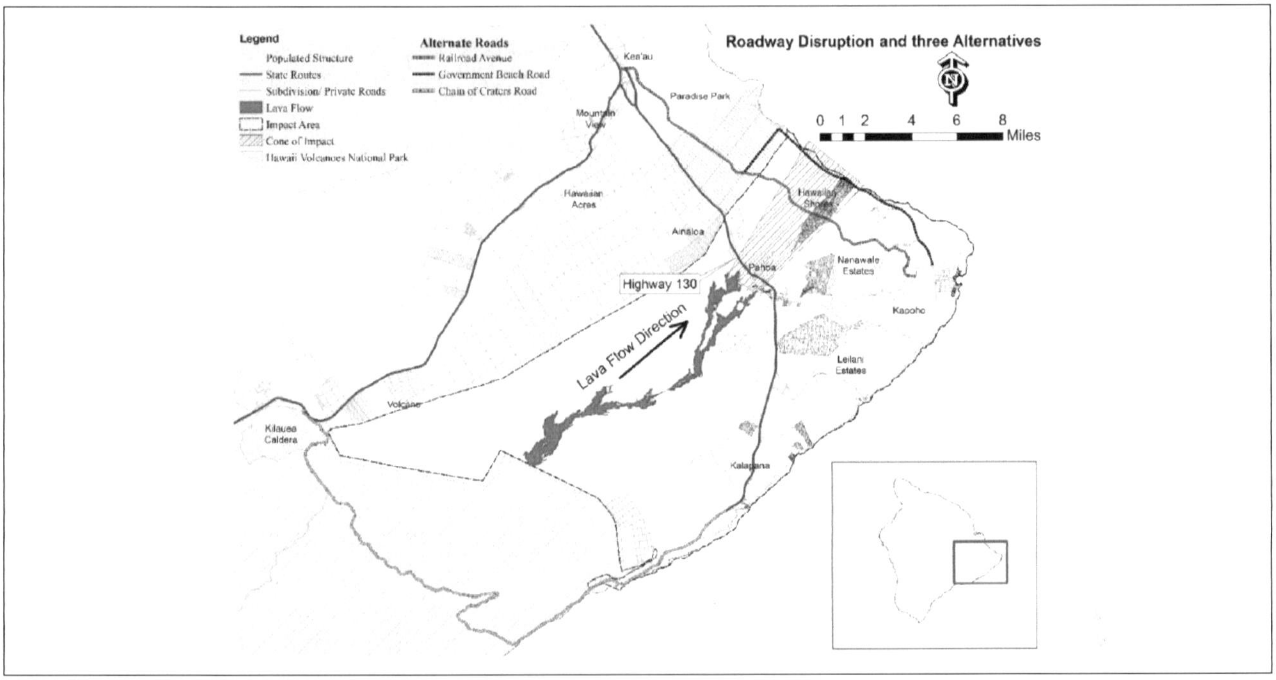

Figure 4. Puna lava flow and alternative transportation routes.

runners were needed until they could be transported back to Boston. While conceptual plans had been discussed, no actual plans had been developed so no shelters were identified, nor staffed, nor were buses available. The incident pointed to the need for evacuation planning for large-scale public events involving temporary shelters, staging of transportation assets, designation of operational protocol, communications, and the assignment of roles and responsibilities (*19*). Contingency planning requires the active involvement of public and private transportation services and needs to be considered part of the emergency response and recovery planning.

Puna Lava Highway Closing

A recent lava flow in Hawaii threatened to close a major state highway. Erupting since 1983, Kilauea volcano is one of the most active in the world. On June 24, 2014, a new lava breakout began on the northwest flank and began moving towards the Puna community. Using drone technology, the University of Hawaii has been monitoring and mapping lava flows in the region (*21*). In September, the governor issued a disaster proclamation. By November, the flow had destroyed a home and covered portions of a cemetery. Businesses and residents began relocating as the lava moved closer to Highway 130. The flow threatened to close the highway affecting approximately 11,000 residents in the area. Three alternative routes were identified including the opening of an abandoned roadway (Railroad Avenue), improving an existing coastal road (Government Beach Road) and reopening a much longer route, (Chain of Craters Road) through the Hawaii Volcanoes National Park (Figure 4). The impacts of these alternatives were estimated and used to support decision-making and contingency planning efforts. The additional travel burden for each of the alternatives as well as the impacts on increased vehicle miles traveled was calculated (*22*).

There was coordination between local, state, and federal emergency management agencies as well as other government agencies including the USGS Hawaiian Volcano Observatory (HVO), transportation agencies, utilities, schools, police and fire and emergency medical services, and other services including solid waste management. Assets such as emergency vehicles, school facilities, and other needed assets were pre-positioned on either side of the expected flow path. After the federal disaster declaration was issued, a JIC was established and numerous community meetings were held to share information and devise appropriate responses and solutions. The University of Hawaii and the National Disaster Preparedness Training Center developed and distributed a just-in-time video along with social media to provide information about the flow and its potential impacts, threats, and hazards. A robust partnership between scientists, emergency managers, government agencies, civic leaders, and members of the community emerged (*22*, *23*).

Although the lava stopped short of crossing the highway, this case provides an example of contingency planning, public engagement, and cross-agency collaboration to manage a slow-moving disaster. In particular, it illustrates the need to recognize uncertainty and how best to manage uncertainty and risk. The lessons may apply to other hazards and communities facing gradual or slow-moving threats from climate change, sea level rise, ground subsidence, infrastructure decay, or other "creeping" disasters.

Superstorm Sandy

Hurricane Sandy began as a low-pressure system south of Jamaica and strengthened into a tropical storm and then became a Category 3 hurricane prior to making landfall in Cuba. As the storm moved north along the southeast coast of the United States, it transitioned into a hybrid, mid-latitude cyclone. A ridge of high pressure to the north and northeast pushed the storm into the heavily urbanized and densely populated areas in the Mid-Atlantic and northeastern regions of the country. This hybrid storm exhibited characteristics of both tropical and mid-latitude cyclones. It brought heavy snow to the Appalachians, rains to the Mid-Atlantic regions, and strong winds and storm surge to New England. In addition to damage from fallen trees, downed power lines, and strong winds, the storm also led to coastal flooding and inundation of populated areas. There were 72 fatalities, many injuries, and over $75 billion in damages. It was the second most costly hurricane after Hurricane Katrina (*9*, *24*, *40*).

Although there have been improvements in forecasting of hurricane tracks, there were differences between the European Center for Medium-Range Weather Forecasts (ECMWF) and the National Weather Service and National Hurricane Center. The models eventually converged but the European forecast came four days earlier than the U.S. forecasts (*9*, *24*, *40*). In Figure 5, the white track is the actual

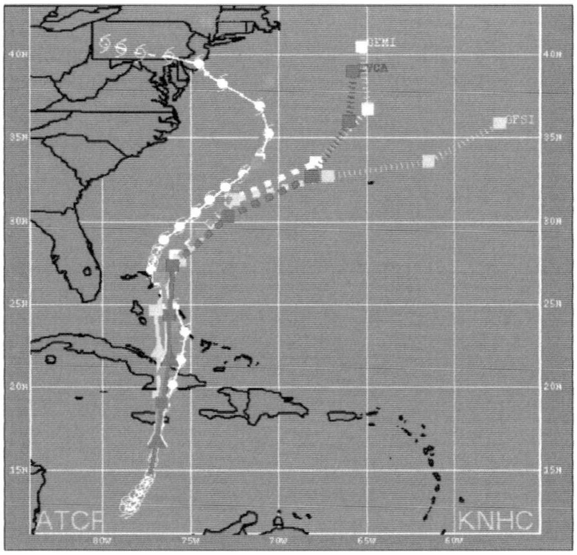

Figure 5. Forecast tracks for Superstorm Sandy.
Source: National Oceanic and Atmospheric Administration, National Hurricane Center.

track of the storm. The European model (in coral) was the first to predict the storm would make a sharp left turn into the U.S. Northeast. The U.S. model (in cyan) and model ensemble (in yellow) were slower to make this prediction.

The U.S. model runs at a coarser spatial resolution (25 kilometers) as opposed to the European model (15 kilometers). The U.S. model takes snapshots of data and builds a global picture of initial conditions, while the European model takes continuous observations and runs its model with real data to produces the forecast. Computationally more advanced, the European model predicted the track earlier than the U.S. model. The official track is in white, the ECMWF global model interpolated is in coral, the Global Forecast System Interpolated (GFSI) is in cyan, the previous cycle GFS ensemble mean interpolated (GEMI) is in yellow, and the Track Variable Consensus Aids (TVCA) model is in red.

As a large storm headed for heavily populated areas, there were concerns regarding evacuation and sheltering. There were cascading effects of power outages, disruption of municipal services, sewage spills, and other impacts beyond wind, rain, flooding, and debris hazards. Because the weather effects varied by location, there were miscommunications about the location of hazards. There were shortages caused by lack of fuel and the inability to effectively restock and manage emergency supplies. The Federal Emergency Management Agency (FEMA) used military aircraft to deliver equipment for clearing roads and restoring critical infrastructure (*25–29*, *41*).

Among the most challenging problems was the evacuation of areas at risk of flooding. Better mapping and modeling of storm surge and flooding, and the impacts on roadways, tunnels, transit and other infrastructure, was needed. Since then, there have been improvements in modeling both storm track and coastal storm surge (*30*, *31*). While Sandy resulted in improvements in the protection of critical transportation infrastructure, it also serves to illustrate the need for strong leadership, backed by sound forecasting and effective communications. Of particular concern are vulnerable, at-risk populations who may not receive warnings nor be able to evacuate. Sandy demonstrated that the transportation needs of the poor, elderly, immigrant and other vulnerable populations need to be central to resilience efforts (*25–29*, *41*). In addition to weather forecasting, it is also necessary to address vulnerabilities and evacuation and sheltering requirements.

Hampton Roads Sea Level Rise Study

Hampton Roads refers to a large body of water connecting the port cities of Virginia Beach, Norfolk, and Newport News. Several rivers empty into Hampton Roads before flowing into the Chesapeake Bay which supports many recreational, fishing, tourism, industrial, and commercial activities and military facilities. Because of a long history of maritime activities, there have been increased concerns regarding climate change, sea level rise, and threats to transportation infrastructure. Significant investment in highways, tunnels, bridges and other transportation infrastructure has occurred. Prepared by the Hampton Roads Transportation Planning Organization in May 2016, the study assesses risks of sea level rise and storms and impacts on transportation infrastructure (*32*). Aligned with the transportation plan time horizon (2045), it includes a methodology for incorporating sea level rise and storm surge into the prioritization of projects as well as design considerations and adaptation strategies. Hampton Roads is second only to New Orleans in terms of vulnerability to sea level rise. The area is experiencing more frequent storm surges and higher tides than normal. Prior research identified risks and threats to military facilities and the local economy. Workshops were conducted with pilot projects to plan for sea level rise. In addition to estimating flood frequencies for interstate, primary/arterial roads, and secondary roadways using 25- and 50-year scenarios, the study identified key evacuation routes and tunnels and areas likely to suffer from poor drainage. Three conditions were considered: 2-foot relative sea level rise; 2-foot relative sea level rise plus a 25-year storm surge; and 2-foot relative sea level rise plus a 50-year storm surge. The at-risk infrastructure was mapped using GIS. This information supported planning and decision-making regarding evacuation route planning, construction of protective barriers, elevating critical infrastructure, and relocating facilities to higher ground.

One of the strongest lessons is the importance of local data and analysis of climate change, sea level rise, and hazard risk exposure. The tools, methods, approaches, and strategies for engagement are useful to other communities interested in building resilience (*32*, *42*, *43*).

Fukushima Disaster

The transportation sector needs to consider requirements for large-scale, complex, cascading disasters. Triggered by an Mw = 9.0 earthquake on March 11, 2011, a tsunami was generated which flooded the nuclear power plant in Fukushima, Japan. More than 18,000 people died. Longer-term health consequences of exposure to radiation are more difficult to ascertain. The Fukushima Nuclear Accident Independent Investigation Commission found that the plant operator (Tokyo Electric Power Company) had failed to conduct an adequate risk assessment and to develop effective safety measures. Tsunami waves (13-meter high) exceeded the height of seawalls and flooded generators needed to run pumps to prevent overheating of reactors. Figure 6 illustrates the multiple levels of protection that were exceeded. Radioactive material was released into the environment. The cleanup will take decades.

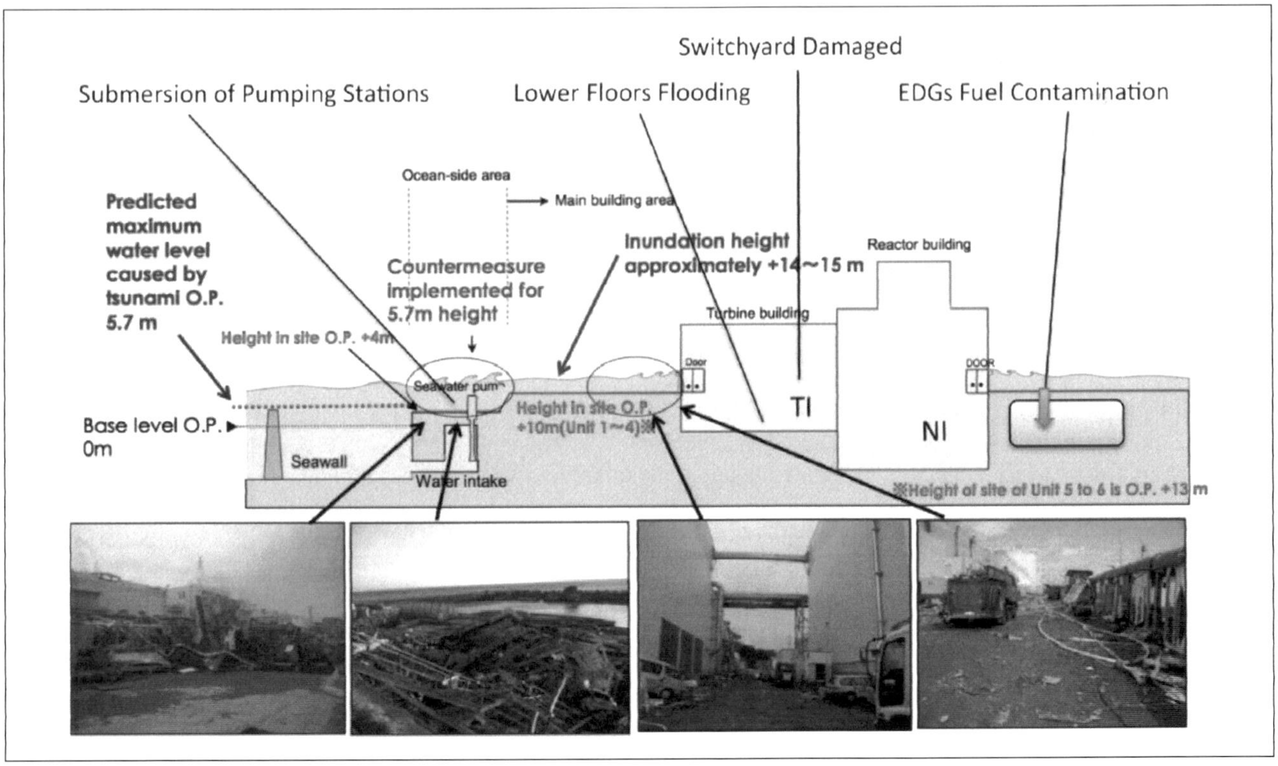

Figure 6. Multiple levels of protection that were exceeded.
Source: Claudio Filippone, Ph.D. and Tokyo Electric Power Company, Curriculum for E0684 Integrating Science into Emergency Management Decision Making and Policy, FEMA Emergency Management Institute.

There are lessons from the disaster. Many people did not evacuate because they believed they were safe because of the sea walls. While the earthquake was quickly detected, the estimated height of the tsunami kept changing as more information became available. People evacuated to shelters which also flooded. Many roadways, railways, and transportation facilities were heavily damaged. The loss of transportation assets slowed recovery. The disaster led to the shutdown of 43 of Japan's 54 nuclear reactors, which also crippled the country's power system and delayed recovery. The disaster disrupted the manufacturing sector, with business closures and impacts to the global supply chain. Organizations were not prepared for a disaster of this magnitude. Few people have sufficient knowledge about nuclear risks and understand requirements related to evacuation and sheltering. The need for training, exercises, and drills are important to building resilience. Coupled natural, technological, and human systems necessitate greater interdisciplinary collaboration to manage uncertainty and complexity. A widening gap between experts and the general public also contributes to the challenges of communicating information regarding hazards, threats, and risks as well as actions to mitigate harm. The human element is critical as large organizations, including transportation agencies, have diverse workforces with divergent skills, responsibilities, and capabilities which influence understanding and management of risk (*44–48*).

Oregon Resilience Plan

A resolution in the Oregon legislature directed the Oregon Seismic Safety Policy Advisory Commission (OSSPAC) to coordinate preparation of a resilience plan that reviews policy options, summarizes relevant reports and studies and makes recommendations to protect lives and keep commerce flowing following a Cascadia earthquake and tsunami. The plan had the following goal: "Oregon citizens will not only be protected from life-threatening physical harm but because of risk reduction measures and pre-disaster planning, communities will recover more quickly and with less continuing vulnerability following a Cascadia subduction zone earthquake and tsunami." Workgroups were charged with determining the likely impacts of a magnitude 9.0 earthquake and tsunami. They estimated the number of fatalities as ranging from 1,250 to 10,000, with 24,000 destroyed buildings and up to 27,000 displaced households with an estimated $30 billion in economic losses. The group estimated that 10 million tons of debris would be generated with power outages in affected areas lasting between one and six months. The plan also focuses on business and workforce continuity, the recovery of damaged communities, and the requirements for rebuilding critical infrastructure. Chapter 5 of the plan focuses on transportation. It identifies goals including effective emergency response and improving access to critical areas. The plan identifies three levels of functionality: minimal, which provides access for emergency responders; functional, which allows for the transportation of needed supplies, fuel, and equipment to heavily damaged areas; and operational (90% capacity) which allows for the resumption of most activities in the affected areas.

The section on transportation includes design considerations for bridges and roadways to improve resilience and contains estimates as to the potential loss of life from landslides, rock fall, and collapsed structures. In addition to highways, shipping, freight, rail, airports, ports, harbors, public transit, and local streets were considered. The analysis identified high priority, backbone roads and ranked them according to Tier 1, 2, and 3 roadways for prioritization of actions. A gap analysis was conducted and estimated time to recovery was also presented in the plan (*49*).

The Oregon Resilience Plan provides a model for states to engage in resilience planning while focusing on specific types of infrastructure as well as economic recovery and rebuilding communities. It illustrates the importance of transportation to response, relief and mass care operations, and recovery. It provides useful information on the process as well as substantive requirements for resilience (*49*).

Summary of Findings

There are common findings as well as specific lessons associated with hazards or community attributes which include:

1. *The centrality of transportation services to all phases of disaster response and recovery.* The cases demonstrated that regardless of whether a natural, technological or man-made hazard, transportation systems are critical to search and rescue, emergency and medical services, communications, power, and critical infrastructure, law and order, and rebuilding and restoring communities following disasters.
2. *Communications are essential to resilience building.* Within and across the organization and with public and private sector organizations and the public, effective communications are essential to response and recovery from disasters; the establishment of a JIC such as in the Hawaii case is useful in communicating uncertainty, risk, and complex scientific information and actions regarding evacuation, and sheltering. Understanding the ways that agencies and the public share information, through conventional media, social media, and technologically-based forms of communications, is critical to improved resilience.
3. *Social Media and Information Technologies Support Resilience.* When Katrina hit Louisiana, Facebook was in its infancy, and Twitter was not available. With Superstorm Sandy, there were 20 million Twitter posts (tweets) related to the storm over a 5-day period (*50*). Following the Boston Marathon Bombings, the Pew Center reported that a quarter of Americans got their information about the bombing

from social media (*51*). Law enforcement and other agencies used social media to communicate in real time (*52*). Social media played a role in the capture of the Boston suspects (*52*, *53*) and the management of transportation assets.

4. *Alignment with ICS and NIMS.* Large disasters bring mutual aid from neighboring jurisdictions and necessitate coordination between state, local, federal, civilian and military, public, private, and non-profit agencies and organizations. ICS and NIMS training improves tactical and operational effectiveness. This was evident in Minnesota and the Boston Marathon cases.
5. *Risk and vulnerability assessment, planning, and pre-*positioning *of assets are essential for effective response and recovery.* Effective risk assessment for severe weather (Sandy), large public gatherings (Boston Marathon), and complex events (Fukushima) is needed to ensure that transportation assets can support response and recovery. The Hampton Roads case demonstrates assessing and planning for sea level rise and storm surge and integrating these hazards and threats into the long-term transportation planning and capital budgeting process.
6. *Transportation officials need to support the integration of science, policy, and operations.* To build resilience, there must be a culture of interdisciplinary knowledge exchange so that new technologies and understanding of complex hazards and threats can be integrated into disaster response, as well as mitigation and adaptation efforts. Hampton Roads highlights the importance of local data and development of climate change adaptation models for transportation assets. Progress has been made in Atlanta and following Hurricane Sandy with regard to weather forecasting and warning and alert systems, but there is still need for more collaboration between scientists, emergency managers, and transportation agencies. This is particularly needed with complex cases such as the Fukushima disaster which has long-term health, environmental, and social consequences.
7. *More attention to understanding and supporting the transportation needs of vulnerable, at-risk populations.* Those with limited mobility, health needs, and other socioeconomic disadvantages are at higher risk of harm from hazards and need to be supported by transportation agencies during disasters. Many of the elderly did not evacuate during the Japan tsunami. This was also the case with Sandy and other disasters.
8. *Transportation assets play a critical role in recovery.* One of the biggest takeaways from the Oregon Resilience Plan is the importance of roads and fuel and transportation assets in rebuilding and restoring communities after a disaster. Transportation systems affect the time to recover and agencies need to prioritize backbone and critical roads needed for the restoration of infrastructure, and critical facilities, as well as jobs and livelihoods (*54*).
9. *Long-term, slow-onset, creeping disasters provide opportunities for innovation and change.* As demonstrated in Hampton Roads and in Oregon, there is value in long-range planning, supported by a realistic assessment of hazards and threats as well as gaps in recovery. Another type of creeping disaster is the lack of investment in repair and maintenance of critical infrastructure such as the I-35W Bridge collapse. Major disasters bring about reform and changes in policies. There is a need for more learning exchange so that communities can implement lessons before disasters strike.

Conclusions

While transportation systems are central to social and economic vitality, they are especially relevant to the resilience of our communities. There must be a strong commitment by transportation industry leaders to build robust, redundant, restorable systems in the event of natural or man-made disasters. Engagement at all levels—executive, management, and operations—is needed along with policy changes and resources to foster resilience. Agencies such as Transport for London have created resilience officers to mainstream resilience thinking into routine operations to reduce the likelihood and impacts of disruption. Resilience requires effective risk assessment and management practices. The adage *"you can't manage what you can't measure"* applies to transportation resilience (*55*). These cases illustrate best practices but there is a need for more research and development as well as training and education on resilience. There are opportunities to integrate the latest science and technology to support effective management and decision-making for transportation officials. These cases highlight the important role of social media, mobile communications devices, and distributed information technologies. New tools such as drones and improvements in data sharing and assimilation support transportation resilience. Improved modeling helps to identify vulnerabilities and exposure, but also the benefits of mitigation and adaptation strategies. The best science is needed to understand the threats from hurricanes, storm surge, sea level rise, earthquakes, tsunamis, hazmat incidents, and nuclear disasters, but also the causes and consequences of terrorism, deferred maintenance, poor oversight of structural engineering and design, and the increased vulnerability and exposure of disadvantaged groups to harm. A cornerstone for building societal resilience begins with planning, designing, engineering and investing in the resilience of our transportation systems.

Acknowledgments

This paper is based on research conducted for NCHRP 20-59(55) Transportation System Resilience: CEO Primer and Engagement Project funded by the National Academies of Sciences, Transportation Research Board. We would like to acknowledge and thank Ms. Deborah Matherly of Louis Berger and Associates, Principal Investigator for NCHRP 20-59(55) as well as other members of the research team including Dr. John Renne (School of Urban and Regional Planning, Florida Atlantic University), Dr. Brian Wolshon (Department of Civil and Environmental Engineering, Louisiana State University), Dr. Pamela Murray-Tuite (Glenn Department of Civil Engineering, Clemson University), and Dr. Anurag Pande (Department of Civil and Environmental Engineering, California Polytechnic State University, San Luis Obispo).

Author Contributions

The authors confirm contribution to the paper as follows: study conception and design: Karl Kim, Oceana Francis, Eric Yamashita; data collection: Karl Kim, Oceana Francis, Eric Yamashita; analysis and interpretation of results: Karl Kim, Oceana Francis, Eric Yamashita; draft manuscript preparation: Karl Kim, Oceana Francis, Eric Yamashita. All authors reviewed the results and approved the final version of the manuscript.

References

1. National Research Council. *Disaster Resilience: A National Imperative*. National Academies Press, Washington, D.C., 2012. https://doi.org/10.17226/13457.
2. Federal Emergency Management Agency. *Emergency Support Function Annexes: Introduction*. https://www.fema.gov/media-library-data/20130726-1825-25045-0604/emergency_support_function_annexes_introduction_2008_.pdf. Accessed May 25, 2017.
3. American Association of State Highway and Transportation Officials. *Fundamental Capabilities of Effective All-Hazards Infrastructure Protection, Resilience, and Emergency Management for State Department of Transportation*. Publication HAZ-1. American Association of State Highway and Transportation Officials, Washington, D.C., 2015.
4. American Association of State Highway and Transportation Officials. *Managing Catastrophic Transportation Emergencies: A Guide for Transportation Executives*. Publication MCTE-1. American Association of State Highway and Transportation Officials, Washington, D.C., 2015.
5. The United States, Department of Transportation. *Adaptation to Climate Change in Transportation Operations and Maintenance, Technical Staff Briefing*. Federal Highway Administration, PowerPoint Presentation, Washington, D.C., 2016.
6. Asam, S., C. Bhat, B. Dix, J. Bauer, and D. Gopalakrishna. *Climate Change Adaptation Guide for Transportation Systems Management, Operations, and Maintenance*. Publication HOP-15-0256. Federal Highway Administration, U.S. Department of Transportation, 2015.
7. Litman, T. Lessons from Katrina and Rita: What Major Disasters Can Teach Transportation Planners. *Journal of Transportation Engineering*, Vol. 132, No. 1, 2006, pp. 11–18.
8. Fatarechi, R., and E. Miller-Hooks Measuring the Performance of Transportation Infrastructure Systems in Disasters: A Comprehensive Review. *Journal of Infrastructure Systems*, Vol. 21, No. 1, 2015, pp. 04014025-1–04014025-15.
9. NDPTC. *Curriculum for E0684 Integrating Science into Emergency Management Decision Making and Policy*. FEMA Emergency Management Institute, 2015.
10. The State of Georgia. *Internal Review and Action Plan: Winter Storm Event*, 2014
11. Sposito, S. *Snow Jam 2014: Timeline of Atlanta Winter Storm*. The Atlanta Journal-Constitution, 2014. http://www.ajc.com/news/snow-jam-2014-timeline-atlanta-winter-storm/cbxR8RX6RhAKVksFxzDEXP/. Accessed May 3, 2017.
12. Burns, R. The Day We Lost Atlanta: How 2 Lousy Inches Paralyzed a Metro Area of 6 Million. *Politico Magazine*, 2014. http://www.politico.com/magazine/story/2014/01/atlanta-snow-storm-102839. Accessed May 3, 2017.
13. CBS News. *Road to Nowhere: Minor Snowstorm brings Atlanta to Standstill*, 2014. http://www.cbsnews.com/news/atlanta-other-parts-of-south-paralyzed-by-ice-snowstorm/. Accessed May 3, 2017.
14. LeClerc, J., and S. Joslym. The Cry World Effect and Weather-Related Decision Making. *Risk Analysis*. Vol. 35, No. 3, 2015, pp. 385–395. http://onlinelibrary.wiley.com/doi/10.1111/risa.12336/full. Accessed May 3, 2017.
15. NTSB. *Collapse of I-35W Highway Bridge Minneapolis*, Minnesota, 2007. https://www.ntsb.gov/investigations/AccidentReports/Reports/HAR0803.pdf. Accessed May 3, 2017.
16. Fielding, E., A. W. Lo, and J. H. Yang. The National Transportation Safety Board: A Model for Systematic Risk Management. *Journal of Investment Management*, Vol. 9, No. 1, 2011, pp. 17–49.
17. Zhu, S., D. Levinson, H. X. Liu, and K. Harder. The Traffic and Behavioral Effects of the I-35W Mississippi River Bridge Collapse. *Transportation Research Part A: Policy and Practice*, Vol. 44, No. 10, 2010, pp. 771–784.
18. Hao, S. I-35W Bridge Collapse. *Journal of Bridge Engineering*. Vol. 15, No. 5, 2010, pp. 608–614.
19. Massachusetts Executive Office of Public Safety. *After Action Report for the 2013 Boston Marathon Bombings*, 2013. http://www.mass.gov/eopss/agencies/mema/news/press-releases/after-action-report-for-the-2013-boston-marathon.html. Accessed May 3, 2017.
20. Walls, R. M., and M. J. Zinner. The Boston Marathon Response: Why Did it Work so Well. *Journal of the American Medical Association*, Vol. 309, No. 23, 2013, pp. 2441–2442. http://jamanetwork.com/journals/jama/fullarticle/1684255. Accessed May 4, 2017.
21. Turner, N. Unmanned Aircraft Applications for Active Volcano Monitoring. *The Geologic Society of America 113th Annual Meeting*, Honolulu, Hawaii, 2017.
22. Kim, K., P. Pant, and E. Yamashita. Managing Uncertainty: Lessons Learned from Volcanic Lava Disruption of Transportation Infrastructure in Puna. *Journal of Emergency Management*, Vol. 16, No. 1, 2018, pp. 29–40.
23. Nakaso, D. *Preparing for Lava Threat Costs Hawaii's Big Island at Least $15 Million*, 2015. http://www.govtech.com/em/disaster/Preparing-Lava-Threat-Costs-Hawaiis-Big-Island-15-Million.html. Accessed May 3, 2017.
24. Blake, E. S., T. B. Kimberlain, R. J. Berg, J. P. Cangialosi, and J. L. Beven, II. *Tropical Cyclone Report: Hurricane Sandy*

(AL182012), 22–29 October 2012, National Hurricane Center, Florida, 2013.
25. Deary, D. S., K. E. Walker, and D. D. Woods. Resilience in the Face of a Superstorm: A Transportation Firm Confronts Hurricane Sandy. *Proc., Human Factors and Ergonomics Society 57th Annual Meeting*, San Diego, CA, 2013.
26Kaufman, S., C. Qing, N. Levenson, and M. Hanson. *Transportation During and After Hurricane Sandy*. Rudin Center for Transportation, NYU Wagner Graduate School of Public Service, 2012.
27. Mark, E. NTSDOT, and Hurricane Sandy: Operational Impacts and Lessons Learned. Presented at 93rd Annual Meeting of the Transportation Research Board, Washington, D.C., 2014.
28.Schmeltz, M. T., K. Sonia, G. Liza Fuentes, A. Kwan, A. Ortega-Williams, and L. Pilar Cowan. Lessons from Hurricane Sandy: A Community Response in Brooklyn, New York. *Journal of Urban Health*, Vol. 90, No. 5, 2013, pp. 799–809.
29.Gibbs, L., and C. Holloway. *Hurricane Sandy After Action Report: Report and Recommendations to Mayor Michael R. Bloomberg*. New York City, NY. 2013.
30. Gall, R., J. Franklin, F. Marks, E. N. Rappaport, and F. Toepfer. The Hurricane Forecast Improvement Project. *Bulletin of the American Meteorological Society*, Vol. 94, No. 3, 2013, pp. 329–343.
31. Bhatia, K. T., D. S. Nolan, A. B. Schumacher, and M. DeMaria. Improving Tropical Cyclone Intensity Forecasts with PRIME. *Weather and Forecasting*, Vol. 32, No. 4, 2017, pp. 1353–1377.
32. Belfield, S. S., and B. McFarlane. *Sea Level Rise and Storm Surge Impacts to Roadways in Hampton Roads*. Hampton Roads Transportation Planning Organization, 2016.
33. Minneapolis DOT. *Mn/DOT's Response to National Transportation Safety Board Recommendations*, 2008. http://www.dot.state.mn.us/i35wbridge/pdf/ntsb-report-response.pdf. Accessed May 3, 2017.
34. Xie, F., and D. Levinson, Evaluating the Effects of the I-35W Bridge Collapse on Road-users in the Twin Cities Metropolitan Region. *Transportation Planning and Technology*, Vol. 34, No. 7, 2011, pp. 691–703.
35. Minneapolis DOT. *Economic Impacts of the I-35W Bridge Collapse*, 2008. http://www.dot.state.mn.us/i35wbridge/rebuild/pdfs/economic-impacts-from-deed.pdf. Accessed May 3, 2017.
36. Kim, K., and J. Davidson. Unmanned Aircraft Systems Used for Disaster Management. *Transportation Research Record: Journal of the Transportation Research Board*, 2015. 2532: 83–90.
37. Hu, Q., C. Connolly Knox, and N. Kapucu. What Have We Learned Since September 11, 2001? A Network Study of the Boston Marathon Bombings Response. *Public Administration Review*. Vol. 74, No. 6, 2014, pp. 698–712.
38. Kellerman, A. L., and K. Peleg. Lessons from Boston. *The New England Journal of Medicine*, Vol. 368, No. 21, 2013, pp. 1956–1957. http://www.nejm.org/doi/full/10.1056/NEJMp1305304#t=article. Accessed May 5, 2017.
39. Biddinger, P. D., A. Baggish, L. Harrington, P. d'Hemecourt, J. Hooley, J. Jones, R. Kue, C. Troyanos, and K. S. Dyer. Be Prepared-The Boston Marathon and Mass-Casualty Events. *The New England Journal of Medicine*, Vol. 368, No. 21, 2013, pp. 1958–1960. http://www.nejm.org/doi/full/10.1056/NEJMp1305480#t=article. Accessed May 5, 2017.
40U.S. Department of Commerce. *Service Assessment: Hurricane/Post-Tropical Cyclone Sandy*, October 22-29, 2012. NOAA, NWS, Silver Spring, MD, 2013
41. Comes, T., and B. Van de Walle. Measuring Disaster Resilience: Impact of Hurricane Sandy on Critical Infrastructure Systems. *Proc., 11th International ISCRAM Conference*, Pennsylvania State University, Pennsylvania, 2014.
42. Kleinosky, L. R., B. Yarnal, and A. Fisher. Vulnerability of Hampton Roads, Virginia to Storm Surge Flooding and Sea Level Rise. *Natural Hazards*, Vol. 40, No. 1, 2006, pp. 43–70.
43. Tompkins, C. F., and C. Deconcini. *Sea-Level Rise and Its Impact on Virginia*, 2014. http://www.wri.org/publication/sea-level-rise-virginia. Accessed May 4, 2017.
44. Amadeo, K. *Japan's 2011 Earthquake, Tsunami and Nuclear Disaster: Economic Impact on Japan and the Rest of the World*, 2017. https://www.thebalance.com/japan-s-2011-earthquake-tsunami-and-nuclear-disaster-3305662. Accessed May 3, 2017.
45. Demetriou, D. *Japan Earthquake, Tsunami, and Fukushima Nuclear Disaster: 2011 Review*, 2011. http://www.telegraph.co.uk/news/worldnews/asia/japan/8953574/Japan-earthquake-tsunami-and-Fukushima-nuclear-disaster-2011-review.html. Accessed May 3, 2017.
46. Koshimura, S., S. Hayashi, and H. Gokon. The Impact of the 2011 Tohoku Earthquake Tsunami Disaster and Implications to the Reconstruction. *Soils and Foundation*, Vol. 55, No. 4, 2014, pp. 560–572.
47. Willauer, D. *Transportation Infrastructure Damage in Japan*, 2011. http://blog.iem.com/2011/03/17/transportation-infrastructure-damage-in-japan/. Accessed May 3, 2017.
48. World Nuclear Association. *Fukushima Accident*, 2017. http://www.world-nuclear.org/information-library/safety-and-security/safety-of-plants/fukushima-accident.aspx. Accessed May 3, 2017.
49.Oregon Seismic Safety Policy Advisory Commission. *The Oregon Resilience Plan: Reducing Risk and Improving Recovery for the Next Cascadia Earthquake and Tsunami*, 2013. http://www.oregon.gov/oem/Documents/Oregon_Resilience_Plan_Final.pdf. Accessed May 1, 2017.
50. Guskin, E., and P. Hitlin. *Hurricane Sandy and Twitter*, Pew Research Center Journalism and Media, 2012. http://www.journalism.org/2012/11/06/hurricane-sandy-and-twitter/. Accessed October 13, 2017.
51Dimock, M., C. Doherty, and A. Tyson. *Most Expect 'Occasional Acts of Terrorism' in the Future*, Pew Research Center U.S. Politics and Policy, 2013. http://www.people-press.org/2013/04/23/most-expect-occasional-acts-of-terrorism-in-the-future/2/. Accessed October 13, 2017.
52. Haddow, G., and K. Haddow. *Social Media and the Boston Marathon Bombings: A Case Study*. Elsevier SciTechConnect: Physical Security and Emergency Management, 2015. http://scitechconnect.elsevier.com/social-marathon/. Accessed October 13, 2017.
53. Presuitti, C. *Voice of America News. "Multi, Social Media Play Huge Role in Solving Boston Bombings"*, 2013. https://

www.voanews.com/a/multi-social-media-play-huge-role-in-solving-boston-bombing/1649774.html. Accessed October 13, 2017.

54. Croope, S., and S. McNeil. Improving Resilience of Critical Infrastructure Systems Postdisaster: Recovery and Mitigation. *Transportation Research Record: Journal of the Transportation Research Board*, 2011. 2234: 3–13.

55. Meany, A., S. Thomas, R. Martins-Tonks, and P. Horne. How Can the Value of the Resilience of Transportation Assets be Assessed? *European Transport Conference 2016. Association for European Transport (AET)*, Barcelona, Spain, 2016.

The Standing Committee on Critical Transportation Infrastructure Protection (ABR10) peer-reviewed this paper (18-00162).

Article

TRR
JOURNAL OF THE TRANSPORTATION RESEARCH BOARD

Transportation Research Record
2018, Vol. 2672(1) 43–53

DOI: 10.1177/0361198118791909
journals.sagepub.com/home/trr
SAGE

Modeling the Effectiveness of Infrastructure and Travel Demand Management Measures to Improve Traffic Congestion during Typhoons

Wisinee Wisetjindawat[1], Sybil Derrible[2], and Amirhassan Kermanshah[3]

Abstract

Many commuters find themselves stranded during natural disasters like typhoons. In the Tokai region in Japan, many road sections become heavily congested during typhoons, with some commuters reporting homebound trips taking more than four times longer than usual because of road flooding at several locations. Although large typhoons are considered extreme events (in terms of magnitude), they occur frequently (i.e., several times per year), substantiating the need for better preparedness. Nonetheless, it is impossible to predict exactly which roads are going to be flooded during a typhoon. As a result, in this study, a stochastic modeling approach was used that assigns a failure probability to each road segment based on climate model outputs for the region. Using this stochastic model, the travel time reliability between any given origin–destination pair can be determined. By applying this model to the road network of the Tokai region, two major measures were identified that could be implemented to reduce severe congestion during a typhoon. First, targeted infrastructure management measures can be implemented to strengthen heavily used roads, thus reducing the failure probability of major roads. Second, travel demand management measures can be implemented, such as asking commuters to leave their workplace or school one or two hours after their normal departure time. Overall, it was found that strengthening heavily used roads has a bigger impact in relieving congestion than delaying departure time, but that combining both strategies achieves the best results.

Building resilience to extreme events has become a priority in many cities around the world. In fact, many cities are revisiting how they plan, design, and build urban infrastructure systems (*1*, *2*). This is notably the case in Japan, which is regularly affected by typhoons and torrential rains. Out of 25.6 typhoons that form in the Pacific on average every year, 11.4 come close to Japan (resulting in heavy rains), and 2.7 actually land in Japan (*3*). When a typhoon reaches its peak during day time, many roads become severely congested during the evening commute, resulting in many stranded commuters. For drivers, traveling under adverse weather conditions is difficult and it involves a high risk of crash. High precipitation notably reduces road capacity and travel speed, thus increasing travel time delay. In addition, degraded road condition caused by flooding forces drivers to make unnecessary U-turns and take detours in order to avoid flood locations, resulting in severe traffic congestion. The focus on typhoons is substantiated by the fact that, of all types of disasters, typhoons are responsible for one of the largest economic losses in Japan. Between 1991 and 2011, typhoons were responsible for an estimated 19.3 trillion yens (approximately US$175 billion), which is more than the losses incurred by the devastating 2011 Great East Japan Earthquake and Tsunami. This is notably as a result of the frequency of typhoons, as they happen more regularly than all other types of disasters (*4*).

The Tokai region is the third largest metropolitan area in Japan and is home to more than 15 million people. The major city in the region is Nagoya. The region is frequently hit by severe typhoons, impacting traffic significantly. For example, Typhoon Roke in September 2011 was one of the most devastating typhoons the

[1]Department of Engineering Mathematics, University of Bristol, Bristol, UK
[2]Civil and Materials Engineering, University of Illinois at Chicago, Chicago, IL
[3]Department of Civil and Environmental Engineering, Vanderbilt University, Nashville, TN

Corresponding Author:
Address Correspondence to Wisinee Wisetjindawat:
w.wisetjindawat@bristol.ac.uk

region ever experienced. Typhoon Roke alone caused an economic loss of more than 1.1 trillion yens (approximately US$10 billion) (4). From the night of September 19 to the evening of September 21, 2011, torrential rains brought by the typhoon affected the entire Tokai region. During the typhoon, public transport service was suspended and several roads were completely flooded. A total of 496 mm of rain fell during the typhoon, with maximum rainfall intensities reaching 78 mm/h (5); as a benchmark, 323 mm or more of rain in a day represents a 200-year rain (i.e., a rain expected to happen once every 200 years) (6). On September 20, when the typhoon hit the region the hardest, the authorities advised the evacuation of a wide area housing approximately 96,000 people, although only 10,320 people were evacuated in the end (5). The typhoon caused a considerable reduction in road capacity; in total seven sections of expressway were closed and 333 locations of normal roads in the region were flooded or obstructed by fallen objects. This event started midday on a weekday, and it affected numerous commuters who had trouble returning home. As train services on several routes were suspended, many commuters could only take taxis to return home, even though some commuters decided instead to spend the night at a hotel, at their workplace, or at a friend's house. Severe traffic congestion was naturally expected because of the significant reduction in road capacity. In addition, the suspension of train services created a sudden increase in traffic demand, thus further worsening road congestion. The congestion started at 12:00, as many companies allowed their workers to leave the workplace early; it was the most severe at 17:00–18:00, and it then continued until 20:00 because many commuters opted to leave their workplace at their usual time (7). As a consequence, many drivers returning home, especially those who left their workplace at their usual time, experienced travel times that were several times longer than normal. For example, a trip from the Central Nagoya to Tajimi, which is located approximately 40 km to the northeast of Central Nagoya, could take as much as four times longer than usual as several roads were damaged, causing drivers to do U-turns and take detours (7).

Typhoons are relatively frequent in the region, and as a result, measures can be taken to reduce the severity of the congestion during a typhoon. In general, measures can be classified into two major groups: (1) infrastructure management and (2) travel demand management. The first group of measures includes, for example, raising or strengthening road structures against flooding, such as using water blockades. This type of measure reduces the likelihood of road failure, and thus it maintains sufficient infrastructure capacity during adverse weather conditions. The second group of measures specifically deals with controlling how the infrastructure is used. Many travel demand management strategies exist; for example, reducing demand or shifting demand away from the peak period can alleviate the severity of congestion. To implement those measures, temporary shelters can be provided to commuters until the peak period has passed, or inexpensive accommodation can be offered to commuters who do not necessarily have to go back home that evening. For these measures, establishing effective communication with commuters is important. Sakamoto and Fujita pointed out that most commuters tend to underestimate the risk of traveling during high-intensity rainfalls (8). For example, many commuters continued with their usual travel plans despite being told that rainfall intensities would reach 50 mm/h.

In this study, a stochastic model is used to test the effectiveness of both groups of measures listed above, specifically to analyze traffic conditions from homebound trips in the Tokai region during a typhoon. The model is calibrated using large-scale climate model outputs and has been previously developed in Wisetjindawat et al. (9) and Kermanshah et al. (10). In the next section, the literature on the main theme of this study is reviewed. Afterwards, the mathematical model that is used is explained. The results of the model are then analyzed, and the effectiveness of the measures listed above are evaluated. Finally, the two groups of measures are discussed and recommendations to improve preparedness for future typhoons are made.

Literature Review

There are two main streams of quantitative analysis for the operation of road networks under extreme events: (1) reliability analysis and (2) vulnerability analysis. In the 1990s, studies focused mainly on travel time reliability for project evaluation (11). Over the last two decades, road network vulnerability analysis has emerged and has received growing attention from numerous researchers around the world (12–16), notably in the wake of the 1995 Kobe Earthquake, the terrorist attack on September 11, 2001 (17), and the 2005 hurricane Katrina in New Orleans.

Reliability analysis focuses on the performance of a road network, considering uncertainties in the road network and their effects on the reliability of the system. There are various sources of uncertainties in a transport network that can be categorized into two main types (18): uncertainties on the demand side (e.g., fluctuations by season, time of day, or day of week), and uncertainties on the supply side (e.g., incidents that lead to reduced capacity such as traffic incidents, bad weather, and natural disasters). These uncertainties can cause severe traffic congestion, thus leading to further variability in travel time.

A variety of measures can be found in reliability studies such as terminal reliability, travel time reliability, and capacity reliability. Terminal reliability, sometimes called connectivity reliability, was the earliest approach to measure the reliability of road networks. The work of Iida and Wakabayashi is an example of terminal reliability (*19*). This approach considers the probability that a given Origin/Destination (OD) pair remained connected in a network. For each OD pair, the network is considered successful if at least one path that connects the OD pair is operational. A disadvantage of this approach is that terminal reliability does not take into account the additional time required as a result of congestion from the increment in travel demand on the remaining operable travel paths. The next development was travel time reliability, which is the most common measure used to evaluate network performance under normal daily congestion (*20*). Travel time reliability is the probability that a trip between a given OD pair can be completed successfully within a specific time interval. In contrast to terminal reliability, it accounts for the congestion on the remaining operable links. Asakura proposed a measure of travel time reliability between an OD pair in a deteriorated road network (*21*). Specifically, the reliability becomes a function of the ratio of travel times between affected and normal conditions. Later, Chen et al. introduced capacity reliability to evaluate the performance of a degradable road system (*20*, *22*). Capacity reliability is defined as the probability that the maximum network capacity is greater than or equal to a required demand level when link capacities are subject to random variations. Using this method, the travel time is compared between two states: with degraded and non-degraded capacities. Thus, this measure of travel time reliability is more suitable for an extreme event when the network is seriously affected and the travel time between an OD pair possibly increases significantly.

Unlike reliability analysis, vulnerability studies focus on identifying the critical elements of a road network, in order to inform mitigation strategies, as the road network structure can have a significant impact on travel demand (*23*). Several analysis methods have been proposed, and Taylor categorized vulnerability studies into four main groups: risk-based inventory analysis; topology-based analysis; serviceability-based analysis; and accessibility-based analysis (*11*). The risk-based approach focuses on the vulnerability of individual components in a network. The topology-based approach, on the other hand, identifies critical locations in a network based on the impacts of their failure on the reduction in network connectivity, such as the work of Kermanshah and colleagues (*10*, *13*), which proposed a vulnerability surface method to quantify the impacts of extreme floods and earthquakes on road networks. The serviceability-based approach analyzes the capability of a transport network to meet certain functional requirements when the network is in some way degraded. Lastly, the accessibility-based approach, which accounts for most vulnerability studies, focuses on the loss in accessibility when one or more elements of a network failed. The work by Jenelius et al. is an example of this approach, as vulnerability is measured as the increase in generalized travel cost when links are closed (*24*). The importance of road segments was measured in terms of accessibility loss caused by network degradation in Sohn (*25*), or in terms of socio-economic impacts in Taylor et al. (*26*, *27*). The impacts of road closures on trip destination and modal choice were considered in Erath (*28*).

Nevertheless, existing vulnerability studies are partially limited. First, few studies account for the simultaneous failure of multiple road segments. Second, many vulnerability studies neglect the decrease in road capacity after a disaster (*17*). In reliability analysis, incidents occurring along roads are modeled using a stochastic method by assuming a probability function to explain the occurrences of the incident, most commonly assumed to follow a Poisson distribution. Similarly, travel time reliability during an extreme event can be assessed if the failure of a road element because of the event is known. Probability of failure was taken into account in several studies. Haghighi et al. presented a probability-based simulation approach with several combinations of link failures, but the damage to links was decided randomly using Monte Carlo simulation (*29*). The purpose of their work was to identify critical links for targeted future improvement; the effectiveness of other types of important measures such as demand management was not tested. In the work of Sumalee and Kurauchi, different traffic regulations were evaluated considering the failure probability of elements in a network after a major disaster (*30*). However, the work was formulated as a network design problem using optimization technique and thus it was limited to only a small test network because of the computational burden.

In this study, thanks to the availability of climate data, the failure probability of each link as a result of torrential rain can be calculated, and, with a stochastic method, the analysis of travel time reliability during an extreme event becomes possible. This assessment is more realistic because: (1) the failure of a road segment is uncertain, (2) multiple cases of joint failures can be assessed, (3) resulting congestion owing to a decrease in road network capacity can be estimated, and (4) several measures, both to improve the current infrastructure and to manage the demand, can be assessed. This study can be categorized as a serviceability-based approach based on the definition provided by Taylor (*11*), in which the measures are evaluated based on the travel time reliability under extreme weather conditions.

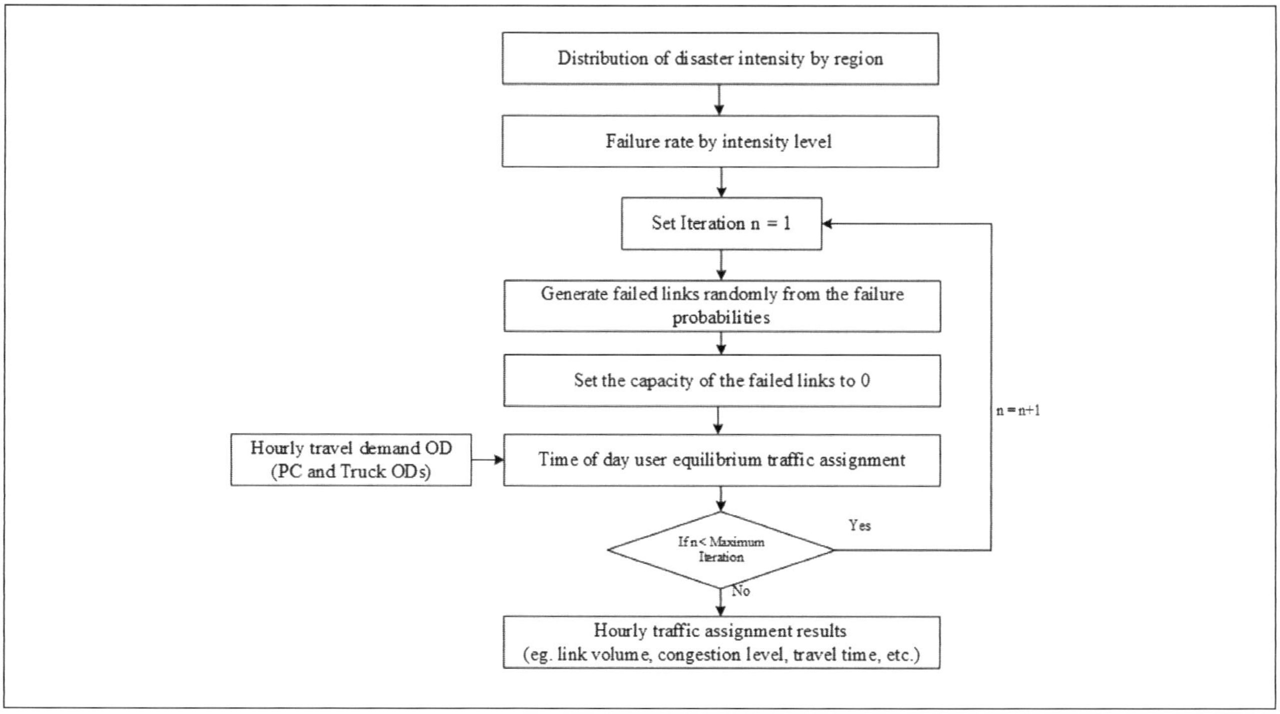

Figure 1. Model framework.

Methodology

Mathematical Model

Here, the stochastic modeling procedure presented in Wisetjindawat et al. (*9*), and shown in Figure 1, is adopted to estimate travel conditions and travel time reliability during an extreme event; the stochastic model assumes that the occurrence of road failure follows a given probabilistic distribution (here, a negative exponential is used). Because it is practically impossible to predict the exact location of road failures during a typhoon, every road segment is given a failure probability based on past climate data and climate projections up to 2100 as presented in Wisetjindawat et al. (*9*). The University of Tsukuba climate model simulations were used in this study in two modes: current (1981–2000) and future projection with RCP 4.5 climate scenario (2081–2100) (*31*).

During the simulation, a road segment is assumed to be either fully operational or to have failed and thus have zero capacity. Although road segments can partially fail (i.e., obstruction of only part of road width), this type of failure is nearly impossible to predict, and during a disaster, it is more likely that the constraint on the capacity of the road network is whether or not a link is usable (*32*, *33*). The failure probability of a link is obtained using a negative exponential distribution, similar to the works by Kroger and Zio (*34*) and Wisetjindawat et al. (*35*). The probability that link *i* fails is defined as follow:

$$p_i = 1 - e^{-\lambda_i \eta_i} \tag{1}$$

where p_i is the failure probability of link *i*; λ_i is a flash flood failure rate; and η_i is the standardized length of link *i* for flash floods. The failure road λ_i of road segment *i* is calculated from the failure rate of the entire climate grid cell, in which the road segment is located, which is in turn calculated as the likelihood of future extreme rainfall considering past extreme rainfall (i.e., beyond 95th percentile) from climate model outputs; note that no hydrologic modeling is used (i.e., runoff modeling), which is why the focus is purely on flash floods. The length of link *i* is then standardized by dividing it by the average length of failed links during Typhoon Roke. Details about the probability assignment method can be found in Wisetjindawat et al. (*9*).

From the simulation procedure in Figure 1, there are two inputs into the model: the road network and the hourly travel demand. The stochastic failure of the road section is used to determine the capacity of the road system during the flash flood. The hourly travel demand, taken from OD matrices of both passenger cars and trucks, are obtained from Fujita et al. (*36*, *37*); they were calibrated from observed link traffic counts. Simultaneous failures of road segments are generated using Monte Carlo analysis. For each iteration (i.e., for each failure scenario), a random number between 0 and 1 is generated for each link and if it is lower than the calculated failure probability, then it is assumed the link has failed. This process is repeated for every link in the road network. The capacity of failed links is set to zero, and the parameters for traffic assignment of other links, such as the number of lanes, capacity, free-flow speed, toll, and other factors, remain the same. Next, the generated road network is input in a traffic assignment model. In this study 50 iterations were run, each iteration taking an hour.

For every iteration, the results from the traffic assignment are obtained and they include link traffic volume, link traverse time, congestion level, link speed, as well as other traffic characteristics. The hourly travel demand matrices are fed into the time-of-day user equilibrium traffic assignment developed by Fujita et al. (*36*, *37*), which uses the standard Bureau of Public Road (BPR) function for link travel time. The parameters for the BPR function and other characteristics of each link are set based on actual values determined for the region. The link travel time function (t_i) of link i is:

$$t_i = t_{f,i}(1 + \alpha(\frac{v_i}{c_i})^{\beta}) \tag{2}$$

where, v_i is the traffic volume of link i; $t_{f,i}$ is the free-flow travel time; α, β are calibrated parameters for the link travel time function; c_i is a random capacity of link i of either 0 (i.e., failed link) or C_{0i} as the capacity of link i under normal conditions.

Determining Travel Time Reliability

In general, travel time reliability is defined as the probability that a trip between an OD pair is completed successfully within a specific time duration. However, this type of definition is only suitable for the evaluation of network performance under normal conditions when the travel time incurred by daily traffic incidents is not much longer than normal conditions. Asakura defined another type of travel time reliability as the ratio of travel times between affected and normal conditions (*21*). This indicator particularly captures the impact of an extreme event when travel time between ODs can become several times longer (*20*); this indicator is adopted in this study. Following the definition by Chen et al. (*20*), given C and C_0 as the vectors of degraded and non-degraded capacities of the network, the corresponding travel time between an OD pair w of the degraded and non-degraded states are defined as $u_w(\mathbf{C})$ and $u_w(\mathbf{C}_0)$, respectively. The travel time reliability is defined as the probability $\tau_w(\theta)$ that the ratio of the travel time between an OD pair w between affected (with capacity vector $\mathbf{C}$) and normal conditions (with capacity vector $\mathbf{C_0}$) is less than a specific level θ, where θ refers to the pre-specified level of service. In this study, these values were set to 2, 4, and 10 to measure the probability that the travel time is twice, four times, and ten times longer than under normal conditions, respectively. The formulation of the travel time reliability is as follows:

$$\tau_w(\theta) = P\left(\frac{u_w(\mathbf{C})}{u_w(\mathbf{C_0})} \leq \theta\right) \tag{3}$$

The results of all cases of failures from the traffic assignment simulation were used to calculate the travel time reliability between a given origin and destination. Travel time reliability is then used as a criterion to evaluate measures against congestion in the following section.

Evaluation of Measures Against Congestion during Typhoons

In this study, the focus is on commuting times between Central Nagoya and Tajimi, as severe traffic congestion occurred between these two areas during past typhoons. Using climate models, it was also identified that the area between the two cities was the most vulnerable to typhoons in the region (*9*). By either managing the infrastructure or travel demand, three cases are evaluated: (1) do-nothing, (2) infrastructure management by strengthening the structure of important roads, and (3) travel demand management by delaying the departure of commuters by one or two hours. The do-nothing case serves to calculate the travel conditions during the flash flood when no measures are implemented. The infrastructure management measures serve to maintain the capacity of important roads during extreme events. Specifically, National Route 19 (i.e., R19) is an important arterial road connecting Central Nagoya to Tajimi, and the first measure tested consists of strengthening it to prevent any capacity reduction during a typhoon; R19 was selected because it passes through a mountainous area, and it was often closed to traffic during past typhoons, which caused severe traffic congestion.

From a modeling point of view, it is assumed the entire road (i.e., all road segments that make up National Route 19) becomes flood proof with a 0% failure probability during a typhoon or torrential rain after the strengthening. Finally, the travel demand management measures shift demand away from the peak period. In this study, this measure assumes the authority recommends commuters delay their usual departure time by one or two hours, either by waiting at their office, school, or at a designated shelter. In the modeled scenario, it is assumed flooding of roads occurs primarily between 16:00 and 20:00, as happened during Typhoon Roke.

Results

The network used in the analysis is the large-scale road network of the Tokai region that consists of 6,734 links

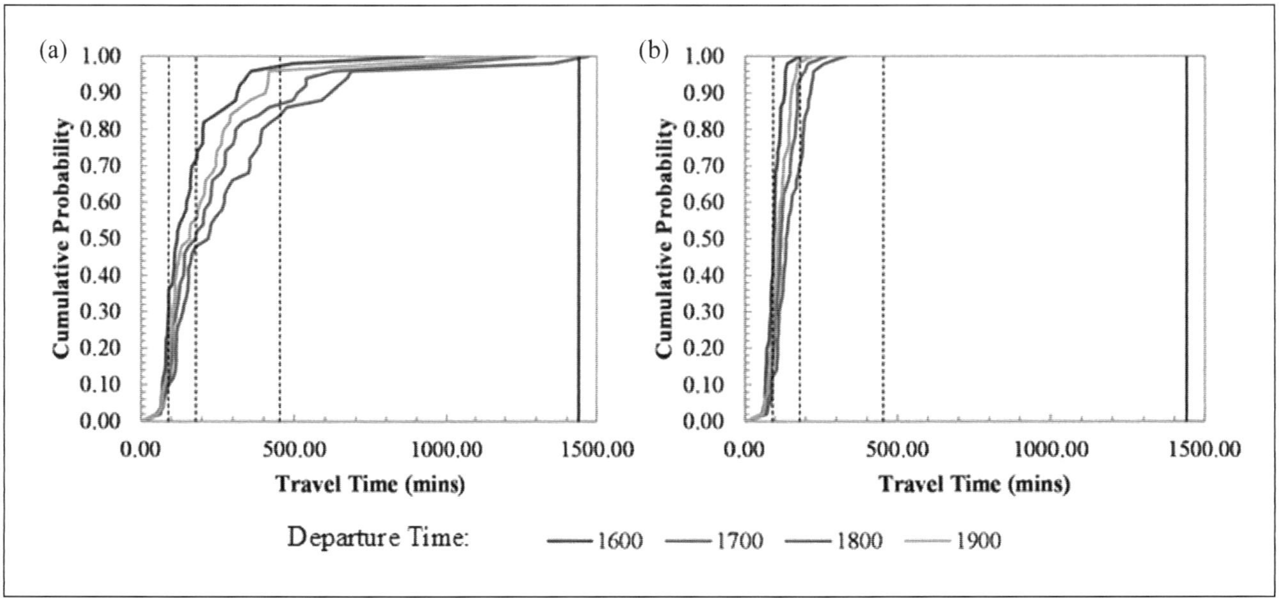

Figure 2. Cumulative probability of travel time reliability from Central Nagoya to Tajimi for do-nothing and strengthening R19 (The dotted lines from left to right represent $\theta = 2$, 4, and 10, respectively. The solid line on the right represents 24 h). (*a*) Do-nothing; (*b*) Strengthening R19.

and 4,323 nodes covering all arterial roads and expressways in the region. The analysis was performed using an hourly travel demand of 452 zones. The number of iterations for each of the three cases was set to 50. Figure 2 compares travel time reliability before and after strengthening National Route 19 at different departure times, starting from 16:00 to 19:00. The x-axis depicts the travel time required to complete a trip between the given OD pair, whereas the y-axis depicts the cumulative probability of completing the trip. The narrower "bends" in the graphs in Figure 2*b* suggest a significant improvement in the reliability of the trips when the road is strengthened.

The travel demand management measure analyzed was an adjustment of departure times for commuters who would normally leave their workplace or school between 16:00 and 19:00. Specifically, the hourly OD matrix of passenger cars during this period was adjusted by shifting the demand for a specific period. Two cases of departure delays were tested: one and two hours. In addition, various percentages of commuters collaborating with the scheme by delaying from 10% to 100% of the demand were tested. Road failure probabilities were left intact. Figure 3 shows the cumulative probability of the trips traveling from Central Nagoya to Tajimi departing at 17:00, 18:00, and 19:00, respectively. Figure 3, *a, b,* and *c* show the travel times when various percentages of commuters delay their departure by an hour. Each graph compares scenarios with four proportions of collaborating commuters: 0%, 50%, 80%, and 100% of commuters who decided to delay their departure. Similarly, Figure 3, *d, e,* and *f* show the travel times that would be experienced when departing at 17:00, 18:00, and 19:00, respectively, if commuters accepted to delay their departure by two hours. Looking at Figure 3*b*, a trip departing at 18:00 can be completed successfully within approximately 45.3 min under normal conditions. In contrast, under flash flood conditions and do-nothing, shown with the blue line in the figure, the trip has a 50% probability to be completed successfully within approximately 3.2 times longer than normal conditions. When the congestion is at its peak at 18:00, as shown in Figure 3*b*, flooded roads increase the severity of the congestion and the reliability of the travel time decreases. By delaying departure by an hour, the peak period shifts to 19:00 because the delayed commuters from the previous time slot added more demand in the next departure hour. Comparing the strategies to delay by one and two hours, the two-hour delay shows further improved travel time reliability at both 17:00 and 18:00 and, a slight improvement to the do-nothing case for 19:00 departures.

Figure 4 shows the average time spent in congestion per vehicle. This value was calculated by weighting by the number of vehicles that actually spend time on congested roads during the affected period (16:00–19:00), at 50% travel time reliability. Mathematically, the average time spent in congestion per vehicle TC_{avg} is defined as:

$$TC_{avg} = \frac{1}{D}\sum_{i} d_i \cdot TT_i \qquad (4)$$

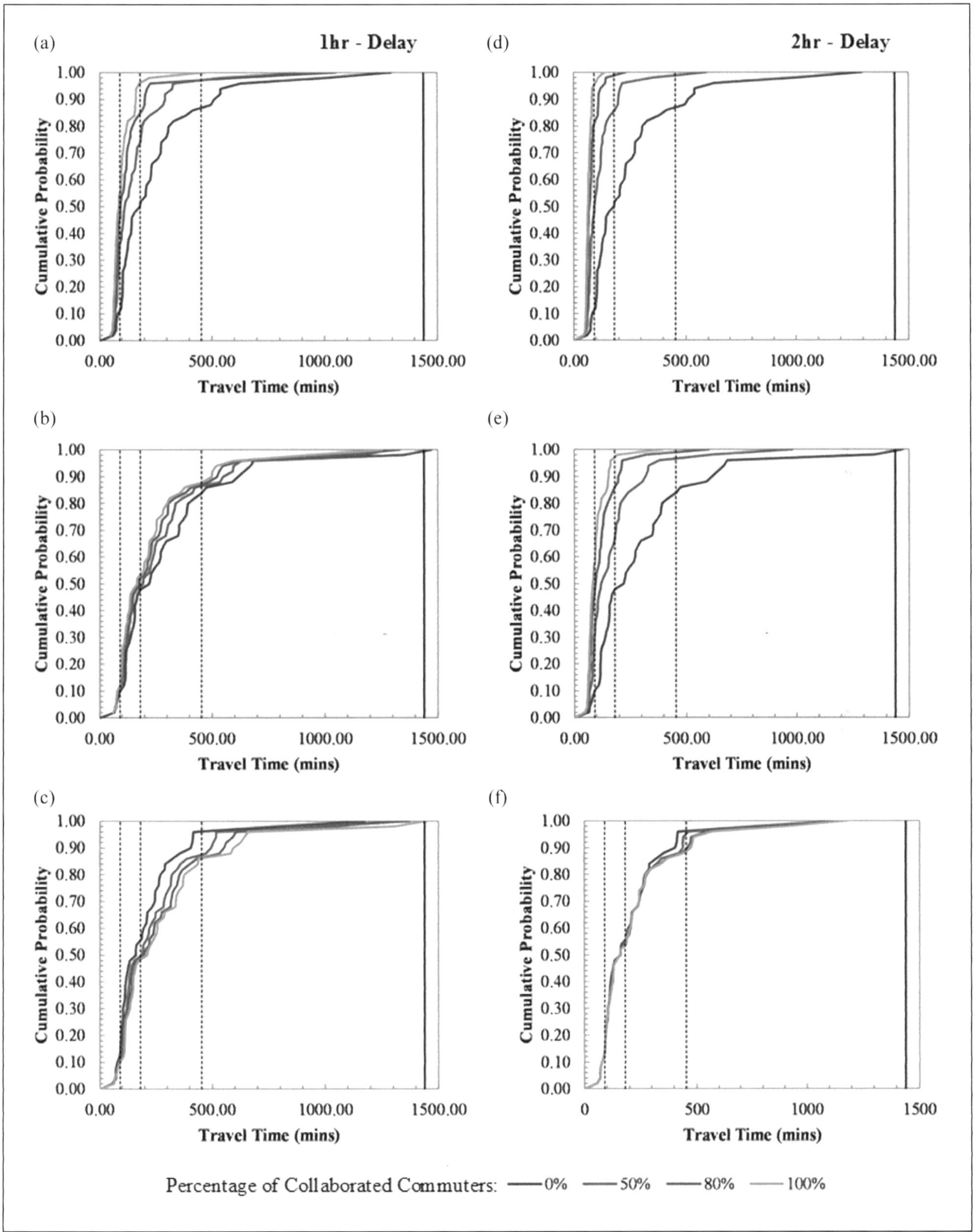

Figure 3. Cumulative probability of travel time reliability from Central Nagoya to Tajimi at different departure times and percentages of commuters collaborating in delaying their departure time. (The dotted lines from left to right represent $\theta = 2$, 4, and 10, respectively. The solid line on the right represents 24 h). (*a*) 17:00 departure, 1 h delay; (*b*) 18:00 departure, 1 h delay; (*c*) 19:00 departure, 1 h delay; (*d*) 17:00 departure, 2 h delay; (*e*) 18:00 departure, 2 h delay; (*f*) 19:00 departure, 2 h delay.

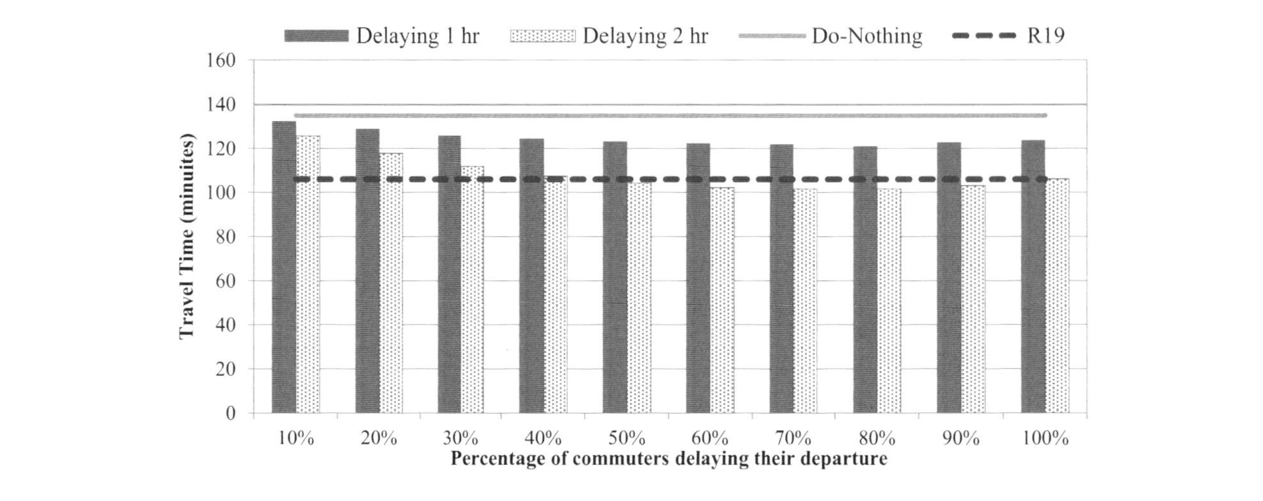

Figure 4. Average time spent in congestion per vehicle at 50% travel time reliability, for different percentages of commuters collaborating in delaying their departure time, comparing with do-nothing and strengthening R19.

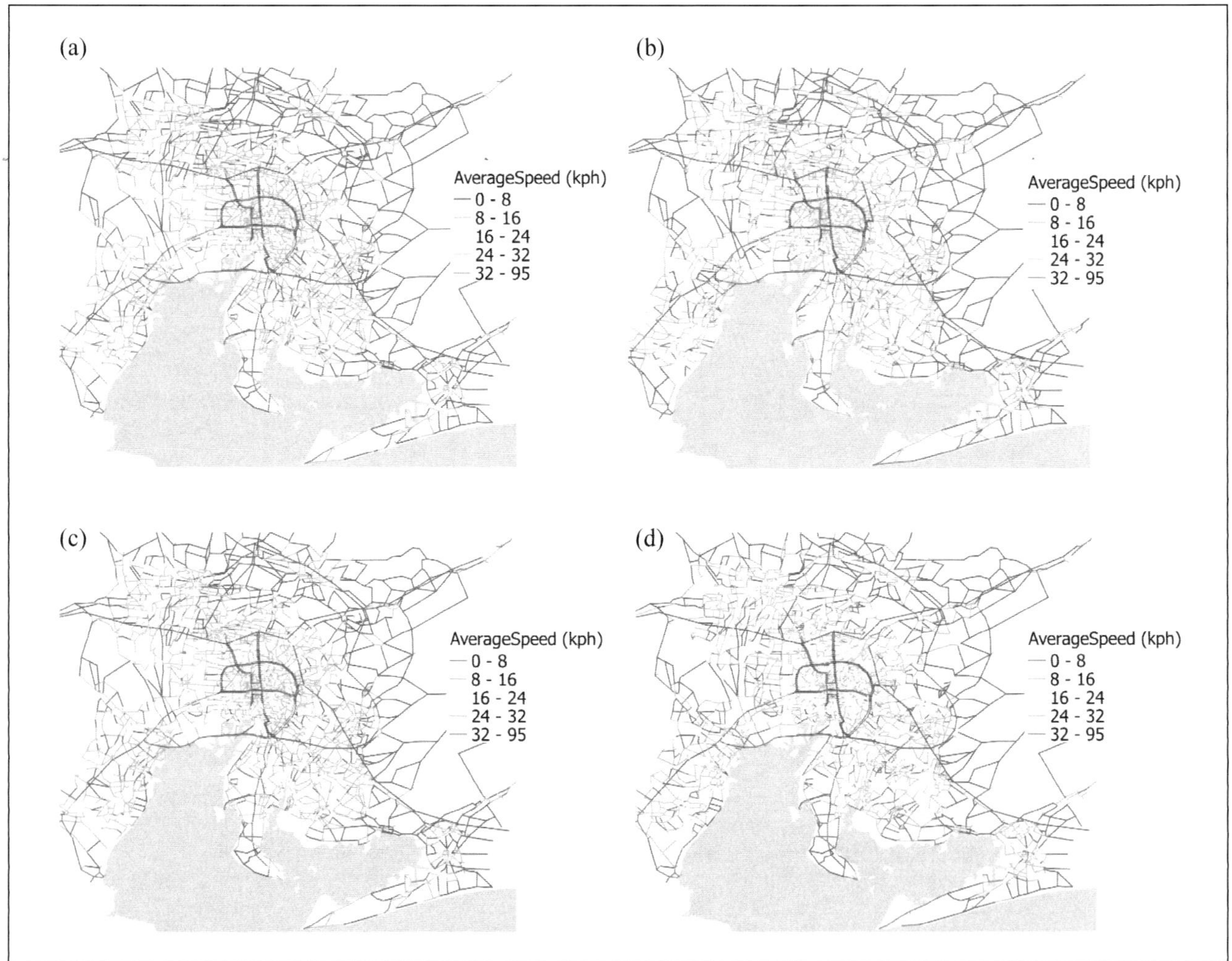

Figure 5. Speeds at 50% travel time reliability at 18:00 under different measures. (*a*) Do-nothing; (*b*) strengthening R19; (*c*) 60% commuters 1 h delay; (*d*) 60% commuters 2 h delay.

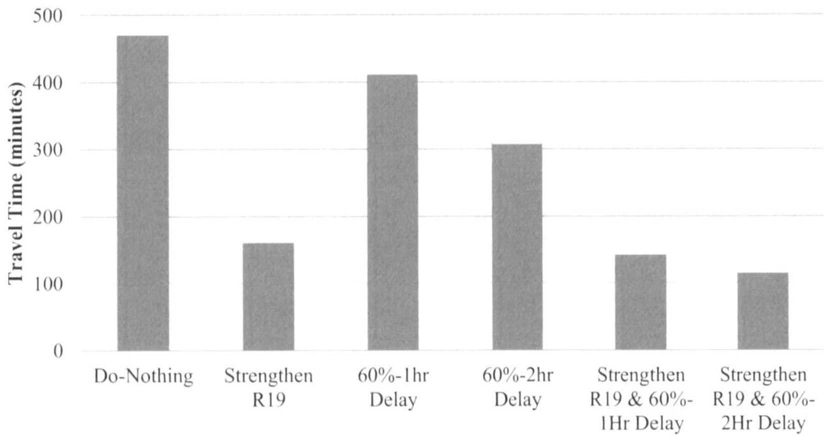

Figure 6. Average time spent in congestion per vehicle at 90% travel time reliability, with different measures.

where D is the total demand (i.e., the sum of all vehicles in the network between 16:00 and 19:00), d_i is the demand at each hour (i.e., $d_{16:00}$ is the number of vehicles leaving at 16:00) and TT_i is the 50% travel time reliability for vehicles leaving at time i (i.e., 16:00).

All measures show an improvement in travel congestion compared with the do-nothing case. The most effective measures are the strengthening of National Route 19 and delaying departures by two hours with at least 40% of commuters collaborating.

Figure 5, *a, b, c,* and *d* shows the 50% travel speed reliability during the evening peak hour with various types of measures: do-nothing, strengthening of National Route 19, and delaying the departure of 60% of commuters by one and two hours, respectively. Travel conditions at 18:00 improved by delaying demand by two hours, compared with other measures.

In order to fully compare the effectiveness of all measures, the average time spent in congestion per vehicle at 90% travel time reliability was also assessed, assuming 60% of commuters collaborate in delaying their departure time. From Figure 6, it can be seen that when measures are applied individually, strengthening the infrastructure is substantially more effective than the two travel demand management measures (although the three measures were more or less as effective at 50% travel time reliability). Combining both groups of measures shows slightly more effective results, especially when commuters are asked to delay their departure by two hours.

Conclusion

This study used a stochastic simulation approach to evaluate the effectiveness of two types of measure for homebound drivers during typhoons in the Tokai region in Japan. The stochastic model outputs a failure probability for each road link during flash floods. During past typhoons, severe traffic congestion was reported, especially on road sections that connect Central Nagoya to Tajimi.

In this study, two main measures were evaluated. The first measure was to strengthen the structure of important roads against flooding so that the infrastructure can maintain its capacity even during typhoons or torrential rains. In this study, National Route 19 was selected, which is an arterial road connecting Central Nagoya to Tajimi. The second measure is to control the travel demand by delaying the departure time of commuters. In past disasters, commuters were often unaware of the severity of the congestion, and many commuters departed at their usual time, resulting in severe traffic congestion. Here, different percentages of collaborating commuters were evaluated. When more commuters collaborate (about 40–80%) in delaying their departure for two hours, travel conditions can improve as well. As expected, successful infrastructure strengthening is more effective in reducing traffic congestion, although in practice, it is rather difficult to achieve completely flood-proof infrastructure. Thus, combining both measures may be preferred to achieve the best results.

In practice, along with infrastructure improvement, establishing good communication with commuters through social media and mobile phones as well as educating commuters to understand the severity of travel conditions during an extreme event is desirable. During typhoons, most companies and schools allow their workers/students to return home earlier than usual; however, this measure was found to be ineffective. Instead, announcements could communicate a suggested departure time by area or workplace/school to alleviate travel demand onto the road network, as well as providing the details of temporary shelters. In addition, affordable accommodation could be provided for those who do not necessarily have to go back home on the day, and workers could be encouraged to work from home.

Although many more measures should be modeled and evaluated, this study showed that successful measures can be implemented, especially in the case of "regular" extreme events like typhoons and torrential rains. Future research could also benefit from including new elements of complexity science, big data, and machine learning (*38–42*).

Author Contributions

The authors confirm contribution to the paper as follows: study conception and design: Wisinee Wisetjindawat; data collection: Wisinee Wisetjindawat; analysis and interpretation of results: Wisinee Wisetjindawat, Sybil Derrible, Amirhassan Kermanshah; draft manuscript preparation: Wisinee Wisetjindawat, Sybil Derrible, Amirhassan Kermanshah. All authors reviewed the results and approved the final version of the manuscript.

References

1. Derrible, S. Urban Infrastructure is Not a Tree: Integrating and Decentralizing Urban Infrastructure Systems. *Environment and Planning B: Urban Analytics and City Science*, Vol. 44, No. 3, 2017, pp. 553–569.
2. Derrible, S. Complexity in Future Cities: The Rise of Networked Infrastructure. *International Journal of Urban Sciences*, Vol. 21, 2017, pp. 68–86.
3. Japan Meteorological Agency. *Average Year Typhoon*. www.data.jma.go.jp/fcd/yoho/typhoon/statistics/average/average.html. Accessed July 24, 2017. (In Japanese)
4. Imakita, E. Issues on Non-Life Insurance. Presented at Workshop on Using Meteorological Information for Industry Advancement, Tokyo, Japan, December 12, 2014. (In Japanese)
5. Fujita, M., M. Ohashi, and J. Sakamoto. Public Transportation User's Relation Between the Degree of Difficulty to Return Home and the Accuracy of Information Acquisition Under Typhoon and Downpour in Tokai Area. *Journal of Japan Society of Civil Engineers*, Vol. 70, No. 5, 2014, pp. I_75–I_83. (In Japanese)
6. Japan Meteorological Agency. Risk Map of Extreme Weather. www.data.jma.go.jp/cpdinfo/riskmap/qt_table.html#area2. Accessed July 24, 2017. (In Japanese)
7. Khan, M. H., W. Wisetjindawat, M. Fujita, and K. Suzuki. An Analysis of Probe Data on Traffic Congestion during Typhoon Using GIS Application. Presented at 10th Conference of Eastern Asia Society for Transportation Studies, Taipei, Taiwan, 2013.
8. Sakamoto, J., and M. Fujita. Empirical Analysis on the Impact of Rain Intensity on Commuters' Departure Decision on Torrential Rain Day. *Journal of the Eastern Asian Society for Transportation Studies*, Vol. 11, 2015, pp. 311–325.
9. Wisetjindawat, W., A. Kermanshah, S. Derrible, and M. Fujita. Stochastic Modeling of Road System Performance During Multihazard Events: Flash Floods, and Earthquakes. *Journal of Infrastructure Systems*, Vol. 23, No. 4, 2017, p. 04017031. https://doi.org/10.1061/(ASCE)IS.1943-555X.0000391
10. Kermanshah, A., S. Derrible, and M. Berkelhammer. Using Climate Models to Estimate Urban Vulnerability to Flash Floods. *Journal of Applied Meteorology and Climatology*, 2008. 2089: 2637–2650.
11. Taylor, M. *Vulnerability Analysis for Transportation Networks*. Elsevier, Cambridge, 2017.
12. Kermanshah, A., and S. Derrible. Robustness of Road Systems to Extreme Flooding: Using Elements of GIS, Travel Demand, and Network Science. *Natural Hazards*, Vol. 86, No. 1, 2017, pp. 151–164.
13. Kermanshah, A., and S. Derrible. A Geographical and Multi-Criteria Vulnerability Assessment of Transportation Networks Against Extreme Earthquakes. *Reliability Engineering & System Safety*, Vol. 153, 2016, pp. 39–49.
14. Kermanshah, A., A. Karduni, F. Peiravian, and S. Derrible. Impact Analysis of Extreme Events on Flows in Spatial Networks. Presented at IEEE International Conference on Big Data (Big Data), Washington, DC, 2014, pp. 29–34.
15. Erath, A., J. Birdsall, K. Axhausen, and R. Hajdin. Vulnerability Assessment Methodology for Swiss Road Network. *Transportation Research Record: Journal of the Transportation Research Board*, 2008. 2137: 118–126.
16. Zhang, X., E. Miller-Hooks, and K. Denny. Assessing the Role of Network Topology in Transportation Network Resilience. *Journal of Transport Geography*, Vol. 43, 2015, pp. 35–45.
17. Chen, X. Z, Q. C. Lu, Z. R. Peng, and J. E. Ash. Analysis of Transportation Network Vulnerability Under Flooding Disaster. *Transportation Research Record: Journal of the Transportation Research Board*, 2015. 2532: 37–44.
18. Lauthep, P. *Stochastic Transport Network Model and Optimization for Reliability and Vulnerability Analysis*. Doctoral thesis. The Hong Kong Polytechnic University, http://ira.lib.polyu.edu.hk/handle/10397/4652, 2011.
19. Iida, Y., and H. Wakabayashi. An Approximation Method of Terminal Reliability of a Road Network Using Partial Minimal Path and Cut Set. Presented at 5th World Conference of Transport Research, Yokohama, 1989, pp. 367–380.
20. Chen, A., H. Yang, H. K. Lo, and W. H. Tang. Capacity Reliability of a Road Network: An Assessment Methodology and Numerical Results. *Transportation Research Part B*, Vol. 36, 2002, pp. 225–252.
21. Asakura, Y. Reliability Measures of an Origin and Destination Pair in a Deteriorated Road Network with Variable Flows. Presented at 4th Meeting of the EURO Working Group in Transportation, Newcastle, UK, 1996.
22. Chen, A., H. Yang, H. K. Lo, and W. A. Tang. Capacity Related Reliability for Transportation Networks. *Journal of Advanced Transportation*, Vol. 33, No. 2, 1999, pp. 183–200.

23. Amini, B., F. Peiravian, M. Mojarraadi, and S. Derrible. Comparative Analysis of Traffic Performance of Urban Transportation Systems. *Transportation Research Record: Journal of the Transportation Research Board*, 2016. 2594: 159–168.
24. Jenelius, E., T. Petersen, and L. G. Mattsson. Importance and Exposure in Road Network Vulnerability Analysis. *Transportation Research Part A*, Vol. 40, 2006, pp. 537–560.
25. Sohn, J. Evaluating the Significance of Highway Network Links under the Flood Damage: An Accessibility Approach. *Transportation Research Part A*, Vol. 40, 2006, pp. 491–506.
26. Taylor, M., S. Sekhar, and G. D'Este. Application of Accessibility Based Methods for Vulnerability Analysis of Strategic Road Networks. *Network and Spatial Economics*, Vol. 6, No. 3–4, 2006, pp. 267–291.
27. Taylor, M. Critical Transport Infrastructure in Urban Areas: Impacts of Traffic Incidents Assessed Using Accessibility-Based Network Vulnerability Analysis. *Growth and Change*, Vol. 39, No. 4, 2008, pp. 593–616.
28. Erath, A. L. *Vulnerability Assessment of Road Transport Infrastructure*. Doctoral dissertation. ETH Zurich, Zurich, 2011.
29. Haghighi, N., K. Fayyaz, X. Liu, and S. Bartlett. Identifying Network-Wide Critical Transportation Links Under Disaster Disruptions: A Multi-Scenario and Probability-Based Simulation Approach. Presented at 96th Transportation Research Board Annual Meeting, Washington, D.C., 2017.
30. Sumalee, A., and F. Kurauchi. Network Capacity Reliability Analysis Considering Traffic Regulation After a Major Disaster. *Network Spatial Economics*, Vol. 6, 2006, pp. 205–219.
31. Suzuki-Parker, A., and H. Kusaka. Future Projection of Thermal Environment in Japan Using WBGT. *Japanese Journal of Biometeorology*, Vol. 52, 2015, pp. 59–72.
32. Rawls, C. G., and M. A. Turnquist. Pre-Positioning of Emergency Supplies for Disaster Response. *Transportation Research Part B*, Vol. 44, 2010, pp. 521–534.
33. Ukkusuri, S. V., and W. F. Yushimito. Location Routing Approach for the Humanitarian Prepositioning Problem. *Transportation Research Record: Journal of the Transportation Research Board*, 2008. 2089: 18–25.
34. Kroger, W., and E. Zio. *Vulnerable Systems*. Springer, London, 2011.
35. Wisetjindawat, W., H. Ito, and M. Fujita. Integrating Stochastic Failure of Road Network and Road Recovery Strategy into Planning of Goods Distribution After a Large-Scale Earthquake. *Transportation Research Record: Journal of the Transportation Research Board*, 2015. 2532: 56–63.
36. Fujita, M., K. Watanabe, and S. Yamada. Hourly OD Flow Estimation from Observed Traffic Flows Based on Tome Coefficient of Daily OD Flow. *JSCE Journal of Traffic Engineering*, Vol. 2, No. 1, 2016, pp. 11–20.
37. Fujita, M., S. Yamada, and S. Murakami. Time Coefficient Estimation for Hourly Origin-Destination Demand from Observed Link Flow Based on Semi-Dynamic Traffic Assignment. *Journal of Advanced Transportation*, Vol. 2017, 2017. https://doi.org/10.1155/2017/6495861
38. Cottrill, C., and S. Derrible. Leveraging Big Data for the Development of Transport Sustainability Indicators. *Journal of Urban Technology*, Vol. 22, No. 1, 2015, pp. 45–64.
39. Derrible, S., and N. Ahmad. Network-Based and Binless Frequency Analyses. *PLoS One*, Vol. 10, No. 11, 2015. https://doi.org/10.1371/journal.pone.0142108
40. Kardani, A., A. Kermanshah, and S. Derrible. A Protocol to Convert Spatial Polyline Data to Network Formats and Applications to World Urban Road Networks. *Scientific Data*, Vol. 3, 2016, p. 160046.
41. Ahmad, N., S. Derrible, T. Eason, and H. Cabezas. Using Fisher Information to Track Stability in Multivariate Systems. *Royal Society Open Science*, Vol. 3, No. 11, 2016, p. 160582.
42. Golshani, N., R. Shabanpour, S. Mahmoudifard, S. Derrible, and A. Mohammadian. Modeling Travel Mode and Timing Decisions: Comparison of Artificial Neural Networks and Copula-based Joint Model. *Travel Behaviour and Society*, Vol. 10, 2018, pp. 21–32.

The Standing Committee on Critical Transportation Infrastructure Protection (ABR10) peer-reviewed this paper (18-02301).

Article

TRR
JOURNAL OF THE TRANSPORTATION RESEARCH BOARD

Transportation Research Record
2018, Vol. 2672(1) 54–65

DOI: 10.1177/0361198118792115
journals.sagepub.com/home/trr
SAGE

Road Network Resilience: How to Identify Critical Links Subject to Day-to-Day Disruptions

Pauline Gauthier[1], Angelo Furno[1], and Nour-Eddin El Faouzi[1,2]

Abstract

Disruptive events occur on road networks on a daily basis and affect traffic flow. Resilience analysis aims to assess the consequences of such disruptions by quantifying the ability of a network to absorb and react to adverse events. In this paper, we advance a methodological approach based on resilience stress testing and a dynamic mesoscopic simulator. We aim to identify and rank the links most critical to the overall performance of the road network, taking into account the dynamic properties of road traffic and focusing on day-to-day disruptions. As a metric to quantify road network performance in the presence of such disruptions, we use the increase in overall travel cost. We then compare our approach with purely topological approaches. We discuss the advantages and drawbacks of the different analyzed metrics. We prove that link ranking varies greatly depending on the metric. Specifically, the proposed stress testing methodology can produce very accurate results by taking into account demand and congestion, but requires computationally intensive simulations, being therefore prohibitive even on medium-sized networks. Conversely, purely static topological metrics can be inaccurate if they do not take into account traffic demand and network dynamics. Our study highlights the need for joining traditional traffic-agnostic topological resilience analysis with demand-aware dynamic stress testing techniques.

Roads are vulnerable to being disrupted by adverse weather conditions, human errors, or technological breakdowns. Such day-to-day disruptions affect traffic flow and may profoundly impair the ability of the transportation network to guarantee basic mobility services as well as management of emergency situations, thus causing fundamental economic and social strains.

Resilience analysis aims to evaluate and predict the consequences of such disruptions, and has become a crucial research concern in recent years (*1*, *2*). In the field of road networks, one possible approach consists in identifying the links that most strongly affect the overall performance. Operators and planners should be aware of the consequences deriving from reduced capacities on links. They should focus their efforts on improving and maintaining such critical links, since they may cause the most severe consequences for traffic operations when disrupted. Thus, a methodological approach and a set of resilience metrics are fundamental to identify critical links.

Similar to other complex systems, a city road transportation network can be modeled as a graph $G = (N, L)$ where road intersections are represented as nodes (N) and roads as links (L). This allows for the use of graph theory and network connectivity analysis to study resilience. Graph theory can make resilience analysis very efficient in terms of computation costs, especially with the widespread adoption of big data technologies and cloud-computing. However, the phenomenon of congestion—dynamic, spatio-temporal, demand dependent—is traditionally not addressed in topological studies on road traffic resilience (*3*, *4*).

In this context, our work aims to answer the following research questions:

- How is it possible to identify the links that are most critical to the operation of the whole road network, especially with respect to day-to-day disruptions?
- Are topological metrics adequate to measure resilience for road networks?

[1]University of Lyon, Lyon, France
[2]Queensland University of Technology, Brisbane, Queensland, Australia

Corresponding Author:
Address correspondence to Angelo Furno: angelo.furno@ifsttar.fr

- Which metrics are the most effective in assessing road network resilience, taking into account their dynamic, spatio-temporal, and demand-dependent properties?

The main contributions of this paper are the following: (a) a methodology based on link-based stress testing and a dynamic mesoscopic simulator is proposed for identifying critical links and quantifying road network robustness; (b) the methodology is compared with multiple topology-based metrics to clearly identify their limitations on a simplified road network; (c) the approach is evaluated on a real-world network to identify and rank the most critical links in a realistic scenario; (d) guidelines are provided to decide when one approach can be preferred to another.

The paper is outlined as follows. The next section deals with the related literature. In the third section, we review different resilience metrics and describe our methodological approach, by presenting the stress testing approach and our dynamic mesoscopic simulator. The fourth section describes our two case studies. The fifth section reports on the evaluation of our approach in the considered case studies. A final discussion is then presented, while some directions for further work are proposed in the final section.

Literature Review

In this section, we firstly introduce the very general concept of resilience based on previous works, by adapting it to the context of road network analysis. We then present relevant road network resilience approaches for identifying critical links.

Resilience Definition

A wide diversity of definitions has been introduced in the literature to characterize resilience. Woods identified four major concepts (*5*): (i) resilience as rebound from trauma and return to equilibrium; (ii) resilience as a synonym for robustness; (iii) resilience as the opposite of brittleness, that is, graceful extensibility when surprise challenges boundaries; and (iv) resilience as network architectures that can sustain the ability to adapt to future surprises as conditions evolve.

Bruneau et al. (*6*) define resilience as the ability to: (i) mitigate hazards (*robustness* or *pre-perturbation resilience*); (ii) contain the effects of disasters during their occurrence (*reactivity*); and (iii) identify disruptions rapidly as they occur and mobilize resources to recover an acceptable traffic flow quickly (*recovery* or *post-perturbation resilience*).

Sullivan et al. define road network robustness as the degree to which the network can function in the presence of capacity disruptions on links. A robust road network can face disruptions on links with only slight increases in overall network-wide travel costs. Conversely, a non-robust road network is subject to substantial increases in costs (*7*).

In the rest of the paper, we tackle the problem of resilience assessment according to the perspective provided by Woods's second concept, that is, *robustness* (*5*), and the definition proposed by Sullivan et al (*7*). The paper proposes an approach based on stress testing for assessing the impact of day-to-day disruptions on network links by measuring overall travel costs, aiming to quantify the ability of the whole road network to absorb such an impact. We remark, as an additional contribution of this paper, that our approach also allows for assessment of resilience under the occurrence of unexpected and possibly very rare events, which corresponds to Woods's third concept that is, graceful extensibility when surprise challenges boundaries (*5*). Our stress testing methodology is also applied for evaluating the impact of sudden variations in travel demand on the overall travel costs.

Approaches to Assessment of Resilience

The majority of the approaches aimed at quantifying resilience are based on topology models and static network connectivity analysis (*3, 4, 8*). Among the plethora of metrics proposed to perform connectivity analysis, betweenness centrality (BC) is traditionally the best choice for traffic network analysis purposes, as it expresses the frequency with which a point falls between pairs of other points on the shortest paths connecting them (*9*). BC has therefore been widely adopted to assess road resilience, by identifying topologically vulnerable links and intersections (*10, 11*). Very recently, approaches based on machine learning and big data solutions (*12*) have been proposed to significantly reduce computation time of BC on several kinds of very large complex networks (*13*).

Even though a large majority of studies focus on static topological features of the network to identify its vulnerabilities, some authors have tried to join network topology features to more dynamic traffic-related information. Augmented definitions of BC take into account time-varying origin–destination (OD) travel demand (*14–17*), as well as travel times (*16, 18, 19*) for shortest paths computation. Such augmented metrics have been adopted for traffic flow analysis and prediction (*14, 15, 18, 19*) and traffic assignment (*16*) as well as network performance monitoring in the presence of extreme events (*17*). To the best of our knowledge, none of the previous approaches has been evaluated with respect to day-to-day disruptions (i.e., reduced link capacity), which represents one of the core aspects of our paper. Also, their applicability to resilience assessment is still neither widespread nor fully understood, these metrics often being inaccurate in highly dynamic environments (*20*) and prohibitive to compute for large-scale networks (*12*).

Another contribution of our paper is a methodology based on stress testing, that is, pushing a system beyond its normal operational capacity and observing how it responds to the applied stress. Stress tests have been widely used in banking (*21*, *22*), medical (*23*, *24*), and hydro-geology domains (*25*). Also referred to as network-disruption analysis, this approach has been occasionally leveraged in the field of transportation to identify critical links in a road network, but it is still at a very early stage (*17*, *26*, *27*). Sullivan et al. (*7*) set different link-based capacity-disruption values for identifying and ranking the most critical links and quantifying road network resilience. Jenelius and Mattsson (*28*) developed the notion of the importance of a link, which is a function of the increase in travel time when the link is disrupted. Their method is demand-aware as they weight travel time by demand. Stress testing allows intra-network comparison; links are ranked based on their contribution to the overall network resilience (*29*, *30*). Other authors have focused on identifying only the most critical nodes to be improved (*30*). Some studies use probability-based models to calculate the likelihood that a network continues functioning after a given stress (*31*, *32*). To the best of our knowledge, the large majority of research works based on stress testing for assessing the resilience of road networks mainly deal with disasters and extreme events (*17*, *33*, *34*), instead of day-to-day disruptions as targeted in this paper.

Assessment of the Literature

The definition and quantification of resilience varies greatly depending on the context and application domain. No universal and totally agreed definition or metric of resilience exists. Despite the plethora of work on resilience in various domains, relatively small numbers of studies have targeted resilience of road networks explicitly, and very few applications on real-world roads have been proposed. Our paper deals instead with a real network in France, in the Paris agglomeration area.

The concept of resilience can be divided into two parts: pre-perturbation resilience (robustness), and post-perturbation resilience. In this study, focused on traffic modeling and analysis, we evaluate pre-perturbation resilience only. We do not account for post-perturbation socio-technical actions. As the aim of this paper is to rank and identify the most critical links in a given network with respect to day-to-day disruptions, we focus on intra-network comparisons.

In the literature, two general approaches are commonly used to quantify resilience (*26*). The first is purely topological and usually demand-insensitive. The second takes into account traffic network performance and includes demand variability via simulation.

It is worth noting the lack of studies combining the stress testing approach with the topological one. We analyze and compare the two different approaches to assess the resilience of road networks, and to determine their advantages and drawbacks in assessing road network resilience. For the first approach, that is, the topological one, we consider and evaluate several metrics from graph theory. As for the second approach, by means of a simulator, we inject perturbations in the network and quantify to what degree it can adjust to them.

Methodology

In the following, we report on the topological metrics considered in our analysis as well as the stress testing technique proposed to perform demand-aware, dynamic disruption analysis.

Graph Theory Metrics

The topological metrics used in this study are based on BC, originally proposed by Freeman (*8*). BC measures the importance of the generic link l of a graph by considering the number of shortest paths that traverse it, and is defined as follows:

$$BC(l) = \sum_{i \neq l \neq j} \frac{\sigma_{ij}(l)}{\sigma_{ij}}, \quad (1)$$

where:

- $\sigma_{ij}(l)$ is the number of shortest paths from node i to node j that traverse link l;
- σ_{ij} is the total number of shortest paths from node i to node j.

In shortest path computation, links can be unweighted or weighted (e.g., in terms of the associated estimated travel time). In this study, we test and compare both cases. We also consider multiple variants of the BC, reported in the following, to model different aspects of a road traffic network.

BC for Entry and Exit Nodes Only. We propose to use an alternative definition of BC consisting in calculating the shortest paths from entry to exit nodes only. (The computation of BC on a subset of nodes [entries and exits] is based on the function *edge_betweenness_centrality_subset* from the *NetworkX* Python library.) This definition introduces two advantages: computation time is reduced, and the definition seems more realistic from a demand-aware perspective, since individuals tend to start and finish their trips over a subset of intersections. This corresponds to the OD representation of the traffic demand. The formula is the same as Equation 1 with some exceptions:

- i is selected from the entry-nodes subset, i.e., intersections used by vehicles to enter the network;
- j is selected from the exit-nodes subset, i.e., intersections used by vehicles to leave the network.

In conclusion, we consider four different formulations of the BC:

- Unweighted BC (BC)
- Travel-time weighted BC (TTWBC)
- Unweighted BC on entry/exit nodes only (BC entries–exits)
- Travel-time weighted BC from entry to exit nodes only (TTWBC entries–exits)

Spatio-temporal traffic properties and phenomena, like demand, congestion and dynamic re-routing, are typically not addressed in graph-based models. Thus, graph-based metrics are usually incapable of capturing these aspects in turn.

Demand-Sensitive Metric

Jenelius et al. (*2*) introduced the demand-aware metric of importance (I) to characterize transportation vulnerability. This metric allows measurement of network performance loss by using travel costs weighted by the traffic demand. Such metric is adequate for our methodological approach as it includes demand and the dynamic phenomenon of congestion (i.e., travel costs increase when traffic is congested). This metric uses a generic notion of travel cost, which can be specified depending on the study context and aim. In this paper, we define travel cost as travel time divided by travel distance (in seconds/kilometer). The importance of a link l is the following:

$$I(l) = \frac{\sum_i \sum_{j \neq i} x_{ij}(c_{ij}^{\delta}(l) - c_{ij}^{0})}{\sum_i \sum_{j \neq i} x_{ij}}, \qquad (2)$$

where:

- x_{ij} is the demand from origin node i to destination node j (number of vehicles);
- $c_{ij}^{\delta}(l)$ is the mean travel cost from origin node i to destination node j when link l is disrupted at level δ;
- c_{ij}^{0} is the mean travel cost from nodes i to j in the base case, i.e., without disruption.

Traffic Model and Algorithms

To model traffic dynamics we use a dynamic mesoscopic simulator based on the Lighthill-Whitham-Richards model (*35*, *36*) and implemented in Matlab by our research group (*37*, *38*). The Lighthill-Whitham-Richards model is formulated in Lagrangian-space coordinates and uses both Lagrangian and Eulerian observations. It represents individual vehicles but only records their transit times at network nodes. A dynamic traffic assignment procedure distributes vehicles along all the possible alternative paths in the network, according to the traffic conditions at the moment the vehicle is generated. More precisely, travel times on all paths are calculated based on traffic flow, and the vehicle chooses the path that requires the smallest travel time. The following parameters have to be specified before running simulations: simulation duration, OD demands, and link capacities. This simulator is adequate for our stress testing approach: as opposed to static topological indicators, it includes traffic dynamic properties such as demand, congestion, traffic-based route assignment, dynamic shortest path computation, and queues. Moreover, travel costs are calculated for each vehicle and can be extracted easily to compute network performance metrics, such as importance from Equation 2.

Stress Testing

The aim of our stress testing methodology is to identify the most critical links in the road network and to assess their resilience by considering the dynamic, spatio-temporal, and demand-dependent properties of network traffic.

Stress testing can be leveraged to quantify the adverse impacts associated with a reduction of capacity on specific links. Disruptive road events such as flooding, obstacles on the road, and traffic accidents are likely to reduce the capacity of a given link and negatively affect network performance. Measuring network performance loss when reducing the capacity of a given link provides the criticality of this link to the operation of the whole network. Therefore, stress testing is an adequate methodological approach to identify and rank the most critical links. It captures the relative importance of the disrupted link to the other links and assess the overall resilience of the whole road network from an intra-network comparison.

Our methodology for road network stress testing is composed of the following steps:

1. *Simulating disruptive road events*:

We propose two strategies to perform this step. In the first one, we simulate day-to-day road disruption as link capacity drops. The *capacity-disruption level* is defined as the reduction in link capacity, expressed as a fraction of the original one. In many studies, the capacity-disruption level is total, that is, 100% of the original value, which means that the capacity of the link is reduced to 0 vehicles per hour (*10*). The link is then completely removed from the road network. However, a 100% capacity-disruption level does not accurately reflect the actual link capacity resulting from frequent day-to-day disruptions or minor events (e.g., number of lane reductions, adverse weather, etc.) that can affect the network. That is why we gradually reduce the capacity to analyze the evolution of the performance depending on the capacity-disruption level. We consider five possible capacity-disruption levels, denoted as δ, on each examined link, that is, $\delta \in \{0\%, 20\%, 40\%, 60\%, 80\%\}$. Therefore, we can consider the following equation to compute the maximum capacity of the generic link l in the presence of a capacity-disruption level δ:

$$\delta = 100 \cdot (1 - \frac{q^{\delta}_{max}(l)}{q^{0}_{max}(l)}), \tag{3}$$

where:

- δ is the capacity-disruption level applied to link l (percentage) with $\delta \in \{0, 20, 40, 60, 80\}$;
- $q^{\delta}_{max}(l)$ is the capacity of link l when it is disrupted at level δ (in vehicles/hour);
- $q^{0}_{max}(l)$ is the capacity of link l in the base case (in vehicles/hour);

As a second strategy to simulate disruptive road events, we consider increases in the traffic demand on specific entry/exit nodes of the network. By this approach, it is possible to simulate exceptional situations like city evacuations following extreme events (e.g., flooding, attacks, etc.) which typically put significant strain on the road infrastructure and result in total congestion. This strategy consists in changing the OD matrix, that is, increasing the traffic flow from given entries, and comparing the stress testing results with another demand level.

Based on the selected strategy for disruptive road events, we set the parameters of our mesoscopic simulator (e.g., link capacity, traffic demand) and we simulate the network in the specific setting. For both strategies, travel costs are collected for all vehicles in order to compute the performance metrics described in the following point.

2. *Computing overall performance loss*:

This step is about quantifying the consequences of the simulated disruptive event on the operation of the whole network. To this purpose, we use the notion of road network performance, measured via the importance metric of Equation 2. Specifically, we consider travel time increase divided by travel distance as a measure of cost. Below, we provide the formulation of the overall performance loss (*PL*), based on the notion of importance, when a link l is disrupted:

$$PL(l, \delta) = \sum_{v=1}^{n} \frac{c^{\delta}_{v}(l) - c^{0}_{v}}{n} \tag{4}$$

where:

- $PL(l, \delta)$ is the overall performance loss when link l is disrupted at level δ (seconds/kilometer);
- $c^{\delta}_{v}(l)$ is the travel cost of vehicle v when link l is disrupted at level δ (seconds/kilometer);
- c^{0}_{v} is the travel cost of vehicle v in the base case (seconds/kilometer);
- n is the number of vehicles in the network.

3. *Analyzing the results*:

For each link we know the performance loss corresponding to the considered capacity-disruption level (i.e., 0%, 20%, 40%, 60%, and 80% of the original link capacity). As an example, $PL(2, 40\%)$ represents the overall performance loss on link 2 when it is disrupted at 40% of its initial capacity. In the evaluation section, we present and discuss the overall performance loss depending on the capacity-disruption level of the link according to two different strategies.

Traffic Robustness Index

The stress testing methodology reported in the previous section allows us to compute the overall performance loss for each link of the network with respect to five different capacity-disruption levels. To compare and identify critical links, we need a unique value of criticality associated to each link. To this purpose, a global metric is required to aggregate the performance loss values in the five different capacity-disruption levels. We propose the stress test criticality (STC) metric, defined as follows, for the generic link l:

$$STC(l) = \int_{\delta} PL(l, \delta), \tag{5}$$

where:

- $STC(l)$ is STC when link l is stress tested (seconds/kilometer)

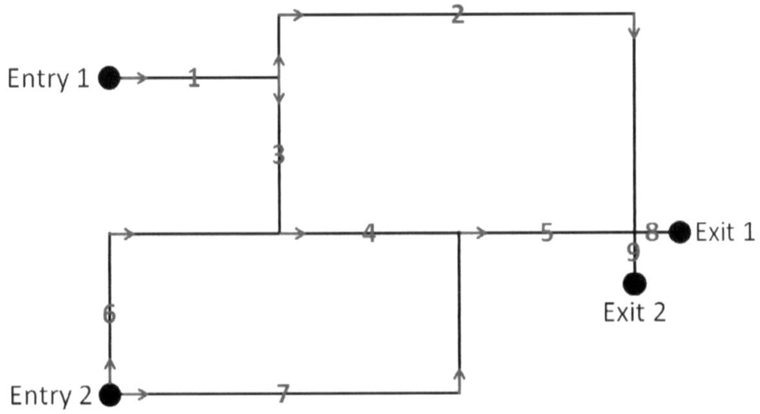

(a) Simple virtual road network

Destination / Origin	Exit 1	Exit 2
Entry 1	375	375
Entry 2	250	1000

(b) OD demand (vehicles/hour) for the simple virtual network (strategy A)

Destination / Origin	Exit 1	Exit 2
Entry 1	500	500
Entry 2	1200	300

(c) OD demand (vehicles/hour) for the simple virtual network in (strategy B)

Figure 1. Simple virtual network. Network structure (*a*) and OD travel demand (*b*, *c*) for the two considered strategies.

- δ is the capacity-disruption level of link l (percentage)
- $PL(l, \delta)$ is the overall performance loss (seconds/kilometer)

We use the trapezoid rule to approximate the integral in Equation 5 from the (five) overall performance values computed on link l.

Case Studies

The methodology and metrics described in the previous sections have been evaluated in two different case studies. The first one is related to a simple virtual network, used as a basic testbed for our approach. The second one is a real road network in France, which we use to confirm the validity of our results in a realistic scenario and to support the discussion on both simulation-based stress testing and topological metrics.

A Simple Virtual Road Network

This network is composed of eight nodes (4 road intersections, 2 entries, and 2 exits) and nine links. The duration of each simulation is fixed to 10 minutes. Figure 1*a* depicts the network structure with numbered links and flow directions. The network has been tested with two different demand levels reported as OD matrices in Figure 1*b* and *c*.

DIRIF: A Real-World Road Network

The DIRIF network is situated in the south of Paris, France, and includes 868 links and 827 nodes (657 intersections, 86 entries, and 84 exits). Its roads are mostly highways. Each simulation duration is fixed to 15 minutes. Since the network is much bigger than the virtual one described above, and traffic demand is extremely low on some links, we specify a higher simulation duration to ensure that enough vehicles travel through the whole network and to collect a significant number of travel cost observations. Simulation is performed with real demand data from 9:00 a.m. to 9:15 a.m., that is, the morning peak-time, in order to increase the probability of observing some performance loss in our stress tests. The network is graphically presented in Figure 3*a* and *b*.

Evaluation

In this section, we present the results of our stress testing methodology and discuss the link ranking derived from the different selected metrics. We show that the ranking

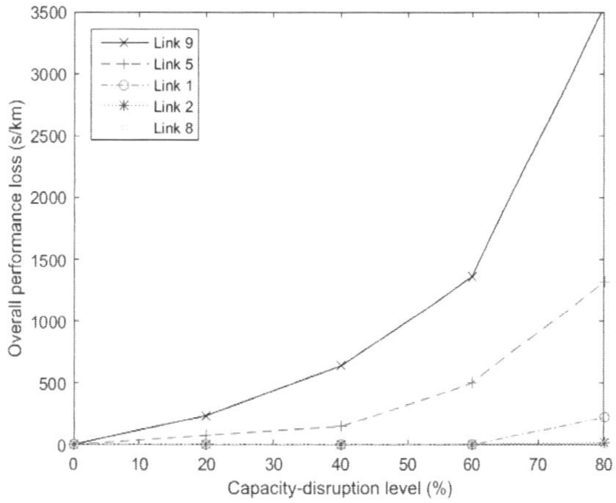

(a) Overall Performance Loss, strategy A (STC A)

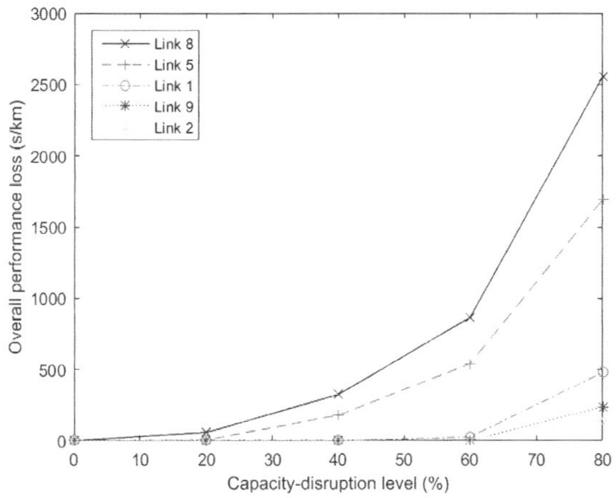

(b) Overall Performance Loss, strategy B (STC B)

Link	STC A	STC B	BC	BC entries-exits	TTWBC	TTWBC entries-exits
9	1st	4th	2nd	1st	4th	3rd
5	2nd	2nd	1st	1st	1st	1st
1	3rd	3rd	2nd	1st	4th	3rd
2	4th	5th	2nd	1st	7th	5th
8	5th	1st	2nd	1st	4th	3rd
6	5th	6th	4th	2th	5th	4th
4	5th	7th	2nd	2th	2nd	2nd
7	5th	8th	3th	1st	6th	4th
3	5th	8th	3th	2th	3rd	3rd

(c) Simple network link rankings generated by the different metrics of criticality

Figure 2. Performance evaluation on the simple virtual network.

of critical links on the same network can vary significantly when different indicators are used, thus proving that simple modifications of one centrality indicator can have a relevant impact on the capacity of the metric to capture different facets of resilience. Moreover, we discuss the advantages and drawbacks of each different approach in assessing road network resilience, and provide guidelines that can be helpful towards the definition of a new enhanced centrality metric.

Application to a Simple Virtual Network

In the scenario of the simple network described in the previous section, we measured STC and all of the proposed topological metrics on all the links. To perform stress testing, we used both strategies described above, that is, link capacity drop (referred to as A in the following) and traffic demand increase (referred to as B). The measures of STC that result from the two strategies above are distinguished as STC A and STC B, respectively. STC A and B are calculated with the same formula, but different parameters are set before stress testing.

First we discuss the results of strategy A. In our simulations, we applied sequentially five capacity-disruption levels (i.e., 0%, 20%, 40%, 60%, 80%) to each link. We then measured the network-wide performance loss (i.e., Equation 4) consequent to the disruption applied to the link. The overall performance loss from our stress tests is reported on the y-axis of Figure 2a, while the corresponding capacity-disruption levels (δ) are reported on the x-axis. Results for different links are depicted with different colors and markers in the figure, using a linear interpolation. For readability, the figure only reports the five most critical links (i.e., those with the highest overall performance loss).

Intuitively, a link capacity drop translates into an increase of network-wide travel cost.

According to the definition provided in Equation 5, the STC of a given link corresponds to the area below its curve. Thus, in strategy A, *link 9* is the most critical to the operation of the whole road network, followed by *links 5, 1, 2*, and 8. By using link ranking from STC as a baseline, we compare in the following the other link rankings as derived from the different selected topological metrics. Figure 2*c* reports on such link rankings for both STC and the whole set of topological metrics.

As a preliminary consideration, it can be observed that the rankings of critical links on the same network may dramatically change depending on the metric, due to the different properties of the network captured by each of them. As an example, *link 5* is on top of all the topological rankings whereas, with STC, it is ranked second, below *link 9*. The top-rank of *link 5* by all the topological metrics can be explained by the large number of shortest paths traversing this link: e.g., paths (*4, 5*), (*7, 5*), (*7, 5, 8*), (*7, 5, 9*), (*6, 4, 5*), (*7, 5, 8*) are all shortest paths.

The different ranking issued by STC A can be easily explained. If *link 5* is disrupted, alternative paths through *link 2* exist for all individuals heading to *exit 1* or *exit 2*. Conversely, when *link 9* is disrupted, no alternative path exists for users willing to travel to *exit 2* from both *entry 1* and *entry 2*, thus resulting in serious congestion and increased travel time for all individuals heading to *exit 2*. Additionally, traffic demand for *exit 2* is very high (see OD matrix in Figure 1*b*). This explains why *link 5* is more critical than *link 9* in terms of topology, but less critical than *link 9* when considering demand data, as made possible by our stress testing methodology (based on dynamic simulations) and captured by the related criticality metric. This simple test clarifies how traditional demand-agnostic approaches may fail in properly ranking edge criticality.

Our simple test also shows that alternative paths may become shortest paths of the network as links are disrupted by adverse event, thus attracting traffic flow previously directed through the disrupted links. This represents another fundamental aspect that is impossible to capture with a static graph-based approach. However, this does not necessarily mean that topological metrics are not good resilience indicators, but rather that road graph modeling should include a dynamic component (e.g., edge weights), and that BC metrics should be rapidly re-computed after relevant network disruptions.

Another striking difference worth analyzing concerns *link 2*: considered as one of the most critical ones according to the BC metric from entries to exits, it is the least critical one for the TTWBC and the TTWBC from entries to exits. The peculiarity of *link 2* is its length; it is the longest one, thus demanding more time to be travelled than the other links. Metrics like BC and BC from entries to exits are not weighted, that is, all links are valued equally, and consequently they are unable to grasp this important aspect. In contrast, links with high travel times are not considered critical by the analyzed weighted approaches, because they are not often part of shortest paths. The same consideration applies to *link 7*, which is the second-longest link of the network.

Finally, it is worth noting that BC values are often clustered. In particular, the BC from entries to exits has many equal values and only two link ranks. Traditional topological metrics appear to have very limited capability to discriminate link criticality at a fine level. In this case, STC does not differentiate all links either, but this is due to capacity-disruption levels. For *links 8, 6, 4, 7*, and *3* which are all ranked in fifth place with the same value, both capacity-disruption and demand levels are not high enough to observe a significant performance loss compared with the base case. For example *link 8* capacity disruption does not affect the overall network performance (see Figure 2*a*). The overall travel cost remains at its base case value. That is why some links have the same criticality value. STC differentiation between links then depends on capacity-disruption and demand levels.

To investigate this aspect further, we use our second stress testing strategy B, that is, with different traffic demand as reported in Figure 1*c*. Results are shown in Figure 2*b* and *c* (i.e., STC B). Link ranking changes significantly when different demand levels are used. As an example, *Link 8* becomes the most critical link, whereas in the previous case it involved no performance loss. This is due to the large increase in demand level associated with *exit 1*, which is directly connected to *link 8* (see Figure 1*a*).

Takeaways: *Critical link ranking is highly variable as different approaches are used. Resilience analysis via topological metrics is limited in the sense that such metrics do not usually take into account traffic demand and network re-configurations following disruptive events. Conversely, the simulation-based stress testing approach is able to capture these aspects thus providing more realistic rankings via the proposed performance loss metric. Stress testing can also be used to compare different road networks and sub-networks by analyzing their response to similar stresses. Travel-time weighted BC produces better estimations of link criticality compared with unweighted BC, which treats all links equally.*

Application to a Real Road Network

To confirm the results of our previous analysis in a realistic scenario, we considered the Paris DIRIF road network, described above. Given the large size of this

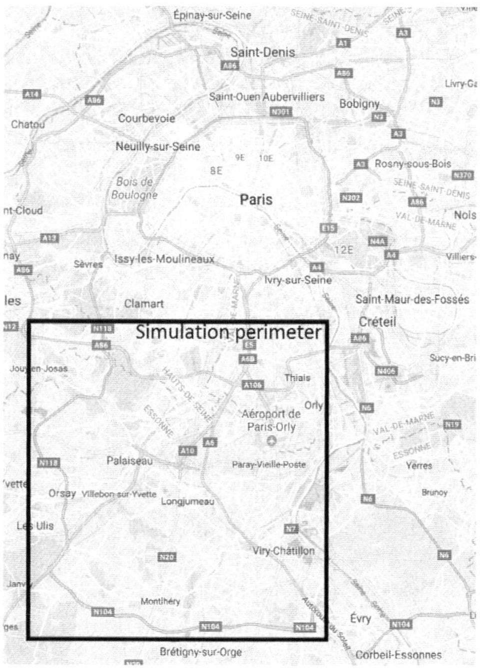

(a) DIRIF, geographic location

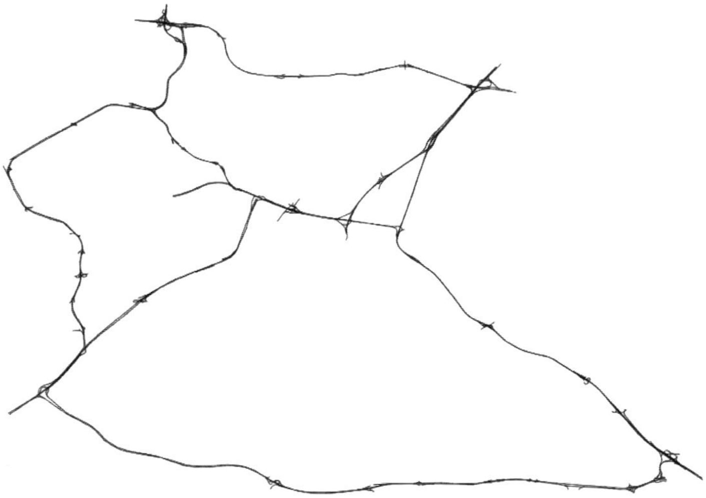

(b) DIRIF network

	Link	STC A	BC	BC entries-exits	TTWBC	TTWBC entries-exits
Highest-demand links	**95**	117	2628	83	2628	83
	93	102	660	83	660	83
	94	101	1974	83	1974	83
Highest-BC links	**802**	42	192497	3029	192717	3039
	803	27	192521	3029	192741	3039
	607	27	192509	3029	192729	3039
	608	44.6	192449	3029	192669	3039
	397	15.8	83164	1139	83164	1139
	672	14.5	10	1	10	1

(c) DIRIF network link values generated by the different metrics of criticality

Figure 3. The DIRIF road network in Paris agglomeration (*a*) and (*b*). Evaluation results (*c*).

network and the high computation time associated with each network simulation (stress testing one link from such a large network with five capacity drops takes more than one hour on an Intel Xeon E3 CPU equipped with 8 GB of RAM running a Matlab implementation of the mesoscopic simulator, properly configured to handle the DIRIF network in Scenarios A and B), it was prohibitive to perform an exhaustive stress test analysis as in the simple network case. Therefore, we performed stress tests on a limited set of representative links: the three links with the highest demand, the three with the highest BC, and three randomly selected edges with BC in three classes of values (high, medium, and low). We discuss in the following only our simulations related to strategy A. (Simulations for strategy B were in line with the results reported in the previous section and are not discussed due to space limitation.) Figure 3*c* reports the actual values of the considered metrics for the analyzed links. Unlike Figure 1, we do not report metric rankings but actual values of the metrics for each analyzed link. This is motivated by the impossibility of obtaining the full ranking for performance loss (i.e., STC).

Consistently with our previous analysis on the simple network, Figure 3*c* shows that rankings of critical links vary significantly. As an example, *links 95, 93, 94* have a very high value of STC A, whereas the topological metrics rate them much less critical than *links 802, 803, 607*. As pointed out in the previous section, the STC A ranking appears to be more realistic since it captures the higher criticality of *links 95, 93, 94* due to the associated higher demand (not reported due to space limitations).

On the small link subset considered in our analysis, taking into account travel times (TTWBC and TTWBC entries–exits) does not significantly change rankings, since link lengths (and therefore the average travel times) happen to be very similar on all considered links. Finally, it is worth noting that in the DIRIF network BC values (especially in the entries/exits variations) tend to cluster significantly themselves (i.e., many edges have very similar values of BC), thus exhibiting a lower discriminant power than in the case of the simple virtual network.

Takeaways: *In a real-world scenario, stress testing proved to be a realistic and reliable approach to evaluate network resilience. Our evaluation confirms the importance of traffic demand and network dynamics for fine-grained ranking of the most vulnerable road network links. Stress testing has however the drawback of requiring high execution times due to computationally intensive network simulations.*

Discussion and Perspectives

From the previous results, we summarize in the following a few guidelines for properly characterizing critical links with respect to day-to-day disruptions, by means of an intra-network approach in different application contexts.

Firstly, if resilience has to be evaluated in a relatively static context (e.g., network maintenance or planning), BC and TTWBC appear to be adequate. In particular, if data about demand and travel times are not available, we recommend BC, BC from entries and exits, and BC on all paths from entries and exits. These indicators do not require special knowledge of network performance and demand data, but only the basic network topology (links and intersections). If traffic demand is the only missing information, STC, TTWBC, and TTWBC from entries to exits should be preferred, since they also take into account travel time information.

If the goal is instead to achieve a more accurate characterization of network resilience, stress testing should be chosen, since it produces reliable results by taking into account traffic demand and congestion phenomena. The drawback is that it requires many computationally intensive simulations, thus being recommended only in application scenarios that allow for larger computation time, or that address small-sized (sub-)networks. Conversely, in domains with very stringent requirements on response time (e.g., on-line vulnerability monitoring), topological indicators could be the only valuable option. However, it is worth remarking that efficient solutions are still required to compute these metrics on very large networks within reasonable computation time. For future work, we are currently working on the switch from a Matlab implementation of our simulation-based tools to a new implementation based on faster programming languages and approaches explicitly designed for big data processing and real-time computation (e.g., Python and Scala on top of the Spark processing framework). In particular, we believe that implementing real-time advanced solutions for data-driven, on-line monitoring of road traffic resilience will constitute a fundamental research problem to investigate.

We advance that, in order to improve road network resilience analysis, future research work is needed that should consider joining graph-based approaches with demand-aware dynamic stress testing techniques. In this context, we believe that a further improvement with topological metrics could be achieved by modeling the road network as a dynamic graph, whose link weights may change over time depending on actual traffic conditions and both structural and performance-related network properties (e.g., road capacity, real-time traffic information, etc.).

Finally, we argue that future work should also consider area-wide disruptions in addition to single link-based disruptions, especially in the light of measuring the impact of extreme events.

Conclusion

Identifying links critical to the overall network performance is part of road network resilience and intra-network analysis. To this purpose, we have analyzed in this paper several topological metrics based on BC and proposed a stress testing approach exploiting a dynamic simulator. Stress testing appears to be a very promising solution for resilience analysis, allowing for measurement of resilience in terms of the overall performance loss of the whole network consequent to simulated link disruptions.

Our analysis shows that link ranking varies greatly when different metrics are used. As opposed to purely topological metrics, the proposed stress testing approach takes into account demand levels and dynamic characteristics of road traffic. However, it requires more computation time and data than traditional graph-based metrics. The choice of a relevant metric for assessing road network resilience should depend on the context and the specific application requirements.

Merging static topological metrics and demand-based approaches could be of further research interest. It could be relevant to adopt dynamic graphs modeling, using link weights to include dynamic information on the network. In such an approach, topological metrics should be computed dynamically by means of efficient quasi real-time solutions.

Author Contributions

The authors confirm contribution to the paper as follows: study conception and design: Gauthier, Furno, El Faouzi; data collection: Gauthier, Furno, El Faouzi; analysis and interpretation of results: Gauthier, Furno, El Faouzi; draft manuscript preparation: Gauthier, Furno, El Faouzi. All authors reviewed the results and approved the final version of the manuscript.

References

1. Berdica, K. An Introduction to Road Vulnerability: What Has Been Done, Is Done and Should Be Done. *Transport Policy*, Vol. 9, No. 2, 2002, pp. 117–127.
2. Jenelius, E., T. Petersen, and L.-G. Mattsson. Importance and Exposure in Road Network Vulnerability Analysis. *Transportation Research Part A: Policy and Practice*, Vol. 40, No. 7, 2006, pp. 537–560.
3. Di Gangi, M., and A. S. Luongo. Measures of Network Vulnerability Indicators for Risk Evaluation and Exposure Reduction. *Environmental Health Risk III*, Vol. 9, 2005, p.1151.
4. Tu, Y., C. Yang, and X. Chen. Methodology for Evaluating and Improving Road Network Topology Vulnerability. In *2010 International Conference on Intelligent Computation Technology and Automation (ICICTA)*, Vol. 2, IEEE, 2010, pp. 664–669.
5. Woods, D. D. Four Concepts for Resilience and the Implications for the Future of Resilience Engineering. *Reliability Engineering & System Safety*, Vol. 141, 2015, pp. 5–9.
6. Bruneau, M., S. E. Chang, R. T. Eguchi, G. C. Lee, T. D. O'Rourke, A. M. Reinhorn, M. Shinozuka, K. Tierney, W. A. Wallace, and D. Von Winterfeldt. A Framework to Quantitatively Assess and Enhance the Seismic Resilience of Communities. *Earthquake Spectra*, Vol. 19, No. 4, 2003, pp. 733–752.
7. Sullivan, J. L., D. C. Novak, L. Aultman-Hall, and D. M. Scott. Identifying Critical Road Segments and Measuring System-wide Robustness in Transportation Networks with Isolating Links: A Link-based Capacity-reduction Approach. *Transportation Research Part A: Policy and Practice*, Vol. 44, No. 5, 2010, pp. 323–336.
8. Freeman, L. C. A Set of Measures of Centrality Based on Betweenness. *Sociometry*, Vol. 40, No. 1, 1977, pp. 35–41.
9. Zhang, Y., X. Wang, P. Zeng, and X. Chen. Centrality Characteristics of Road Network Patterns of Traffic Analysis Zones. *Transportation Research Record: Journal of the Transportation Research Board*, 2011. 2256: 16–24.
10. King, D., A. Shalaby, and P. Eng. Performance Metrics and Analysis of Transit Network Resilience in Toronto. Presented at 95th Annual Meeting of the Transportation Research Board, Washington, D.C., 2016.
11. Berche, B., C. von Ferber, T. Holovatch, and Y. Holovatch. Resilience of Public Transport Networks Against Attacks. *The European Physical Journal B*, Vol. 71, No. 1, 2009, pp. 125–137.
12. Furno, A., N. E. El Faouzi, R. Sharma, and E. Zimeo. Two-level Clustering Fast Betweenness Centrality Computation for Requirement-driven Approximation. *IEEE Big Data 2017 Conference*, 2017.
13. Furno, A., N. E. El Faouzi, R. Sharma, and E. Zimeo. Reducing Pivots of Approximated Betweenness Computation by Hierarchically Clustering Complex Networks. In *International Conference on Complex Networks and their Applications*, Springer, 2017, pp. 65–77.
14. Jayasinghe, A., K. Sano, and H. Nishiuchi. Explaining Traffic Flow Patterns Using Centrality Measures. *International Journal for Traffic & Transport Engineering*, Vol. 5, No. 2, 2015, pp. 134–149.
15. Kazerani, A., and S. Winter. *Modified Betweenness Centrality for Predicting Traffic Flow*, Vol. 2, 2009.
16. Puzis, R., Y. Altshuler, Y. Elovici, S. Bekhor, Y. Shiftan, and A. Pentland. Augmented Betweenness Centrality for Environmentally Aware Traffic Monitoring in Transportation Networks. *Journal of Intelligent Transportation Systems*, Vol. 17, No. 1, 2013, pp. 91–105.
17. Wisetjindawat, W., A. Kermanshah, S. Derrible, and M. Fujita. Stochastic Modeling of Road System Performance During Multihazard Events: Flash Floods and Earthquakes. *Journal of Infrastructure Systems*, Vol. 23, No. 4, 2017, 04017031.
18. Altshuler, Y., R. Puzis, Y. Elovici, S. Bekhor, and A. B. Pentland. Augmented Betweenness Centrality for Mobility Prediction in Transportation Networks. *Finding Patterns of Human Behaviors in Network and Mobility Data (NEMO)*, 2011.
19. Cheng, Y.-Y., R. Ka-Wei Lee, E.-P. Lim, and F. Zhu. Measuring Centralities for Transportation Networks Beyond Structures. *Applications of Social Media and Social Network Analysis*, Springer, 2015, pp. 23–39.
20. Duan, Y., and F. Lu. Robustness of City Road Networks at Different Granularities. *Physica A: Statistical Mechanics and its Applications*, Vol. 411, 2014, pp. 21–34.
21. Haldane, A. Why Banks Failed the Stress Test. *BIS Review*, Vol. 18, 2009, 2009p.
22. Schuermann, T. Stress Testing Banks. *International Journal of Forecasting*, Vol. 30, No. 3, 2014, pp. 717–728.
23. Goldschlager, N., A. Selzer, and K. Cohn. Treadmill Stress Tests as Indicators of Presence and Severity of Coronary Artery Disease. *Annals of Internal Medicine*, Vol. 85, No. 3, 1976, pp. 277–286.
24. Baertschi, S. W., K. M. Alsante, and R. A. Reed. *Pharmaceutical Stress Testing: Predicting Drug Degradation*. CRC Press, 2016.

25. Zhang, L. M., L. Gao, S. Y. Zhou, R. W. M. Cheung, and S. Lacasse. Stress Testing Framework for Managing Landslide Risks Under Extreme Storms. *Workshop on World Landslide Forum*, Springer, 2017, pp. 17–32.
26. Mattsson, L.-G., and E. Jenelius. Vulnerability and Resilience of Transport Systems – A Discussion of Recent Research. *Transportation Research Part A: Policy and Practice*, Vol. 81, 2015, pp. 16–34.
27. Clarke, J., and E. O'Brien. A Multi-hazard Risk Assessment Methodology, Stress Test Framework and Decision Support Tool for Transport Infrastructure Networks. *Transportation Research Procedia*, Vol. 14, 2016, pp. 1355–1363.
28. Jenelius, E., and L.-G. Mattsson. Road Network Vulnerability Analysis of Area-covering Disruptions: A Grid-based Approach with Case Study. *Transportation Research Part A: Policy and Practice*, Vol. 46, No. 5, 2012, pp. 746–760.
29. Scott, D. M., D. C. Novak, L. Aultman-Hall, and F. Guo. Network Robustness Index: A New Method for Identifying Critical Links and Evaluating the Performance of Transportation Networks. *Journal of Transport Geography*, Vol. 14, No. 3, 2006, pp. 215–227.
30. Berdica, K., and M. Lars-Göran. Vulnerability: A Model-based Case Study of the Road Network in Stockholm. In: Murray A.T., Grubesic T.H. (eds) *Critical Infrastructure. Advances in Spatial Science*. Springer, Berlin, Heidelberg, 2007, pp. 81–106.
31. Zhen-Ping, D., and A. Nicholson. Degradable Transportation Systems: Sensitivity and Reliability Analysis. *Transportation Research Part B: Methodological*, Vol. 31, No. 3, 1997, pp. 225–237.
32. Church, R., and M. P. Scaparra. Analysis of Facility Systems Reliability When Subject to Attack or a Natural Disaster. In: Murray A.T., Grubesic T.H. (eds) *Critical Infrastructure. Advances in Spatial Science*. Springer, Berlin, Heidelberg, 2007, pp. 221–241.
33. Donovan, B., and D. B. Work. Using Coarse GPS Data to Quantify City-scale Transportation System Resilience to Extreme Events. *arXiv preprint arXiv:1507.06011*, 2015.
34. Mayada, O., A. Mostashari, and R. Nilchiani. Assessing Resilience in a Regional Road-based Transportation Network. *International Journal of Industrial and Systems Engineering*, Vol. 13, No. 4, 2013, pp. 389–408.
35. Lighthill, M. J., and G. B. Whitham. On Kinematic Waves. II. A Theory of Traffic Flow on Long Crowded Roads. In *Proceedings of the Royal Society of London A: Mathematical, Physical and Engineering Sciences*, Vol. 229, The Royal Society, 1955, pp. 317–345.
36. Richards, P. I. Shock Waves on the Highway. *Operations Research*, Vol. 4, No. 1, 1956, pp. 42–51.
37. Duret, A., L. Leclercq, and N.-E. El Faouzi. Data Assimilation Using a Mesoscopic Lighthill–Whitham–Richards Model and Loop Detector Data: Methodology and Large-scale Network Application. *Transportation Research Record: Journal of the Transportation Research Board*, 2016. 2560: 26–36.
38. Laval, J. A., and L. Leclercq. The Hamilton–Jacobi Partial Differential Equation and the Three Representations of Traffic Flow. *Transportation Research Part B: Methodological*, Vol. 52, 2013, pp. 17–30.

The Standing Committee on Critical Transportation Infrastructure Protection (ABR10) peer-reviewed this paper (18-03204).

Article

TRR
JOURNAL OF THE TRANSPORTATION RESEARCH BOARD

Transportation Research Record
2018, Vol. 2672(1) 66–75

DOI: 10.1177/0361198118794057
journals.sagepub.com/home/trr
SAGE

Spatially Clustered Autonomous Vehicle Malware: Producing New Urban Geographies of Inequity

Evan W. Vassallo[1] and Kevin Manaugh[2]

Abstract

Malicious software (malware) is both a hurdle to autonomous vehicle (AV) adoption and a serious threat to AV occupant safety. Yet, to date, the topic of how subnational transportation decision makers should handle this cybersecurity threat has been underexplored. This paper describes how malware can spread through vehicle-to-X networks and infect AVs to a non-technical planning audience. It then explains how AV malware can cluster spatially in linguistic, socioeconomic, and political enclaves of American cities. Since AVs can identify new indicators of safety imperceptible to humans, malware clusters can produce new geographies of accessibility, mobility, economic, and environmental inequity across a city. In provisioning the inherently localized technical malware prevention tools available to them, subnational transportation planners have the capacity to reproduce or overcome historical and current transportation system inequities.

When malicious software (malware) infects an ordinary personal computer, data is in danger. When malware (*1*) infects a light autonomous passenger vehicle (here denoted as AV), *human lives* are in danger (*2*). Vehicular malware can be programmed to take control of vehicle components (such as the brakes and the engine) and cause a collision (*3*). Driverless AVs may provide many benefits including improved traffic flow (*4*), more leisure time for would-be drivers (*5*), and reduced vehicle ownership due to increased car sharing (*5*). Among these, saving human lives through preventing collisions caused by human error is one of the most significant (*6*). To achieve this safety improvement, AVs require massive amounts of information from on-board image processing, radar, and other sensory technologies (*7*).

To improve traffic safety, American, European, and Asian regulators plan to require AVs to be capable of disseminating some of this information through vehicle-to-X (V2X) networks (*8*). The foremost types of V2X networks are vehicle-to-vehicle (V2V) networks (which facilitate communication between vehicles) and vehicle-to-infrastructure (V2I) networks (which connect vehicles to roadway infrastructure devices, or roadside units, RSUs). While some researchers (e.g., [*9*, *10*]) define V2X networks as being distinct from V2V networks, the definition this paper uses appears to be more common—especially in corporate communications and among regulators (*8*). V2X regulations would necessitate the construction of V2X network infrastructure to function.

Like all computer networks that circulate information, V2X networks can also enable malware transmission and infection (*11*). As such, planners should consider preventing V2X network-facilitated vehicular malware infection in order to increase road safety. As manufacturers may begin to commercialize AVs within the next decade by some estimates (*4*, *12*) and V2X regulations may come into force even sooner (*8*), the need to start planning for vehicular cybersecurity threats is urgent.

AV malware infection may prevent the equitable distribution of AV benefits across a city. The targeted, non-generalizable nature of malware threats—combined with the localized, immobile infection points that comprise V2I networks—can yield spatial clusters of malware. Human-driven vehicles could also connect to these networks and be infected by vehicular malware, but AV computers will be better equipped to perceive, predict, and avoid malware-prone areas than humans are. As a result, the physical spaces these clusters occupy could receive a disproportionate share of the accessibility, mobility, economic, and environmental benefits AVs may produce.

[1]Department of Geography, McGill University, Montreal, QC, Canada
[2]Department of Geography and McGill School of Environment, McGill University, Montreal, QC, Canada

Corresponding Author:
Address correspondence to Kevin Manaugh: kevin.manaugh@mcgill.ca

Fortunately, several tools that can prevent vehicular malware infection are available to state- and municipal-level transportation planners. The most viable of these tools, though, are only effective within a set distance away from where planners deploy them. The deployment of malware prevention measures can reproduce or overcome historical and current spatial inequities in urban accessibility and mobility.

Scope

AVs are vulnerable to malicious attacks through many channels. For example, attackers may physically tamper with a vehicle's hardware, intercept a vehicle's communication signals, or hijack a vehicle's connection to a centralized server. This paper, however, will focus primarily on AV-compatible malware transmitted through V2I, and to a lesser extent, V2V networks.

The geographically stratifying effects of AV malware infection would become more pronounced when some minimum density of AVs that consistently make routing decisions, choose driving destinations, and execute entire trips themselves begins to operate in a city. SAE International's J3016 vehicular automation taxonomy (*13*), adopted by the National Highway Transportation Safety Association (NHTSA) as of October 2016 (*14*), would classify a vehicle with these abilities as Level 4 (on a 6-point scale from Level 0 to Level 5). Though it is difficult to determine an exact threshold AV market share, the arguments presented here will assume a widespread, mainstream adoption of Level-4 automation (a.k.a, "Conditional Automation" [*13*]) AVs.

Both national and subnational policy affect the malware threats AV (and related infrastructure) adoption pose. As this paper will examine the role of metropolitan- and state-level planners in preventing AV malware, focusing on a specific country will allow for more concrete analysis of subnational-level malware prevention tools. Unlike many other countries, American federal (*15*) and state (*16*, *17*) entities have already proposed regulations governing Level-4 AVs. This early insight into the American AV regulatory landscape, combined with a rich set of American transportation equity case studies, provides a solid literature base to work from. Therefore, this paper will confine its analysis to American urban areas.

With the exception of some limited-access highways, most urban roads (and, by extension, their supporting infrastructure) are under the control of state- and municipal-level transportation planning departments (*18*). Therefore, this paper will only discuss malware prevention measures that would reasonably fall within the jurisdiction of these departments.

To improve accessibility for non-technical planning audiences, this paper will only describe technical malware prevention measures in terms of broad categories of physical devices and very high-level technical architecture. Other, lower-level technological variables (such as network architecture and cybersecurity design) may also increase or decrease the extent to which malware spatially clusters across a city; however, a discussion of these variables falls outside the scope of this paper.

Literature Review

A wide-ranging literature on vehicular cybersecurity threats and countermeasures already exists (*19*, *20*). This paper will focus on malware infection and prevention, only two aspects of the broader AV cybersecurity domain. Kim et al. have described how information (including malware) can propagate through a V2X network (*21*). Basaras et al. and Zhang et al. both proposed distributed, network-based measures to prevent malware from propagating further once detected (*11*, *22*, *23*). The technical details of their solutions will appear in the Malware Prevention section. Each of these authors also discusses the challenges and limitations of implementing a malware prevention solution. Like the specific articles discussed thus far, virtually all AV malware prevention research the authors found is directed at a technical audience. This paper will describe a portion of these malware threats and prevention measures to a non-technical, transportation planning audience.

Despite this abundance of vehicular cybersecurity research, a comparative dearth of work examines how public policy and transportation planners should address cybersecurity issues. Litman, and Schoettle and Sivak briefly list cybersecurity threats as roadblocks to AV adoption and AV benefit fulfillment, respectively (*5*, *24*). Fagnant and Kockelman express a need for focused AV cybersecurity policy research (*4*). This knowledge gap held by transportation policy researchers also applies to planning practitioners. As recently as 2015, only 2 out of 25 selected American metropolitan planning organizations included AVs in their long-term regional transportation plans (RTPs) (*25*), let alone AV malware prevention strategies.

Researchers like Mezzour et al. and Anthe et al. found that rates of computer malware infection vary widely between countries (*26*, *27*); however, the authors are not aware of any studies that attempted to detect spatial malware variance across a single city. Nor could they find any literature that discusses the spatially stratifying effect of both AV malware infection and prevention in urban environments.

Methodology

The authors conducted a literature review spanning several fields, examining peer-reviewed work on the nature of malware, AV-specific malware threats, and how malware propagates through V2X networks. Another line of review uncovered articles about AV malware prevention efforts, especially those that subnational planning departments could implement. Yet another prong focused on how spatially clustered demographic groups have disproportionately experienced either accessibility gains or losses through infrastructure construction or planning policy. Finally, the authors reviewed articles that suggested certain AV characteristics (such as manufacturer and language setting) could cluster spatially.

AV Malware Threats

In order to infect a computer, malware requires a conduit. Specifically, a computer or a set of computers (such as an AV) must have some digital connection to the world around it to be vulnerable to malware attacks. Such a connection could be a physical peripheral outlet (such as a USB port) or a wireless connection (over a protocol such as Bluetooth or Wi-Fi). As vehicles form digital connections to V2X networks, these networks can serve as malware infection conduits. Figure 1 illustrates how information (including malware) can propagate through V2I and V2V networks.

All passenger automobiles on the market today have some kind of digital connection to the outside world, with an ever-growing share of them maintaining persistent Internet connections while on the road (*29*). Through these connections, attackers may tamper with on-board computers and remotely hijack vehicle functions such as braking and speed control (*3*). Despite these vulnerabilities, it appears that no neighborhood-scale spatial clustering of malware exists in American cities (although no published work describes an attempt to find or even theorize one). The introduction of AVs and V2X networks, though, may create the potential for spatial malware clustering. By removing the need to pay attention to the road, AVs will allow drivers to conduct Internet browsing activity that increases malware infection risk (like pirating media or viewing pornography) (*30*) that falls outside the limited browsing options vehicle infotainment systems offer today (*31*). Further, the deployment of RSUs to form V2I networks, though designed to improve safety, will introduce immobile information broadcast points that could infect vehicles driving within RSU signal range.

Malware Threat Variability

Attackers, by necessity, write malware to target specific software characteristics. Just as no one piece of software can run on every computer, no one piece of malware can successfully infect all computers. One can transmit malware to an AV, and that AV can attempt to open that malware, but it may not actually infect the AV depending on the AV's software characteristics. Although attackers often target characteristics like operating systems (e.g., Windows, Linux) and installed applications (e.g., navigation or mapping tools), they also target traits like software manufacturer and system software locale (i.e., language and date settings) (*27*).

Certain AV software characteristics can cluster spatially within a city around demographic group and land use clusters (see Figure 2 for an illustration). For example, malware targeting a locale corresponding to one of a city's minority languages could infect RSUs located inside that city's associated linguistic enclave. Infected RSUs could then infect vehicles both passing through and parked in the enclave. By targeting particular AV software manufacturers, attackers could also indirectly induce malware clustering in socioeconomic enclaves.

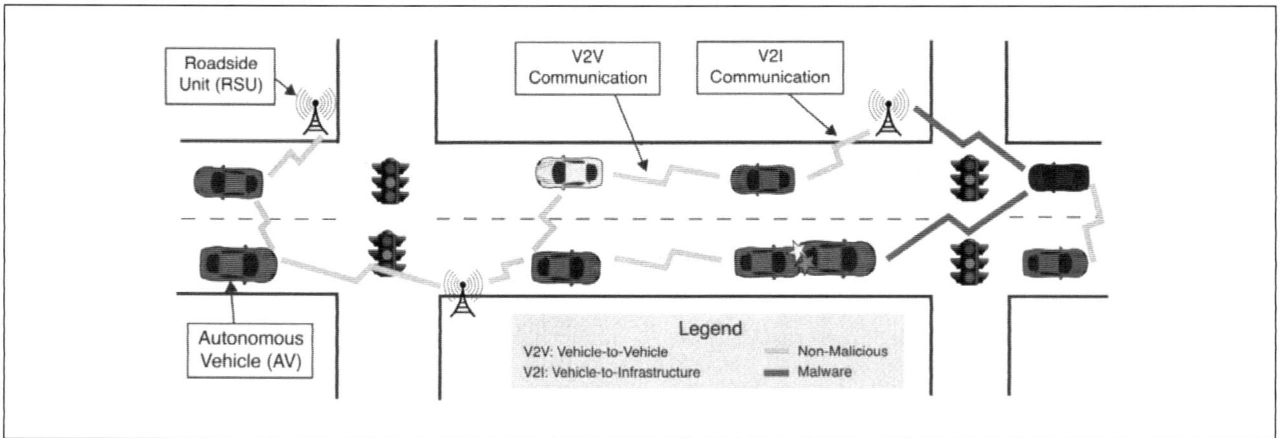

Figure 1. A figure adapted from Al-Qutayri et al. (*28*) depicting how both malware and non-malicious information can travel through vehicle-to-vehicle (V2V) and vehicle-to-infrastructure (V2I) networks comprised of autonomous vehicles (AVs) and roadside units (RSUs).

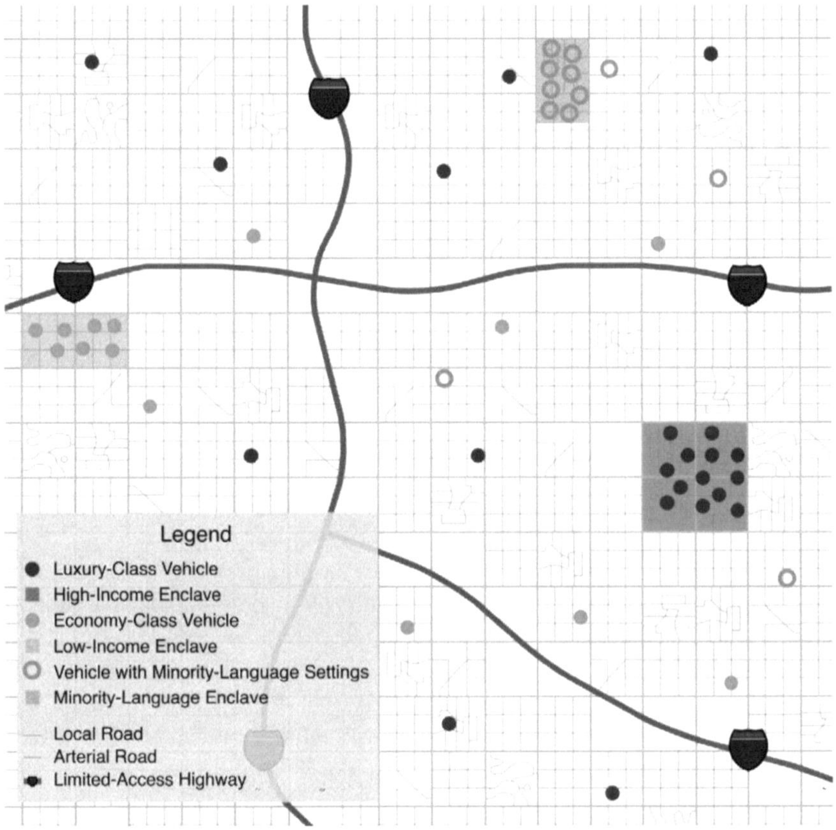

Figure 2. A fictional, schematic map illustrating how certain autonomous vehicle (AV) brands and other AV characteristics could overlap demographic group enclaves.

For example, they could target a manufacturer that exclusively produces luxury- or economy-class vehicles. This same phenomenon can occur in areas where residents holding similar political ideologies cluster. Sexton and Sexton found that every 1% of Green Party voter share (as a proxy for environmental protection support) in a Colorado zip code area added 1.01% to the Toyota Prius's market share in that area (*32*). They theorized that people in favor of strong environmental protections tend to purchase Priuses in part to broadcast their political values and in turn bolster their social status within their communities. Should attackers target a specific brand, manufacturer, or other AV software characteristic, the physical enclaves hosting clusters of that AV characteristic could also contain an associated malware cluster.

Even if attackers target all AV characteristics at the same frequency in aggregate, the temporal and scalar unevenness of these attacks can still produce spatial malware clustering. Should a certain AV manufacturer face several malware crises in rapid succession, it could attract significant public attention and thereby decrease public trust in that brand and manufacturer (*33*). AV routing algorithms could then reasonably factor public mistrust of a manufacturer into their decision making. Similarly, the scale and the type of damage inflicted by malware can vary across attacks. For instance, the impact of malware that suddenly steers the vehicle into an object would be more severe, and attract more public attention, than the impact of malware that simply adjusts an outdoor temperature reading. Much like malware's *temporal* unevenness, its *scalar* unevenness can also change public perception of safety and, by extension, AV perception of safety. In short, the spatial clustering of certain AV brands and characteristics can cause certain demographic groups (and the areas where they cluster) to experience disproportionate rates of AV malware infection.

AV Risk Aversion and the Transportation System

Many researchers, including Sparrow, and Howard and Millard-Ball, believe AVs will be more risk-averse than human-driven vehicles (*6*, *34*). In order to assuage pedestrian and general public skepticism about AV safety, they argue, AVs must prioritize safety above most other considerations when driving. It is unclear from the literature and from manufacturer statements which risks AVs would avert most. For instance, Gogoll and Müller, among others, have debated whose safety AVs should prioritize in a crash scenario: its passengers or pedestrians and other motorists around it (*35*, *36*). As it is unclear which safety risks AVs would prioritize and how AVs would mitigate those risks, it is possible that AVs would avoid malware-prone areas when planning a route or suggesting a destination of interest (e.g., a gas station).

Acting on Perceived Geographies of Malware Risk

The widespread adoption of AVs offers vehicle manufacturers an unprecedented opportunity to track the spatial and temporal parameters of malware infection. To navigate (and thus be useful), AVs will require *near-constant* access to global positioning systems and other location-tracking tools (*17*). AVs could then theoretically build a location history (as one emerging AV manufacturer already does openly on smartphones [*37*]), either onboard or on a centralized server. This history could be cross-referenced with the download initiation times of files contemporaneously or retroactively recognized as malware. In turn, vehicle manufacturers could develop machine-readable maps of malware infection. Using these infection maps, AVs could then avoid areas they deem unsafe or less safe than others.

Just as it could for AVs, perception of spatially distributed crime and safety plays into human transportation decisions (*38*) and where humans choose to locate generally. Humans use some combination of crime indicators, varying from qualitative (such as anecdotes) to quantitative (such as crime statistics), to gauge the safety of an area (*39*). Ordinary citizens moving through or stationary in real-world space generally would not have access to sufficient data, would not think to access this data, or would not be able to process this data quickly enough to accurately determine where such indicators locate in the space around them at all times. For example, a human-readable map of spatial vehicular malware infection incidents will likely not be available in the near future, especially as vehicle manufacturers have an interest in keeping the requisite data private. Despite wanting to base decisions in part on the location of pre-determined indicators, human deficiencies in recalling and processing data cause humans to misconstrue the geography of crime surrounding them. In other words, they develop mental maps that do not necessarily reflect real-world conditions (*40*), skewing their route plan and destination choices.

Inaccurate cognition of vehicular malware infection in a city could create new or bolster existing neighborhood safety reputations. Fear of areas perceived as malware-ridden (crime-ridden) can then reproduce both measurable economic and social outcomes of oppressed racial and class groups and reinforce negative stereotypes held against them (*41*). The same phenomenon could occur with enclaves unjustly perceived as "dirty" or "disease-ridden" (*42*) as the public often conflates cybersecurity threats with biological ones (*43*, *44*). Acute variations of this bias could arise from spurts of media-sustained malware panic (*44*) induced by sudden malware attacks (*43*). Even if AV malware attacks are very limited in scale as Fagnant and Kockelman suggest, humankind's historically reactionary response to even the specter of malware infection can seep into AV systems and drive spatial differentiation of malware (*4*).

Through their designers, AV decision-making systems can co-opt (at times irrational) human safety judgment methods. Artificial intelligence (AI) and machine learning algorithms, the technologies that power AVs, imitate existing human pattern-matching and reasoning techniques (*45*). An AI agent learns "optimal" behavior by running human-sourced, and consequently human-biased, training data through its human-developed learning algorithms (*45*, *46*). As a result, a handful of AV system designers that choose AV reasoning mechanisms and training data can effectively dictate how a significant portion of all transportation actors will plan their routes: which data they should consider, which patterns they should recognize from it, and which human biases they should perpetuate with algorithmic precision and consistency. Unlike human-driven vehicles, which rely on their human driver's own unique personal experience and social conditioning to navigate, fleets of AVs will move through space informed by a few opaque, monolithic viewpoints held by a small subset of people removed from the vehicle's operating environment—whether those viewpoints are accurate, just, or neither. This dynamic could intensify the avoidance of enclaves occupied by marginalized groups.

Even accurate representations of malware geographies held by AVs can yield skewed route plan and destination choices. When AVs become transportation decision makers, they can leverage their advanced computing power and AI technology to uncover and process new crime geographies. First, they can readily perceive new indicators of criminal activity (such as malware infection rates) that humans cannot. Second, AV computer systems can determine their position in space, compare their position with available malware information, and act accordingly in real time—all much more quickly than humans can. Each AV, by accessing a pool of sensory information generated by a network of AVs, can perceive a large portion of all outdoor urban space at all times. These systems, like any computer, can precisely select which spatial indicators will influence their decisions and optimize their decision-making around these indicators. Given the disastrous potential of malware infection on immediate passenger safety (*2*) and existing public skepticism of AV safety (*24*), AV manufacturers have reason to consider safety factors like malware infection probability when choosing a route or a destination. Once manufacturers do not expect human occupants to control the vehicle (as SAE automation Level 4 requires), AVs can fully express their malware-skewed safety geography in their decisions, in turn inducing real-world transportation geography changes.

Malware Risk Aversion and Transportation Inequities

Reduced trips to and through an area caused by real or perceived malware concentration can clearly produce and reproduce some spatial transportation inequities. As a benefit to malware-prone areas, lower vehicle density would likely result in fewer local pollutants (which can be inequitably distributed across a city [*47*]) and fewer adverse health effects for area residents (*48*, *49*). Less demand for road space (more available road capacity) could temporarily increase mobility in the area (*50*). As a disadvantage, an area that receives fewer trips could receive reduced transportation infrastructure investment. As this investment decreases, accessibility from and to an area may also decrease with it. Public and government official resentment of oppressed enclaves, bolstered by increased malware infection rates, could render public funding for these areas less politically feasible (*51*). Even groups that have generally benefited from past inequitable transportation planning decisions, such as wealthy, white residents (*52*), could lose accessibility benefits if they live in enclaves at high risk of malware infection.

Other potential impacts resulting from malware concentration are less clear. The mobility and accessibility changes that can occur in malware-prone areas (especially abrupt changes resulting from malware's temporal and scalar unevenness) could have complicated, time-variant, and location-dependent impacts on land use, induced traffic demand, and economic activity that are difficult to predict (*53–55*). A thorough discussion of these more complex consequences falls outside the scope of this paper. The extent to which AV malware will decrease or increase AV or general vehicular travel overall (and thus mitigate or exacerbate intra-city inequity) is also unclear.

The risk of AV malware infection shares elements with some of contemporary society's environmental risks as framed by Beck's "risk society" framework (*56*). Smog and nuclear pollution exposure, Beck argues, are only partially related to class and social status. He writes that although the wealthy can afford to distance themselves somewhat from these environmental hazards, they ultimately affect everyone across a wide region. Similarly, there is no evidence that the wealthy could better protect themselves from malware by buying a more expensive AV. Even if this were possible, it is difficult to imagine a more expensive AV never passing near (and thereby communicating with) a less expensive (and hypothetically more vulnerable to malware) one on any given trip. It is primarily by shaping the environment these vehicles operate in (in this case, the RSUs that comprise V2I networks) that malware risk may be selectively mitigated. A careful RSU deployment process, therefore, could help offset traditional inequalities (*56*).

Malware Prevention

Like malware infection, malware prevention measures can produce spatially uneven outcomes. Human error is the most critical cause of malware infection (*57*). As such, cybersecurity education is a crucial malware prevention tactic, especially for emerging technologies like AVs (*58*). By necessity, local authorities would need to choose which people in which areas of a city would receive this education. Authorities are at risk of mistakenly targeting groups that are most prone to *physical, pathological* infection instead of those most prone to malware infection (*44*). Such erroneous education targeting could even further expose groups already at higher malware infection risk.

Authorities may also deploy technical prevention measures. Vehicle manufacturers and software developers would implement many such measures directly into the AVs they create (*20*, *59*). These in-vehicle solutions, however, would likely fall under the authority of the federal NHTSA (*58*). Basaras et al. and Zhang et al. outline the flaws of relying solely on in-vehicle malware infection mechanisms (*11*, *23*). Chiefly, they argue, maintaining an up-to-date database of all possible malware threats on each vehicle's computer is infeasible due to storage limitations and network lag. As an alternative, they propose storing a large database of known malware threats on a centralized server ("the cloud"). Under this cloud-based plan, all V2I traffic would route through the cloud, which would in turn flag any malware-laden communication. Routing all traffic this way could lead to unacceptable lag and delay times (*60*) and would introduce a single point of failure into the system.

To address technical routing flaws, RSUs could store a cached (duplicate) version of the threat database. RSUs would then periodically synchronize their on-board database with the cloud. Since storing this cache would require a substantial amount of hard disk space (*23*), regional transportation authorities may need to selectively invest in storage-enhanced RSUs. Alternatively,

Basaras et al. (*11*) propose a scheme in which so-called specialized hardware—devices similar to but functionally distinct from RSUs—would isolate infected vehicles from a V2V network. This solution would also require selective, spatially intentional, deployment decisions. The most viable AV malware prevention schemes for subnational planners may require planning authorities, especially where monetary resources are scarce, to selectively upgrade or construct new roadside communication devices.

The inherently spatial nature of the human- and technical-focused malware prevention efforts discussed above have a real-world impact on how the benefits of widespread AV adoption are distributed across a city. As described earlier, pockets of malware concentration can result in an uneven distribution of the benefits AV adoption can provide. Therefore, planners should be aware of how their deployment of malware prevention resources can reproduce or overcome historical and current inequities in transportation systems.

Discussion and Conclusion

This paper helps fill a knowledge gap of how AV security issues may affect policy and planning outcomes. In particular, it affirms that planners and policymakers should consider the perverse outcomes of infrastructure investment decisions, especially when these decisions concern emerging technologies. Without taking proper precautions, the widespread safety, mobility, and accessibility benefits promised by AV technology may be unevenly distributed.

The authors hope this paper will inspire ongoing and long-term discussion among planners on how provisioning AI-powered services can yield inequities in urban systems. Planners need to better prepare for and adapt to technology change (*12*), and the adoption of automation services is no different. Steps like discussing AI implications in professional planning programs (both in school and continuing education) and instituting policies that mandate the consideration of emerging technology in planning documents could help spur future-minded dialogue among planners.

Although the full vision this paper presents may not be fulfilled immediately, subtle impacts of AV malware may arise as vehicles gradually become more automated. Even though the spatial differentiation of malware may be an AV-specific phenomenon, human-driven vehicles may still be infected by malware, especially if V2X networks are rolled out before the commercial AV market opens in earnest. There is urgency, therefore, for planners to start considering and discussing vehicular cybersecurity solutions with each other.

Future Work

A city network simulation comprising AVs, RSUs, and malware infection could help predict the scale and distribution of spatial vehicular malware infection. This technique could help planners to provision malware prevention measures. As connected vehicles (vehicles that communicate with each other but are not necessarily autonomous) and AVs are introduced in the coming years, researchers may have the opportunity to detect malware infection rates across a city and the spatial patterns vehicular malware infection can create.

This emerging problem of AV malware clustering opens up several new research questions related to regulation, planning, and shared vehicle provisioning. Which institutions and organizations (both private and public) have the most suitable competencies and most appropriate powers to regulate AV cybersecurity? Should the selection of cybersecurity and route planning-related AI algorithms and training data be regulated to foster socially just AV decision-making, and if so, what regulations should be put in place? How should land use and transport models, such as traditional four-step travel demand models, incorporate AV malware clustering? Further, how should planning models treat previously imperceptible geographies made more salient by the adoption of mobile, always-on, AI-driven devices?

When spatially provisioning malware prevention services, planners could institute several different guiding strategies. For instance, they could focus investment on areas they predict will be most targeted in malware attacks, they could organize investment to primarily reconcile past inequitable planning decisions made in historically marginalized areas, or they could take an egalitarian approach by provisioning resources evenly. Additional research is needed into how planners should choose a guiding strategy, the advantages and disadvantages of each guiding strategy, and the technical feasibility of a malware infection prediction-based strategy.

More broadly, researchers should consider what other new geographies the decision-making and computational power of AVs (and emplaced AI generally) can uncover in a city by identifying and focusing on new spatial indicators. Some of these geographies may directly contribute to transportation decisions, like geographies of leisure and traffic flow. Others, like political and social geographies, are more abstract and contribute to urban life generally. Saving human lives is important, but human life in the city encompasses much more than physical safety. AVs and advanced, emplaced AI, with their capacity to consider all space at all times, could affect not just the transportation system but the entire American urban experience. Researchers should approach this incoming technical revolution just as holistically.

Acknowledgments

The authors acknowledge Andrew Dwyer of Oxford University, who inspired them to explore the potential for uneven malware propagation across a city.

Author Contributions

The authors confirm contribution to the paper as follows: study conception and design: EV, KM; data collection: EV; analysis and interpretation of results: EV; draft manuscript preparation: EV, KM. All authors reviewed the results and approved the final version of the manuscript.

References

1. Crandall, J. R., R. Ensafi, S. Forrest, J. Ladau, and B. Shebaro. The Ecology of Malware. *Proc., 2008 Workshop on New Security Paradigms*. The Association for Computing Machinery, New York, NY, 2009. pp. 99-106.
2. Tyagi, P., and D. Dembla. Investigating the Security Threats in Vehicular Ad Hoc Networks (VANETs): Towards Security Engineering for Safer On-Road Transportation. *2014 International Conference on Advances in Computing, Communications and Informatics,* Research Publishing Services, 2014, pp. 2084–2090.
3. Koscher, K., A. Czeskis, F. Roesner, S. Patel, T. Kohno, S. Checkoway, D. McCoy, B. Kantor, D. Anderson, H. Shacham, and S. Savage. Experimental Security Analysis of a Modern Automobile. *2010 IEEE Symposium on Security and Privacy*, IEEE Computer Society Conference Publishing Services, 2010, pp. 447–462.
4. Fagnant, D. J., and K. Kockelman. Preparing a Nation for Autonomous Vehicles: Opportunities, Barriers and Policy Recommendations. *Transportation Research Part A: Policy and Practice*, Vol. 77, 2015, pp. 167–181.
5. Litman, T. *Autonomous Vehicle Implementation Predictions: Implications for Transport Planning*. Victoria Transport Policy Institute, Victoria, BC, Canada, 2017.
6. Sparrow, R., and M. Howard. When Human Beings are like Drunk Robots: Driverless Vehicles, Ethics, and the Future of Transport. *Transportation Research Part C: Emerging Technologies*, Vol. 80, 2017, pp. 206–215.
7. Leonard, J., J. How, S. Teller, M. Berger, S. Campbell, G. Fiore, L. Fletcher, E. Frazzoli, A. Huang, S. Karaman, O. Koch, Y. Kuwata, D. Moore, E. Olson, S. Peters, J. Teo, R. Truax, M. Walter, D. Barrett, A. Epstein, K. Maheloni, K. Moyer, T. Jones, R. Buckley, M. Antone, R. Galejs, S. Krishnamurthy, and J. Williams. A Perception-Driven Autonomous Urban Vehicle. *Journal of Field Robotics*, Vol. 25, No. 10, 2008, pp. 727–774.
8. National Highway Traffic Safety Administration, Department of Transportation. Federal Motor Vehicle Safety Standards; V2V Communications. *Federal Register*, Vol. 82, No. 8, 2017.
9. Kanchanasut, K., S. Boonsiripant, A. Tunpan, H. K. Kim, and M. Ekpanyapong. Internet of Cars through Commodity V2V and V2X Mobile Routers: Applications for Developing Countries. *KSCE Journal of Civil Engineering*, Vol. 19, No. 6, 2015, pp. 1897–1904.
10. Ibáñez, J. A. G., S. Zeadally, and J. Contreras-Castillo. Integration Challenges of Intelligent Transportation Systems with Connected Vehicle, Cloud Computing, and Internet of Things Technologies. *IEEE Wireless Communications*, Vol. 22, No. 6, 2015, pp. 122–128.
11. Basaras, P., I. Belikaidis, L. Maglaras, and D. Katsaros. Blocking Epidemic Propagation in Vehicular Networks. *Proc., 12th Annual Conference on Wireless On-Demand Network Systems and Services (WONS)*, IEEE, New York, NY, 2016, pp. 1–8.
12. Dorr, A. Technology Blindness and Temporal Imprecision: Rethinking the Long Term in an Era of Accelerating Technological Change. *Foresight*, Vol. 18, No. 4, 2016, pp. 391–413.
13. SAE International. *Taxonomy and Definitions for Terms Related to Driving Automation Systems for On-Road Motor Vehicles*. No. J3016_201609. 2016.
14. SAE International. *NHTSA Adopts SAE International Standard Defining Autonomous Vehicles; SAE Releases New Version for Free - J3016 States and Defines Six Levels of Automation in On-Road Motor Vehicles*. https://perma.cc/C4EH-75Y9. Accessed July 19, 2017.
15. *Federal Automated Vehicles Policy: Accelerating the Next Revolution in Roadway Safety*. Publication DOT HS 812 329. National Highway Traffic Safety Administration, U.S. Department of Transportation, Washington, D.C., 2016.
16. Brodsky, J. S. Autonomous Vehicle Regulation: How an Uncertain Legal Landscape May Hit The Brakes on Self-Driving Cars. *Berkeley Technology Law Journal*, Vol. 31, No. 2, 2016, pp. 851–877.
17. Anderson, J. M., N. Kalra, K. D. Stanley, P. Sorensen, C. Samaras, and O. A. Oluwatola. *Autonomous Vehicle Technology: A Guide for Policymakers*. RAND Corporation, Santa Monica, CA, 2014.
18. Weiner, E. *Urban Transportation Planning in the United States: An Historical Overview*. Praeger, Westport, CT, 1999.
19. Haddrell, M. *Towards an Autonomous Vehicle Enabled Society: Cyber Attacks and Countermeasures*. Information Security Group, Royal Holloway University of London, London, UK, 2016.
20. Bertschy, K. *Vehicle Computer and Network Security: Vulnerabilities and Recommendations*. Utica College, Utica, NY, 2015.
21. Kim, Y. H., S. Peeta, and X. He. An Analytical Model to Characterize the Spatiotemporal Propagation of Information Under Vehicle-to-Vehicle Communications. *IEEE Transactions on Intelligent Transportation Systems*, Vol. PP, No. 99, 2017, pp. 1–10.
22. Basaras, P., D. Katsaros, and L. Tassiulas. Dynamically Blocking Contagions in Complex Networks by Cutting Vital Connections.In *2015 IEEE International Conference on Communications (ICC)*, 2015, pp. 1170–1175.
23. Zhang, T., H. Antunes, and S. Aggarwal. Defending Connected Vehicles Against Malware: Challenges and a Solution Framework. *IEEE Internet of Things Journal*, Vol. 1, No. 1, 2014, pp. 10–21.
24. Schoettle, B., and M. Sivak. *Public Opinion about Self-Driving Vehicles in China, India, Japan, the US, the UK, and*

Australia. University of Michigan Transportation Research Institute, Ann Arbor, MI, 2014.
25. Guerra, E. Planning for Cars That Drive Themselves. *Journal of Planning Education and Research*, Vol. 36, No. 2, 2016, pp. 210–224.
26. Mezzour, G., K. M. Carley, and L. R. Carley. An Empirical Study of Global Malware Encounters. *Proc., 2015 Symposium and Bootcamp on the Science of Security*, The Association for Computing Machinery, Inc., New York, NY, 2015.
27. Anthe, C., E. Argyle, E. Douglas, S. Fender, E. Florio, C. Foster, R. Gowrishankar, V. Grebennikov, P. Henry, A. Hulett, I. Ivanov, M. Johnson, J. Jones, T. Kerk, M. Letourneau, M. Mallen, M. Miller, C. Mills, N. Ng, H. O'Dea, J. P. Dee, S. Pavithran, D. Pecelj, F. Plazo, T. Rains, P. Rebriy, K. Selvaraj, T. Shinder, N. Sood, T. Teller, and V. Thakur. *Microsoft Security Intelligence*. Report. No. 21. Microsoft Corporation, Redmond, WA, 2016.
28. Al-Qutayri, M., C. Yeun, and F. Al-Hawi. Security and Privacy of Intelligent VANETs. *Computational Intelligence and Modern Heuristics*, InTech, 2010, pp. 191–218.
29. Baker, E. H., D. Crusius, M. Fischer, W. Gerling, K. Gnaserakan, H. Kerstan, F. Kuhnert, J. Kusber, J. Mohs, M. Schulte, J. Seyfferth, J. Stephan, and T. Warnke. *Connected Car Report 2016: Opportunities, Risk, and Turmoil on the Road to Autonomous Vehicles*. Connected Car Report, Strategy, New York, NY, 2016.
30. Holt, T. J., and A. M. Bossler. Examining the Relationship Between Routine Activities and Malware Infection Indicators. *Journal of Contemporary Criminal Justice*, Vol. 29, No. 4, 2013, pp. 420–436.
31. Holstein, T., M. Wallmyr, J. Wietzke, and R. Land. Current Challenges in Compositing Heterogeneous User Interfaces for Automotive Purposes. *Human-Computer Interaction: Interaction Technologies*, Springer, Cham, Switzerland, 2015, pp. 531–542.
32. Sexton, S. E., and A. L. Sexton. Conspicuous Conservation: The Prius Halo and Willingness to Pay for Environmental Bona Fides. *Journal of Environmental Economics and Management*, Vol. 67, No. 3, 2014, pp. 303–317.
33. Nestas, L., and K. Hole. Building and Maintaining Trust in Internet Voting. *Computer*, Vol. 45, No. 5, 2012, pp. 74–80.
34. Millard-Ball, A. Pedestrians, Autonomous Vehicles, and Cities. *Journal of Planning Education and Research*, 2016, pp. 1–7.
35. Gogoll, J., and J. F. Müller. Autonomous Cars: In Favor of a Mandatory Ethics Setting. *Science and Engineering Ethics*, 2016, pp. 1–20.
36. Bonnefon, J.-F., A. Shariff, and I. Rahwan. The Social Dilemma of Autonomous Vehicles. *Science*, Vol. 352, No. 6293, 2016, pp. 1573–1576.
37. Elliott, M. *Is Google Is Tracking You? Find Out Here*. CNET. https://perma.cc/LP4N-JF9W. Accessed July 19, 2017.
38. Cao, X., P. L. Mokhtarian, and S. L. Handy. The Relationship between the Built Environment and Nonwork Travel: A Case Study of Northern California. *Transportation Research Part A: Policy and Practice*, Vol. 43, No. 5, 2009, pp. 548–559.
39. Lopez, N., and C. Lukinbeal. Comparing Police and Residents' Perceptions of Crime in a Phoenix Neighborhood Using Mental Maps in GIS. *Yearbook of the Association of Pacific Coast Geographers*, Vol. 72, No. 1, 2010.
40. Egenhofer, M. J., and D. M. Mark. Naive Geography. *Spatial Information Theory: A Theoretical Basis for GIS*, Springer, Berlin, Germany, 1995. pp. 1–15.
41. Drakulich, K. M. Strangers, Neighbors, and Race. *Race and Justice*, Vol. 2, No. 4, 2012, pp. 322–355.
42. Keil, R., and H. Ali. Multiculturalism, Racism and Infectious Disease in the Global City: The Experience of the 2003 SARS Outbreak in Toronto. *Topia: Canadian Journal of Cultural Studies*, No. 16, 2006, pp. 23–49.
43. Lupton, D. Panic Computing: The Viral Metaphor and Computer Technology. *Cultural Studies*, Vol. 8, No. 3, 1994, pp. 556–568.
44. Sandywell, B. Monsters in Cyberspace: Cyberphobia and Cultural Panic in the Information Age. *Information, Communication & Society*, Vol. 9, No. 1, 2006, pp. 39–61.
45. Russell, S. J., P. Norvig, and E. Davis. *Artificial Intelligence: A Modern Approach*. Prentice Hall, Upper Saddle River, NJ, 2010.
46. Schmidhuber, J. Deep Learning in Neural Networks: An Overview. *Neural Networks*, Vol. 61, 2015, pp. 85–117.
47. Sider, T., M. Hatzopoulou, N. Eluru, G. Goulet-Langlois, and K. Manaugh. Smog and Socioeconomics: An Evaluation of Equity in Traffic-Related Air Pollution Generation and Exposure. *Environment and Planning B: Planning and Design*, Vol. 42, No. 5, 2015, pp. 870–887.
48. English, P., R. Neutra, R. Scalf, M. Sullivan, L. Waller, and L. Zhu. Examining Associations between Childhood Asthma and Traffic Flow Using a Geographic Information System. *Environmental Health Perspectives*, Vol. 107, No. 9, 1999, pp. 761–767.
49. Wilhelm, M., and B. Ritz. Residential Proximity to Traffic and Adverse Birth Outcomes in Los Angeles County, California, 1994-1996. *Environmental Health Perspectives*, Vol. 111, No. 2, 2003, pp. 207–216.
50. Douglass B. Lee, J., L. A. Klein, and G. Camus. Induced Traffic and Induced Demand. *Transportation Research Record: Journal of the Transportation Research Board*, 1999. 1659: 68–75.
51. Bullard, R. D., G. S. Johnson, and A. O. Torres. *Highway Robbery: Transportation Racism and New Routes to Equity*. South End Press, Cambridge, MA, 2004.
52. Golub, A., R. A. Marcantonio, and T. W. Sanchez. Race, Space, and Struggles for Mobility: Transportation Impacts on African Americans in Oakland and the East Bay. *Urban Geography*, Vol. 34, No. 5, 2013, pp. 699–728.
53. Handy, S. Smart Growth and the Transportation-Land Use Connection: What Does the Research Tell Us? *International Regional Science Review*, Vol. 28, No. 2, 2005, pp. 146–167.
54. Noland, R. B. Relationships between Highway Capacity and Induced Vehicle Travel. *Transportation Research Part A: Policy and Practice*, Vol. 35, No. 1, 2001, pp. 47–72.

55. Weisbrod, G., D. Vary, and G. Treyz. Measuring Economic Costs of Urban Traffic Congestion to Business. *Transportation Research Record: Journal of the Transportation Research Board*, 2003. 1839: 98–106.
56. Beck, U. *Risk Society: Towards a New Modernity*. Sage Publications, London, UK, 1992.
57. Colella, A., and C. Colombini. Security Paradigm in Ubiquitous Computing. *2012 Sixth International Conference on Innovative Mobile and Internet Services in Ubiquitous Computing*, IEEE Computer Society Conference Publishing Services, 2012, pp. 634–638.
58. National Highway Traffic Safety Administration. *Cybersecurity Best Practices for Modern Vehicles*. Publication DOT HS 812 333, National Highway Traffic Safety Administration, U.S. Department of Transportation, Washington, D.C., 2016.
59. Blommendaal, C. *Information Security Risks for Car Manufacturers Based on the In-Vehicle Network*. University of Twente, Enschede, The Netherlands, 2015.
60. Eiza, M. H., and Q. Ni. Driving with Sharks: Rethinking Connected Vehicles with Vehicle Cyber Security. *IEEE Vehicular Technology Magazine*, Vol. 12, No. 2, 2017, pp. 45–51.

The Standing Committee on Critical Transportation Infrastructure Protection (ABR10) peer-reviewed this paper (18-04199).

Article

TRR
JOURNAL OF THE TRANSPORTATION RESEARCH BOARD

Transportation Research Record
2018, Vol. 2672(1) 76–89

DOI: 10.1177/0361198118784378
journals.sagepub.com/home/trr
SAGE

Modeling Cyber Attacks at Intelligent Traffic Signals

Gurcan Comert[1], Jacquan Pollard[1], David M. Nicol[2], Kartik Palani[3], and Babu Vignesh[3]

Abstract

Transportation networks are considered one of the critical physical infrastructures for resilient cities (cyber-physical systems). In efforts to minimize adverse effects that come with the advancement of vehicular technologies, various governmental agencies, such as the U.S. Department of Homeland Security and the National Highway Traffic Safety Administration (NHTSA), work together. This paper develops belief-network-based attack modeling at signalized traffic networks under connected vehicle and intelligent signals frameworks. For different types of cyber attacks, defined in the literature, risk areas and impacts of attacks are evaluated. Vulnerability scores, technically based on the selected metrics, are calculated for signal controllers. In addition, the effect of having redundant traffic sensing systems on intersection performance measures is demonstrated in terms of average queue length differences.

Resilience of critical infrastructures is defined as their ability to withstand an upsetting event, deliver essential levels of service during it, and recover quickly after it. With the increase of connected systems, cyber attacks that can target critical infrastructure systems are becoming more troubling. Transportation networks are considered one of the critical physical infrastructures for resilient cities (cyber-physical systems [CPSs] [*1*]). According to recent reports (*2–4*), several benefits are foreseen from upcoming technologies such as connected and autonomous vehicles (CAVs), including up to 80% reduction in fatalities from multi-vehicle crashes and prevention of the majority of human-error-related incidents, which takes out about 94% of all incidents. These intelligent applications, however, come at a price; for example, in 2015 alone 1.5 million vehicles were recalled because of cybersecurity vulnerabilities. NHTSA's current research focuses on CAVs that are heavily involved in secure implementations which will enable the field and its technology experts to harness efficient, reliable, and secure system design (*3*). Some of these topics can be listed as anomaly-based intrusion detection systems, cybersecurity of firmware updates, cybersecurity on heavy vehicles, vehicle-to-vehicle (V2V) communication interfaces, and trusted vehicle-to-everything (V2X) communications (*5*). The main goals for any critical infrastructure are quick detection of attacks and rapid mitigation efforts (*6*). There are many attack types, of which some can be resolved via detection and some require redundant systems and sensors. In intelligent transportation systems (ITS), to increase security and resiliency in case of possible attacks or benign system errors during different events, research is likewise needed into detection using various sensors and data types. Research is also necessary to enhance confidence in sensor readings by checking consistency with other sensors and information sources as well as validating control system commands (*7*). This paper investigates attack modeling and impacts on intelligent signals. For different types of cyber attacks defined in the literature (e.g., [*8*]). risks are calculated via probabilistic graphical models (e.g., the Bayesian network [BN] [*9*]) and quantified by system simulations.

In the literature, related studies with analytical graphical models include factor graphs, BNs, and Markov random fields (*10*). Two detailed works on current attack types, survey studies on CPS security, list possible attacks and discuss their mitigation techniques (*11*, *12*). In these

[1]Department of Physics and Engineering, Benedict College, Columbia, SC
[2]Department of Electrical and Computer Engineering, University of Illinois, Urbana-Champaign, IL
[3]Coordinated Science Laboratory, University of Illinois, Urbana-Champaign, IL

Corresponding Author:
Address correspondence to Gurcan Comert: comertg@benedict.edu

studies, abstract cyber-physical models for smart cars are presented. Possible attacks are listed as criminal, privacy, tracking, profiling, political threats with different structures replay, command (message) injection, eavesdropping, and denial of service (DOS).

Existing research efforts on vehicular communications discuss possible attacks and their mitigation methods (*5*, *8*). A recent study on current signal cabinets presents the idea that the possible impact of attacks could be unmanageable queues (*13*). Overall, ITS applications require protocols that conflict with anonymity and privacy requirements. The research looks at quantifying such risks and at traffic control in the case of either lost or faulty communications. In sum, studies on the quick detection of such cases and possible redundant data resources for cost effective control are needed for improving the resiliency of transportation networks.

Moreover, model-based attacks, usually for power grids, are investigated by researchers (*14*). Intrusion models for different control systems and proper modeling for moving systems, as in vehicular or mobile ad hoc network (VANET/MANET) cases, are well-investigated in (*11*), in which reputation management in vehicular networks is recommended. Possible revoking or blacklisting of information contributors is also recognized in a similar survey study specifically on cooperative ITS (*8*). Driver privacy and safety critical applications are also starting to be investigated by the researchers (*15*).

The particular contributions of this paper can be listed as:

- Development of a connected vehicle-based signal system model with cyber-physical representation
- Application of a probabilistic expert system (BN) for modeling anomalies and attacks (malicious messages or benign failures)
- Presentation of the impact of redundancy systems (sensors) and a number of different traffic states (from parameters: market penetration rate, and vehicle–pedestrian traffic composition versus control reliability)

The rest of the paper is organized as follows. Section 2 introduces the problem and the approach. Section 3 explains the analytical modeling and BNs. Section 4 presents simulations from BNs. The section also describes analysis with and without sensors at an isolated traffic signal. Finally, section 5 summarizes findings and addresses possible future research directions.

Methodology

The main objective of this paper is to develop analytical graphical models for quantification of possible attack impacts on intelligent traffic signal components through controller states. First, a BN is generated based on the physical model of an intelligent traffic signal given in Figure 1 which is adopted from (*16*). Second, different attack surfaces, scenarios, and possible impacts for risks are obtained from (*5*). Finally, the two previous steps are combined and standard vulnerability scores from (*17*) are calculated.

In Figure 1, possible intelligent signal components (i.e., physical devices) are depicted. The signal controller unit has a processor, memory, and connections (e.g., ethernet, wireless, or other ports). Although simplified, the figure is part of the field test architecture of connected adaptive signals (*18*). In an arterial setting, each intersection would have a communication device and a controller overseen by a single traffic management system. For communication among the devices, the security certificate authority (e.g., Security Credential Management System [SCMS]) is interfaced to the roadside equipment (RSE) and used to provide security certificates to trusted on-board equipment (OBE).

In the figure, various modes of passenger vehicles, emergency vehicles, freight, transit, motorcycles, and other motorized travelers are represented as vehicles. Pedestrians represent any other nonmotorized modes such as bicycles. These vehicles and pedestrians can also be equipped; that is, they could have an OBE or portable device which can communicate traffic control system (via RSE). In this study, the traffic control system consists of the signal controller, other surveillance technologies and environmental sensors, and an RSE. The RSE radio manages the 5.9 GHz Dedicated Short Range Communications (DSRC) between vehicles and the infrastructure. Specific channels are designated for vehicles to broadcast basic safety messages (BSM-channel 172) and RSE to broadcast the map message (MAP), the signal phase and timing messages (SPaTs), and signal status messages (SSM-channel 182).

In a broader view of the traffic network, signal controllers can be positioned as part of a traffic management system in which different control devices can be connected and coordinated for various transportation modes. These modes can be sensed by different detection technologies (fixed-inductive loops, video, infrared, radar, magnetic plates, pedestrian buttons or solely mobile-GPS, and cell phones, etc.). These detectors provide information to the signal control algorithms for presence to change or extend phases. These communications can be direct (pedestrian push button) or based on inference (video cameras–queues).

The functions of the signal controller can be listed as processing traffic sensor data, pedestrian protection, conflict monitoring, and giving indicative output data for roads, other signals, or users. It also monitors RSE operation and provides a device interface for field

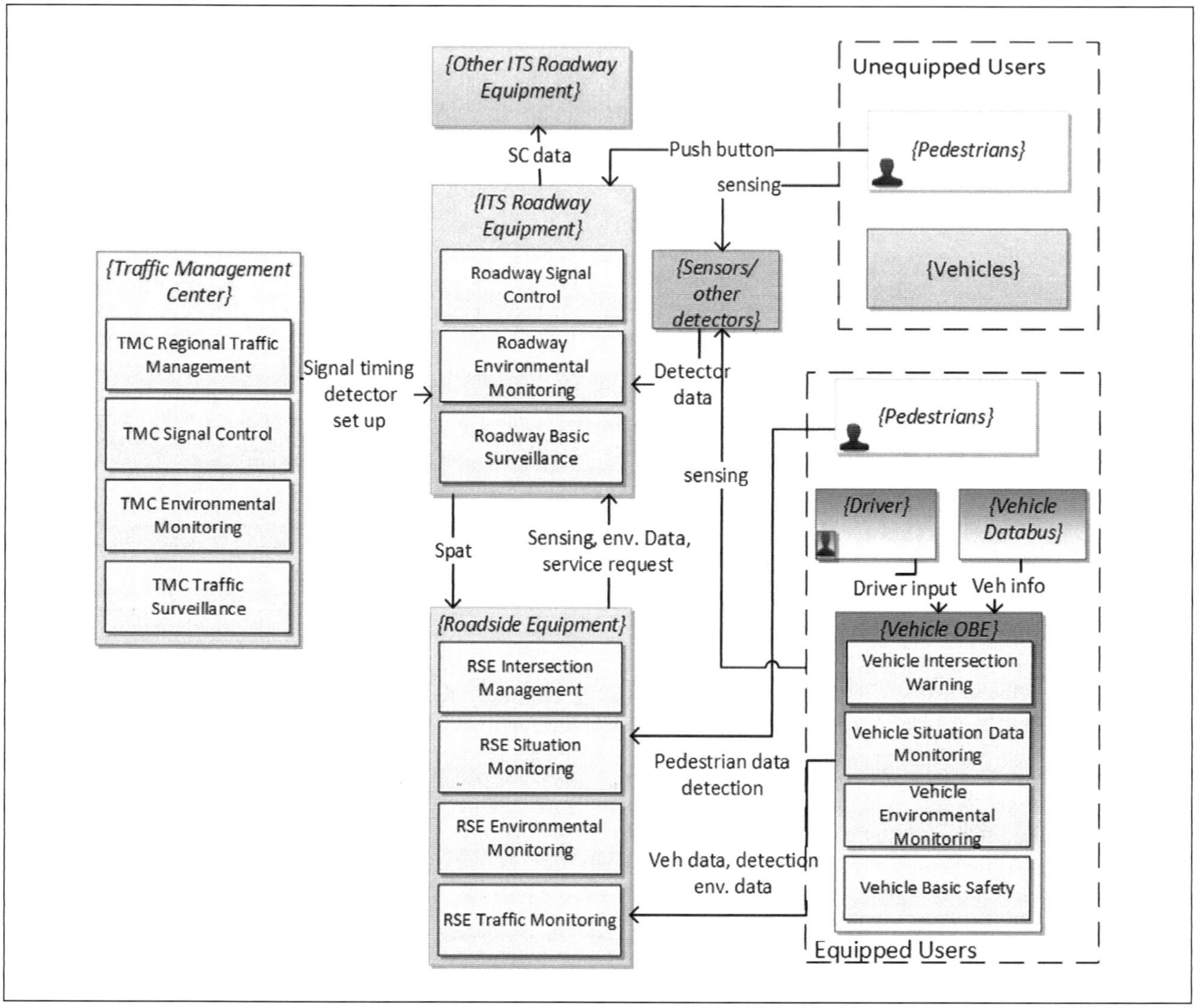

Figure 1. A possible intelligent signal adopted from (*16*).

management stations. The impacts on applications can be listed as data distribution for eco-signaling, eco-transit, emergency vehicle preemption, freight signal priority, pedestrian mobility, red light violation warning, and transit priority. In this study, for simplicity, the impacts of cyber attacks are quantified as changes in service times, specifically average queue lengths (*13*).

Bayesian Networks

Bayesian (belief) networks are stochastic graphical expert systems that depict conditional (in)-dependences of probabilities of interest. Basically, they are expressions of complex joint distributions with a large number of variables involving levels of conditional independences. They can be used for finding probability distributions of a variable with multiple states which can be expressed as replications of such a variable with unobserved states (i.e., hidden Markov models); some variables can also be assumed as parameters connecting different realizations, therefore calculating the posterior probabilities of such parameters.

BNs in a cybersecurity context are utilized to model hypothesized communications and to generate risk and reliability in the presence of possible attacks (*9*, *19*, *20*). Example cases on safety can be found in (*21*, *22*). The literature involving simulation and analytical models also addresses different attack types. From a cybersecurity viewpoint, in general they include message injection, replay, data manipulation, eavesdropping, routing, DOS, and coordinated multiple attacks. However, in this study they are revised for a connected vehicle (CV) framework (*5*). The physical model Figure 1 introduced above is modified with the addition of multiple modes, detection

Table 1. Attack Surfaces for Equipped Vehicles and Pedestrians

Target	Type	Feasibility	Metric	Detection	Metric
SA	LTC	low	$[h\,l\,l]$	med	$[l\,h\,l]^T \times [h\,l\,l]$
	CRL	med	$[l\,h\,l]$	med	$[l\,h\,l]^T \times [l\,h\,l]$
	PC	med	$[l\,h\,l]$	med	$[l\,h\,l]^T \times [l\,h\,l]$
RSE	WSA	high	$[l\,l\,h]$	low	$[h\,l\,l]^T \times [l\,l\,h]$
	DB	high	$[l\,l\,h]$	med	$[l\,h\,l]^T \times [l\,l\,h]$
	DOSR	high	$[l\,l\,h]$	high	$[l\,l\,h]^T \times [l\,l\,h]$
	SD	low	$[h\,l\,l]$	high	$[l\,l\,h]^T \times [h\,l\,l]$
VEHs	CB	low	$[h\,l\,l]$	low	$[h\,l\,l]^T \times [l\,h\,l]$
	BLK	med	$[l\,h\,l]$	low	$[h\,l\,l]^T \times [l\,h\,l]$
	RB	low	$[h\,l\,l]$	low	$[h\,l\,l]^T \times [h\,l\,l]$
	BSM	high	$[l\,l\,h]$	low	$[h\,l\,l]^T \times [l\,l\,h]$
	DOS	high	$[l\,l\,h]$	low	$[h\,l\,l]^T \times [l\,l\,h]$
	MP	high	$[l\,l\,h]$	med-high	$[l\,m\,h]^T \times [l\,l\,h]$
	DCC	med	$[l\,h\,l]$	med	$[h\,l\,l]^T \times [l\,h\,l]$
	LC	med	$[l\,h\,l]$	low	$[h\,l\,l]^T \times [h\,l\,l]$

Note: vehs = vehicles; pedes = pedestrians.

interfaces, and a series of other signals in simulation. The critical issue, though, when data are obtained only from CVs, is that any error could result in significant costs. Therefore, the impacts of back up surveillance systems and attacks in a cyber-physical context are investigated.

In the study, multiple BNs are generated to see the impact of having redundant detection systems which can be treated as additional costs. BNs are utilized to calculate conditional probability distributions and expected vulnerabilities and their standard deviations via Monte Carlo simulations. Some predictions are also provided given assumed evidences for different scenarios. Thus, a mixture of various BN usages is incorporated into this paper. First, variables (nodes) and their discrete states are defined. Note that these are assumptions mostly compiled from (*5*, *17*). Second, nodes and respective edges are formed from Figure 1. These nodes are mainly variables that are affecting signal control states. Parental nodes (attack surfaces) along with their vulnerability scores are given in Table 1. Metrics from (*17*), (low, med, high) and (none, low, med, high, cri), represent likelihoods of attacks, states of a node after an attack, and detection probabilities. Standard values for such metrics are also described below. The last column of the table, for example, shows a conditional probability matrix of $p(\mathrm{SDS}|\mathrm{LTC})$ in multiplication of $[l\,h\,l]^T_{(1\times 3)}$ and $[h\,l\,l]_{(1\times 3)}$ matrices. It basically denotes the probability of detection given a fake long-term certificate (LTC) attack. The marginal distribution of LTC is also given as $[h\,l\,l]$ because feasibility is low.

The model's system constraints or assumed variables, and their descriptions and metrics (states) are given in Table 2. For an example description, a signal controller box is denoted by (SC). Traffic composition (TC) consists of equipped and unequipped roadway users that are denoted as pedestrians (EP), cars (CR), emergency vehicles (EM), freight trucks (FR), transit (TR), unequipped pedestrians (UP), and unequipped vehicles (UV). This is by no means an exhaustive list of possible attacks, and extensions can easily be incorporated into the developed networks. For instance, satellite systems or time source manipulations can be incorporated on multiple surfaces along with detailed feasibility and impact of such attacks. If available, more complex metrics or probability distributions can be included for feasibility of the attacks.

Figure 2 shows designed BNs under assumed dependencies among the nodes given in Table 2. Figure 2*a* is a BN with single signal controller including a sensor (additional detection) node. This is critical as the reliability of the SC could increase via additional ground truth checks as well as control via this sensor which could be quantified using simulation in subsection. However, this would involve additional cost. Similarly, detection nodes can be seen as a cost for monitoring and delay of communications. In the figure, SC is connected on pedestrian push buttons (it is assumed that there is no impact on the SC state), other ITS which can be other signals, TMC, S, and most importantly RSE. Transition of the risk at an SC to series of other SCs (i.e., ten controllers) is demonstrated by the BN in Figure 2*b*.

Now, suppose all the nodes in Table 2 $N = \{\mathrm{SC}, \mathrm{SA}, \ldots, \mathrm{DCC}, \mathrm{LC}\}$. Given:

$$p(\mathrm{SC}) = \sum_{N/\mathrm{SC}} \prod_{N/\mathrm{SC}} p(\mathrm{SC}|\mathrm{pa}(\mathrm{SC})) \qquad (1)$$

where pa(SC) represents the parental variables of SC. Simply when nodes are incorporated, Equation 1 follows:

Table 2. Nodes on the Designed BNs with Assumed Metrics

Node	Description	States
SC	signal controller	none, low, med, high, cri
SA	certificate (security) authority	low, med, high
RSE	roadside equipment	low, med, high
TC	traffic composition	UP, EP, UV, CR, EM, FR, TR
UP	unequipped pedestrians	none, low, med, high, cri
EP	equipped pedestrians	none, low, med, high, cri
UV	unequipped vehicles	none, low, med, high, cri
CR	regular equipped vehicles	none, low, med, high, cri
EM	emergency vehicles	none, low, med, high, cri
FR	freight trucks	none, low, med, high, cri
TR	transit vehicles	none, low, med, high, cri
ITS	other ITS devices	low, med, high
TMC	traffic management center	low, med, high
S	sensor	true, false
DD	detection by driver	low, med, high
DP	detection by pedestrian	low, med, high
DS	detection by system (vehs)	low, med, high
PDS	detection by system (pedes)	low, med, high
RDS	detection by system (RSE)	low, med, high
SDS	detection by system (SA)	low, med, high
LTC	fake long-term certificate	low, med, high
CRL	fake certificate revocation list	low, med, high
PC	pseudonym certificate	low, med, high
WSA	wrong safety warning and signal phasing	low, med, high
DB	database and map poisoning	low, med, high
DOSR	denial of service (RSE)	low, med, high
SD	device shutdown	low, med, high
CB	sending channel busy	low, med, high
BLK	block pseudonym change	low, med, high
RB	remote reboot	low, med, high
BSM	send/intrusion/fake basic safety message	low, med, high
DOS	denial of service (vehs, pedes)	low, med, high
MP	database and map poisoning (vehs)	low, med, high
DCC	distributed congestion control mechanisms	low, med, high
LC	location tracking	low, med, high

Note: vehs = vehicles; pedes = pedestrians.

$$p(\mathrm{SC}) = \sum_{(S,\,\mathrm{RSE},\ldots,\,\mathrm{WSA})} p(\mathrm{SC, S, RSE,\ldots, WSA}) \\ = p(\mathrm{SC}|\mathrm{S, RSE,\ldots, WSA})p(\mathrm{RSE}|\mathrm{RDS},\ldots) \\ p(\mathrm{S}|\mathrm{UV, UP, EP,\ldots, TR})\ldots p(\mathrm{WSA}) \tag{2}$$

Probabilities in Equation (2) from BNs are propagated (posteriors) by the junction tree algorithm with $O(n^3)$ time complexity for marginal distributions. The algorithm actually yields the multiplication of conditional probabilities including the independences, a simplified version. The algorithm, on directed acyclic graphs, first marries parent nodes, then triangulates and removes unnecessary links through generation of cliques and assignments, and reports message propagation (*22*). Note that parent nodes $p(\mathrm{WSA})$ require probability assumptions for uncertain evidence setting. Scales are adopted from (*5*) and assigned from (*17*). Specifically state vectors (levels) for RSE (low, med, high) and SC (none, low, med, high, cri) are assigned with maximum probabilities (0.25, 0.50, 0.75) and (0.01, 0.39, 0.69, 0.89, 1.0) respectively. Note that for RSE these probabilities represent low (affected by a serious attack, failure like), med (less affected), and high (working) states. For SC, the state probabilities are none (not affected), low (working state), med (still working, but service times would be affected), and high and cri, which are highly affected states in which traffic is disrupted with long delays (e.g., hazardous conditions for emergency vehicles). To address the sensitivity, these probabilities are generated randomly from a uniform distribution and the resulting posteriors are obtained as distributions for each state rather than a single value. Moreover, the resulting expected utilities for vulnerabilities are calculated on a scale of 0–10, on which up to 0.1 shows no risk or impact and 10 means cri vulnerability. BNs are coded and developed using gRain package in R (*23*). Examples of

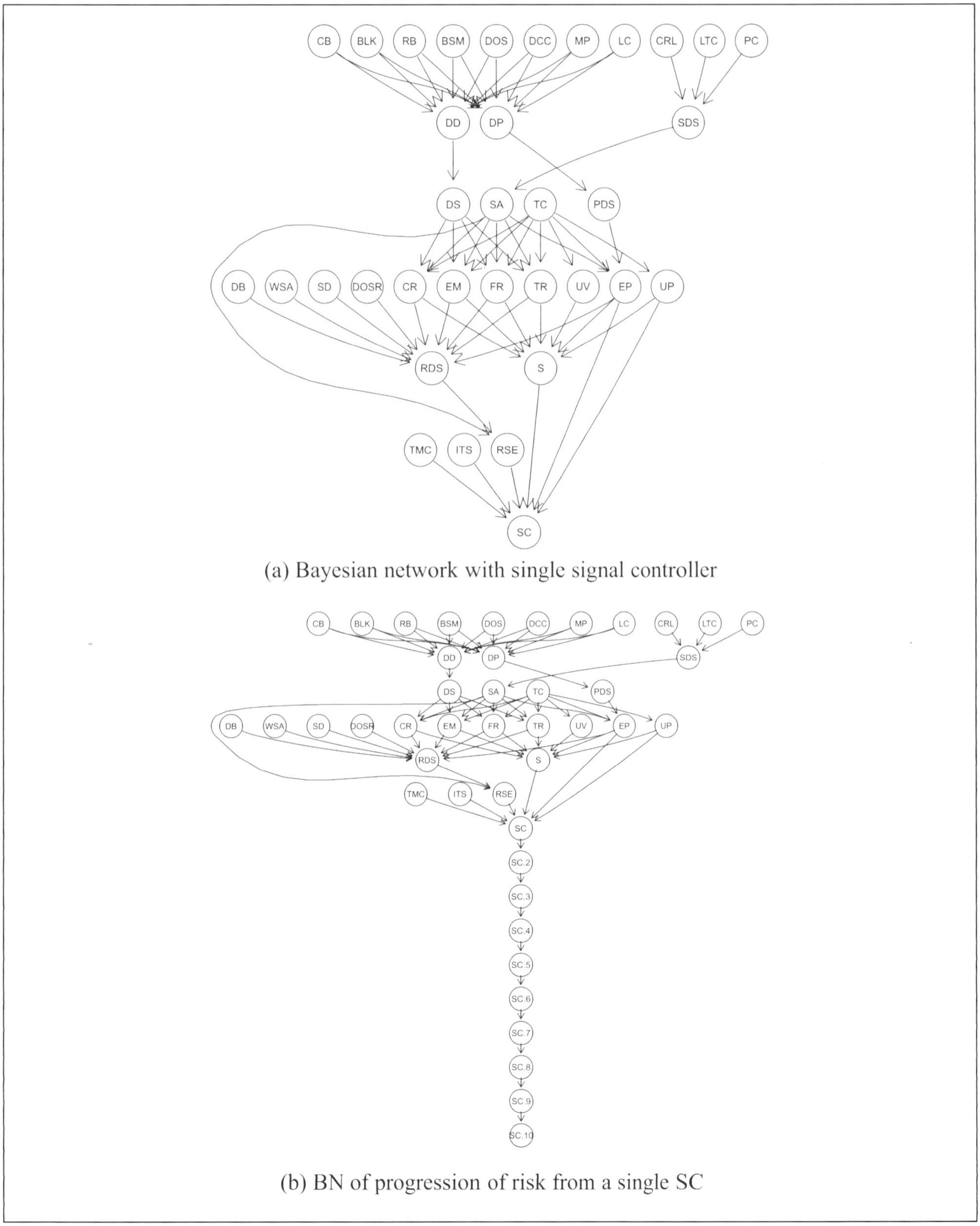

(a) Bayesian network with single signal controller

(b) BN of progression of risk from a single SC

Figure 2. Designed BNs.

Table 3.

Propagated p(SC|RSE), p(SC|S), and p(RSE|RDS) with sensor.

		SC						RSE		
		none	low	med	high	cri		low	med	high
RSE	low	0.923	0.076	0.00002	0.00003	0.00002	RDS	0.007	0.198	0.794
	med	0.923	0.076	0.0001	0.0001	0.00004		0.890	0.057	0.053
	high	0.923	0.076	0.00005	0.0002	0.0001		0.999	0.00003	0.00003
S	true	0.937	0.062	0.00006	0.00008	0.00005				
	false	0.703	0.296	0.00031	0.00041	0.00024				

Propagated p(SC.10|SC), p(SC.10|S), and p(SC.10|RSE) with sensor.

		SC.10				
		none	low	med	high	cri
SC	none	0.372	0.340	0.195	0.059	0.033
	low	0.279	0.328	0.254	0.088	0.049
	med	0.172	0.291	0.320	0.132	0.084
	high	0.142	0.268	0.330	0.153	0.106
	cri	0.116	0.251	0.341	0.169	0.122
RSE	low	0.366	0.339	0.199	0.062	0.034
	med	0.360	0.338	0.202	0.064	0.035
	high	0.358	0.337	0.203	0.065	0.036
S	true	0.366	0.339	0.199	0.062	0.034
	false	0.347	0.337	0.211	0.068	0.037

propagated conditional distributions are shown in Table 3. Given different states of RSE, probabilities for SC states are calculated. Notice that they are all high for the none state as states of other nodes are also affecting the SC. If S is false, though, p(SC = none) decreases. Similar probabilities can be calculated easily for other higher nodes. For instance, detection state transition to RSE can be calculated as p(RSE|RDS). Reasonably, from the table, if the detection state is low, RSE can be at high state, else RSE is at failing low state. Likewise, the propagation of risks to a series of other SCs (from Figure 2*b*) is also represented. For instance, at SC.10, probabilities of p(SC = low) and p(SC = med) are significantly higher. Note that impacts at attacked SC's RSE and S states do not play an important role. This can be seen from the equal probabilities across different states.

Simulations

Types of Simulation

Analyzing the cyber-physical model, impact on an SC is propagated through service times (see Figure 2*a*). Primary objectives are quantification and identification of possible problems. These would yield risk areas, impacts, and possible solutions for more resilient systems. Monte Carlo simulations (MCSs) are carried out on BNs, including a series of intersections as well as replications of complete BNs. MCSs generate probability distributions and expected utility as well as dispersion so that confidence intervals can be generated. A second approach generates a Bayesian framework for posterior probability of a critical node common among these signals, such as SA, considering it as a parameter.

This section presents numerical results for BNs with and without redundant surveillance systems. MCSs are run for 10,000 cycles which result in state probabilities, and in expected, and standard deviation of, vulnerabilities. Simulations are repeated 200 times to obtain distributions of these measures for CV market penetrations of p = (0.01, 0.05, 0.10, 0.20, 0.50, 0.75, 0.90, 1.00), which is constituted by (CR = 0.80, EM = 0.03, FR = 0.07, TR = 0.10). Vehicle and pedestrian traffic are 95% and 5%, respectively. In Figure 3*a*, probabilities, expected value, and standard deviation of vulnerability of SC are presented. Boxplots are used to show the ranges within 200 replications, as metrics are assigned randomly from uniform distribution. Probabilities are used to calculate the moments. The probability of SC being at the no effect level is about 93% and about 6% at low impact. These values vary in the ranges 0.80–0.99 and 0–0.20, respectively, changing the central tendency and the dispersion of vulnerabilities. The expected value certainly falls mostly into the low impact (> 0.1) region. The median of standard deviations is about 0.9. Thus a simple confidence interval for mean level (0.37) with mean

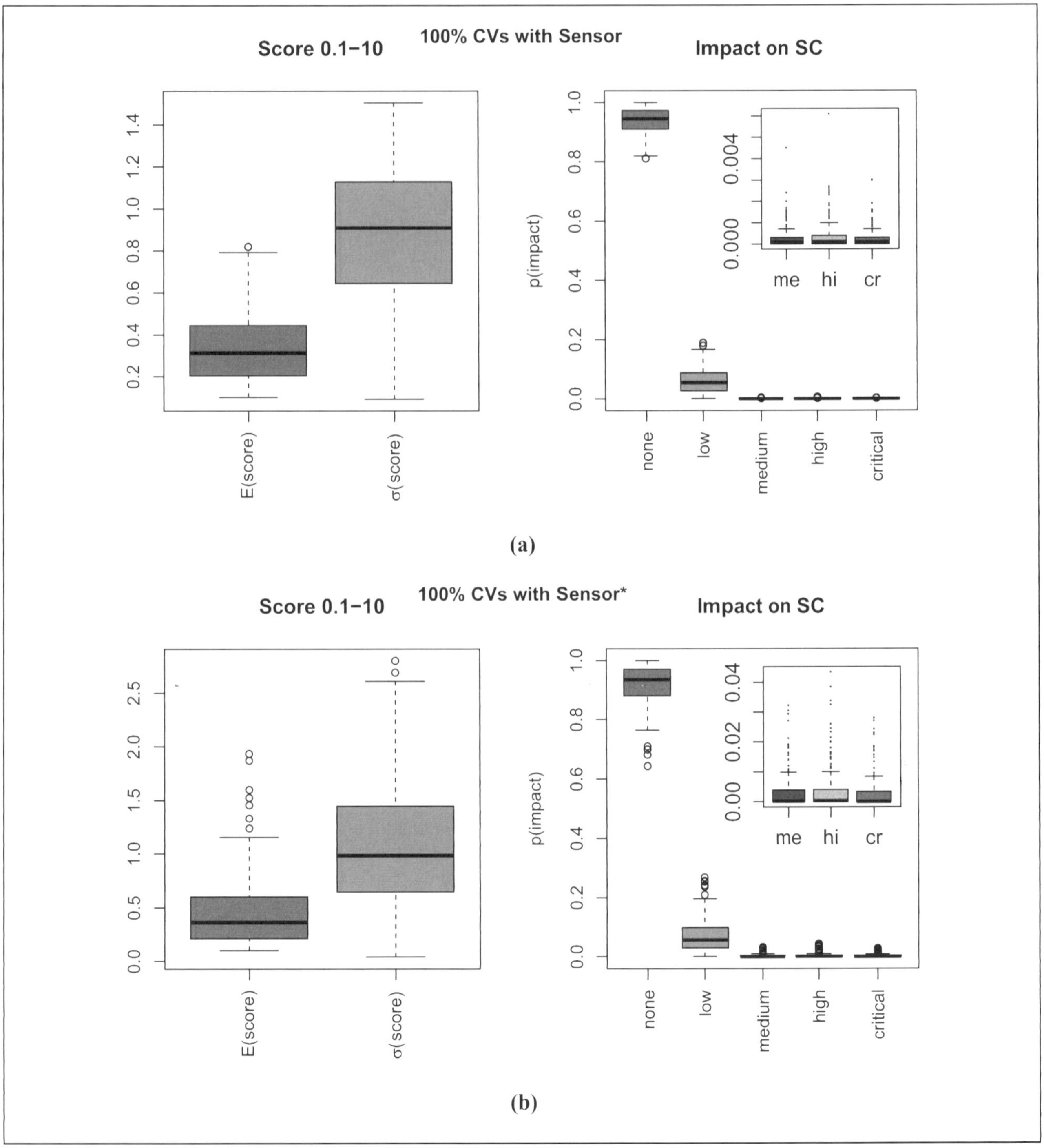

Figure 3. Simulated score values from BN with sensors.

standard deviation (0.93) $\mu \pm 2\sigma$ would be within (0, 2.23), which is still in the low impact scale.

Further simulations are carried out to check the effect of randomly generating all metric values. These values are only assigned for 200 replications but not for each node on BNs. Figure 3*b* shows somewhat higher expected values; however, much higher standard deviations with risk likelihoods vary more. Still, median values are very close; with the mean values (0.57, 1.32), the same confidence interval in this case would be (0, 3.20), very close to medium risk level. This could be due to not exploring enough of the metric ranges and resulting higher dispersions. Therefore, a completely random approach would give more robust results. Similar

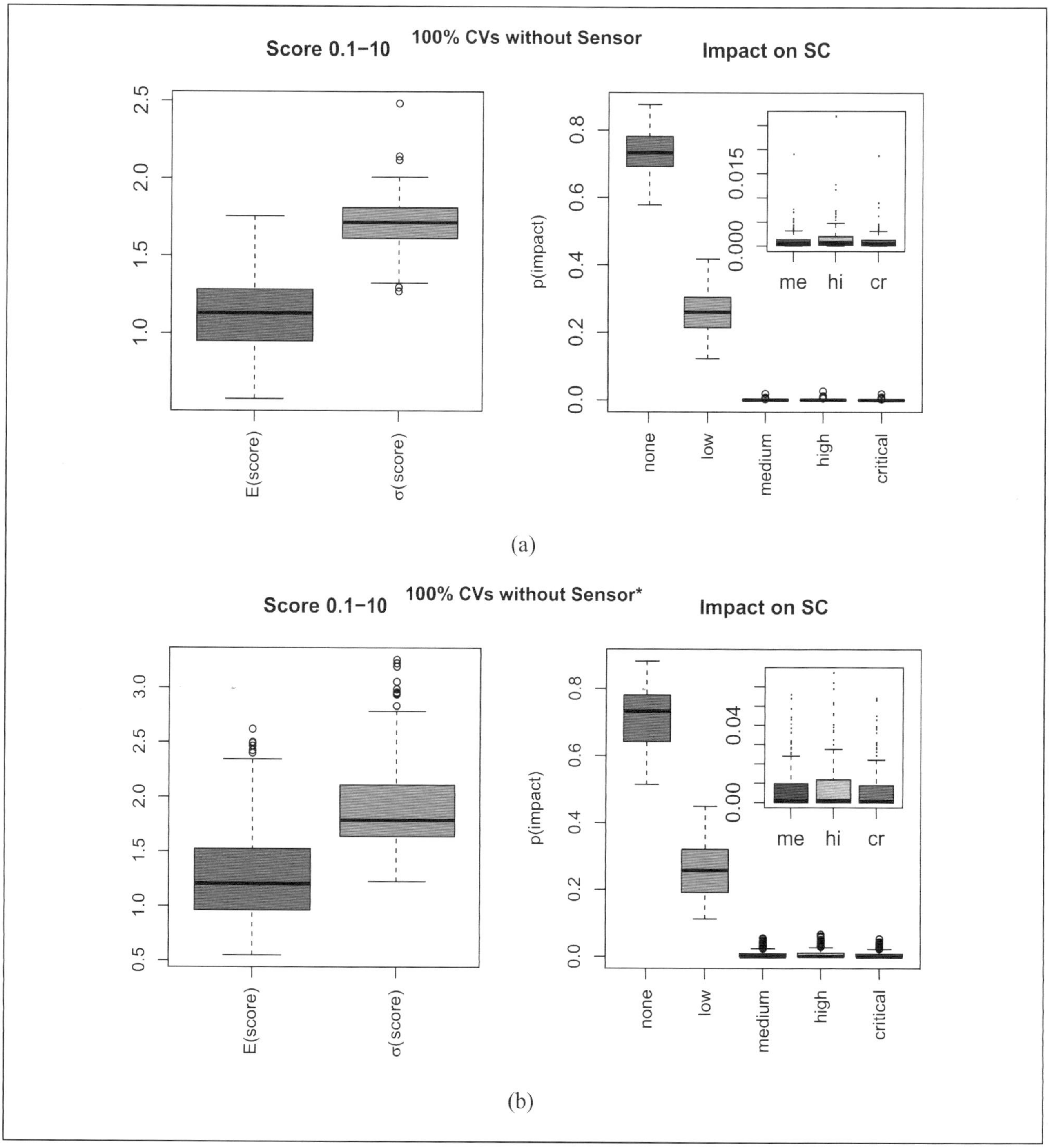

Figure 4. Simulated score values from BN without sensors.

results are demonstrated in Figure 4 for the without sensor case. They reveal much higher risk levels and variations. Risk likelihood is still mainly within low level; however, the low state is more probable than none. The confidence interval in this case would be (0,4.81), with mean and standard deviation values of 1.19 and 1.81, respectively. In this case vulnerability falls into the much more serious med level. As before, the less random version yields higher dispersion, with confidence interval (0, 5.03) with means 1.28 and 1.87. Furthermore, basic progression of the risk present on signal control to the next 10 signals is represented in Figure 2*b*. The vulnerability of SC and SC.10 signal controllers is presented in Figure 5. From the figure, it

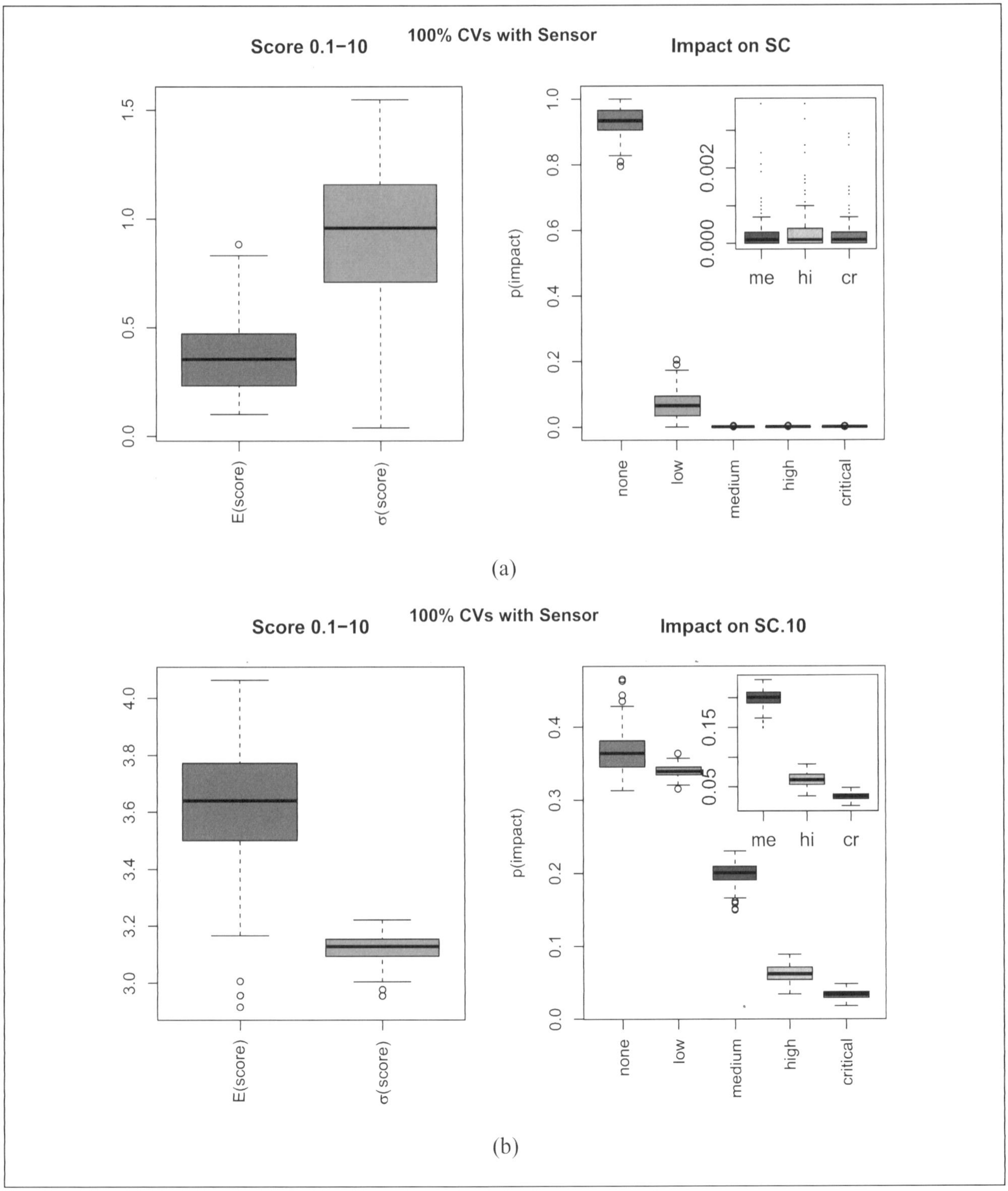

Figure 5. Simulated score values from BN with sensors.

is observed that the risk, on moving from SC to SC.10, increases in a serial scheme based on defined transition probabilities, which is intuitive. The average score standard deviation is about 3.1, with mean level 3.65. Thus, the vulnerability score is already very close to med level and confidence intervals would fall into all serious levels, med to high impact scales (0, 9.85). In this system, however, dependency is only through SC nodes, not

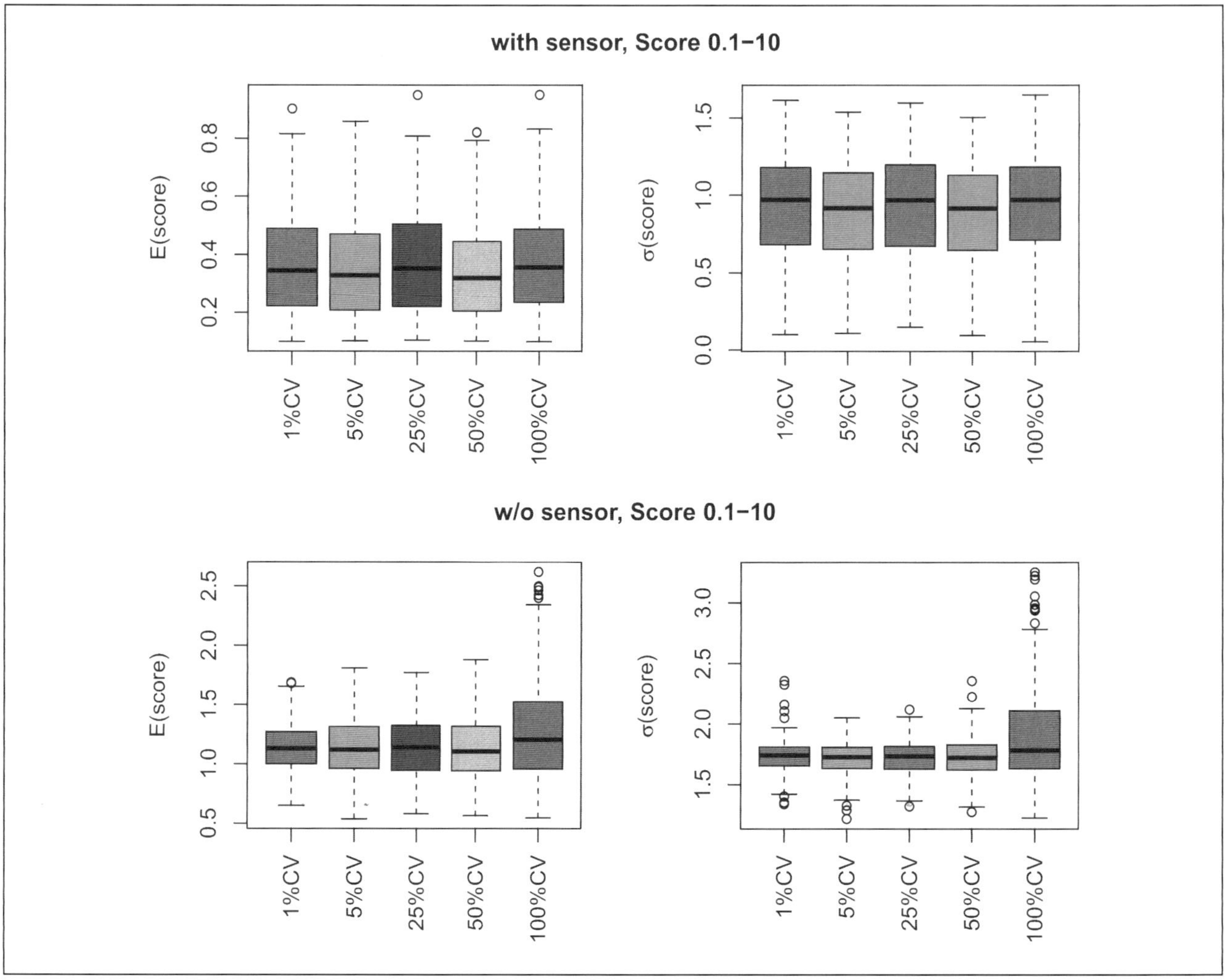

Figure 6. Risk values for SC at different %CV levels.

others. Thus, overall risk may change if other nodes are included.

Figure 6 is given to demonstrate the impact of CV market penetration. This is incorporated as increasing likelihood of CVs in TC by multiplying unequipped modes with $(1 - p)$ and weighing CVs with p. Thus, as p increases, CV population will increase, resulting in more communications with RSE and SC. However, based on this method, CV percentage does not have a significant effect on SC vulnerability level. Certainly, it should be noted that a different incorporation (i.e., network design) might show otherwise.

Traffic Process Simulations

A possible quantification of the impact of state probabilities on an isolated signalized intersection is shown via traffic queue simulations using a ProModel Process Simulator. For simplicity, the signal control method is assumed to be fixed and communications are not included. Changes are incorporated only through service times. Average queue lengths are demonstrated for regular and affected scenarios with and without sensors. The intersection is designed with approaches that can accommodate a maximum of 130 average-length cars per approach (Figure 7*a*). For the simulation set up, first the saturation level is found. To compare with the results from microsimulations (*24*), an isolated intersection with two one-way approaches is designed with 45 s green for each approach and volumes = (500, 600, 700, ..., 2000) vehicles per hour (vph) per approach. Similar results are observed with saturation at about 1000 vph per approach. Simulations are run for 1 h (40 cycles) and 100 replications. Figure 7*b* below shows volume versus average queue lengths as number of vehicles (v) and volume-to-capacity ratios (utilization). From the figure, the saturation point can be seen at close to 2000 vph with capacity of 23 vehicles per cycle and 1.95 seconds per

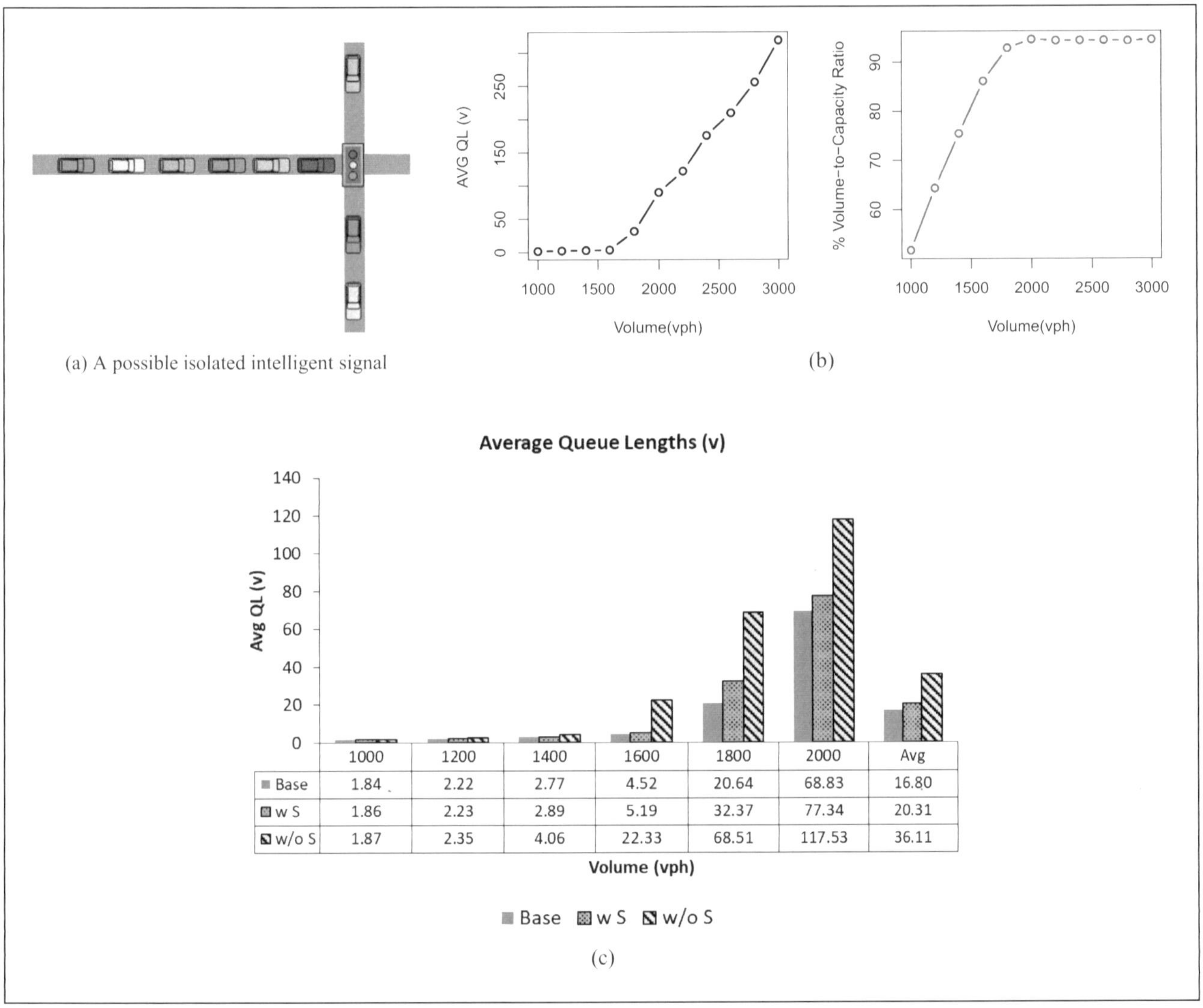

Figure 7. Process traffic simulation results.

vehicle (spv) discharge rate. Using the same volume rates, this time up to 2200 vph per intersection, possible interruptions are generated from the probability distribution BNs. Impacts are reflected on 1.95 spv, with (none, low, med, high, cri) = (1.95, 2.60, 3.90, 9.75, 19.5) spv and average probabilities, with sensors, of (0.930, 0.069, 0.0005, 0.0005, 0.0005) and, without sensors, of (0.730, 0.260, 0.003, 0.004, 0.003), respectively. Figure 7*c* demonstrates an overall summary of the impact of risk levels on queue lengths that are obtained from simulations which are medium level realistic between microscopic and point queue simulations. The figure provides the overall evolution of base performance measures for with sensor and without sensor scenarios. As would be expected, impact increases as volume gets higher. The impact on a single intersection is much higher, about 15% average queue length (QL) increase with a sensor. Without any sensor, changes are much higher with average 125% of average queue length across all volume levels.

Conclusions

In this paper, models are developed based on BNs for possible attack risks on intelligent signals in a CV framework. From these models, risk probabilities and expected utilities (impacts) are deduced for SCs with and without redundant traffic surveillance systems. The impact of risks at an isolated signalized traffic intersection is quantified via simulations as average queue length differences, which can be interpreted as meaning higher cost, higher fuel consumption, and higher emissions. From simple simulations with and without redundant systems, the impact of risk is estimated as an increase of queues by averages of 15% and 125%, respectively. Further improvements are possible via:

- Quantification of impacts of derived risks at signalized traffic networks via simulations as different performance measures
- Microscopic traffic simulation including communications with more realistic vehicle movements
- Expression of the systems as a flow network for possible attack paths to optimize sensor deployment and minimize communication delays
- Inclusion of alternative CV market penetration incorporation

A continuation study would aim to generate data using part real equipment and part simulations. Then, it would utilize this data to develop crucial detection models for critical components and generate mitigation efforts. It is essential to have optimum deployment as unnecessary control can result in communication delays, leading to problems for safety critical applications such as collision warning. As a systematic approach, similar methodology would be adopted for different ITS applications.

Acknowledgments

This research is supported by the U.S. Department of Homeland Security Summer Research Team Program and was conducted at the Critical Infrastructure Resilience Institute, University of Illinois, Urbana-Champaign. It was managed by ORAU. The research is also partially supported by USDOT Regional University Transportation Center for Connected Multimodal Mobility and NSF Grant Nos. 1719501 and 1400991.

Author Contributions

The authors confirm contribution to the paper as follows: study conception and design: GC; data collection: GC; analysis and interpretation of results: GC; draft manuscript preparation: GC. The author reviewed the results and approved the final version of the manuscript.

References

1. Cárdenas, A. A., S. Amin, Z.-S. Lin, Y.-L. Huang, C.-Y. Huang, and S. Sastry. Attacks Against Process Control Systems: Risk Assessment, Detection, and Response. *Proc., 6th ACM Symposium on Information, Computer and Communications Security*, ACM, 2011, pp. 355–366.
2. Kaiser, L. *Transportation Industrial Control System (ICS) Cybersecurity Standards Strategy 2013–2023*. Technical Report. National Highway Traffic Safety Administration, 2013.
3. National Highway Traffic Safety Administration. *Cybersecurity Best Practices for Modern Vehicles*. Technical Report No. DOT HS 812 333. National Highway Traffic Safety Administration, USDOT, 2016.
4. Beck, K. *Smart Security? Evaluating Security Resiliency in the U. S. Department of Transportation's Smart City Challenge*, 2017.
5. Petit, J., and S. E. Shladover. Potential Cyberattacks on Automated Vehicles. *IEEE Transactions on Intelligent Transportation Systems*, Vol. 16, No. 2, 2015, pp. 546–556.
6. Nicol, M. D. *Critical Infrastructure Resilience Institute*, 2016, talk, USCG.
7. Peisert, S., J. Margulies, D. M. Nicol, H. Khurana, and C. Sawall, Designed-In Security for Cyber-Physical Systems. *IEEE Security & Privacy*, Vol. 12, No. 5, 2014, pp. 9–12.
8. van der Heijden, R. W., S. Dietzel, T. Leinmüller, and F. Kargl. Survey on Misbehavior Detection in Cooperative Intelligent Transportation Systems. *arXiv preprint arXiv:1610.06810*, 2016.
9. Al-Sultan, S., A. H. Al-Bayatti, and H. Zedan. Context-Aware Driver Behavior Detection System in Intelligent Transportation Systems. *IEEE Transactions on Vehicular Technology*, Vol. 62, No. 9, 2013, pp. 4264–4275.
10. Cao, P., E. Badger, Z. Kalbarczyk, R. Iyer, and A. Slagell. Preemptive Intrusion Detection: Theoretical Framework and Real-World Measurements. *Proc., 2015 Symposium and Bootcamp on the Science of Security*, ACM, 2015, p. 5.
11. Mitchell, R., and I.-R. Chen. A Survey of Intrusion Detection Techniques for Cyber-Physical Systems. *ACM Computing Surveys (CSUR)*, Vol. 46, No. 4, 2014, p. 55.
12. Humayed, A., J. Lin, F. Li, and B. Luo. Cyber-Physical Systems Security—A Survey. *arXiv preprint arXiv: 1701.04525, 2017.*
13. Ernst, J. M., and A. J. Michaels. Framework for Evaluating the Severity of Cybervulnerability of a Traffic Cabinet. *Transportation Research Record: Journal of the Transportation Research Board*, 2017. 2619: 55–63.
14. Sridhar, S., and M. Govindarasu. Model-Based Attack Detection and Mitigation for Automatic Generation Control. *IEEE Transactions on Smart Grid*, Vol. 5, No. 2, 2014, pp. 580–591.
15. Sucasas, V., G. Mantas, F. B. Saghezchi, A. Radwan, and J. Rodriguez. An Autonomous Privacy-Preserving Authentication Scheme for Intelligent Transportation Systems. *Computers & Security*, Vol. 60, 2016, pp. 193–205.
16. CVRIA. *Physical Diagram of Intelligent Signal System*, 2015. http://local.iteris.com/cvria/html/applications. Accessed June 26, 2017.
17. NIST. *Common Vulnerability Scoring System*, 2017. https://nvd.nist.gov/vuln-metrics/cvss/v3-calculator. Accessed June 30, 2017.
18. Head, L., W. B. Zhang, G. Duncan, E. Raamot, and R. Jose. *MultiModal Intelligent Traffic Signal System Phase II: System Development, Deployment and Field Test-MMITSS*. Technical Report. Center for Transportation Studies, University of Virginia, 2016.
19. Huang, J., and D. M. Nicol. Evidence-Based Trust Reasoning. *Proc., 2014 Symposium and Bootcamp on the Science of Security*, ACM, 2014, p. 17.
20. Wan, J., D. Zhang, S. Zhao, L. Yang, and J. Lloret. Context-Aware Vehicular Cyber-Physical Systems with Cloud

Support: Architecture, Challenges, and Solutions. *IEEE Communications Magazine*, Vol. 52, No. 8, 2014, pp. 106–113.

21. MacKay, D. J. *Information Theory, Inference and Learning Algorithms*. Cambridge University Press, 2003.
22. Barber, D. *Bayesian Reasoning and Machine Learning*. Cambridge University Press, 2012.
23. Højsgaard, S. Graphical Independence Networks with the gRain Package for R. *Journal of Statistical Software*, Vol. 46, No. 10, 2012, pp. 1–26.
24. Comert, G. Simple Analytical Models for Estimating the Queue Lengths from Probe Vehicles at Traffic Signals. *Transportation Research Part B: Methodological*, Vol. 55, 2013, pp. 59–74.

The Standing Committee on Critical Transportation Infrastructure Protection (ABR10) peer-reviewed this paper (18-06284).

Article

TRR
JOURNAL OF THE TRANSPORTATION RESEARCH BOARD

Transportation Research Record
2018, Vol. 2672(1) 90–100

DOI: 10.1177/0361198118789953
journals.sagepub.com/home/trr
SAGE

Transit Information Utilization during an Extreme Weather Event: An Analysis of Smartphone App Data

Coline Remy[1], Candace Brakewood[2], Niloofar Ghahramani[3], Eun Jin Kwak[4], and Jonathan Peters[4]

Abstract

Extreme weather events such as heavy snow can severely disrupt urban transportation systems. When this occurs, travelers often seek information about the status of transportation services. This study aims to assess information utilization during an extreme weather event by analyzing data from a smartphone application ("app") called Transit, which provides real-time transit and shared mobility information in many cities. This research focuses on a snowstorm that hit the northeastern United States in January 2016 and severely disrupted transit and shared mobility services. An analysis of Transit app data is conducted in four parts for New York City, Philadelphia, and Washington, D.C. First, hourly app utilization during the snowstorm was compared with mean hourly app utilization prior to the storm. Second, the rate of app usage was calculated by dividing hourly utilization during the storm by the mean hourly volume before the storm. Third, an ordinary least squares regression model of hourly app usage was estimated for each city. Last, a feature within the app used to request Uber vehicles was examined. The results of the first three analyses reveal that overall app usage decreased during the snowstorm in all three cities; after the storm, New York experienced a significant increase in overall app use during the first Monday commuting period. The analysis of Uber data reveals that app users continued to search for ridehailing services during the snowstorm, despite travel bans. These findings are important for transportation operators and app developers to understand how travelers use information during extreme weather events.

Extreme weather events such as heavy snow can severely disrupt urban transportation systems. One recent example occurred in January 2016, when Winter Storm Jonas dumped historic amounts of snow throughout the northeastern region of the United States. Jonas was the single largest snowstorm on record for numerous locations in the northeast, including New York City's LaGuardia Airport and Baltimore–Washington International Airport (*1*). This historic snowfall deeply affected the transportation systems in New York City, Philadelphia, and Washington, D.C. The local governments declared travel bans, public transit systems were shut down, and the period of recovery in some areas lasted for nearly a week after the snowstorm.

When extreme weather events such as Winter Storm Jonas occur, travelers often seek information about the status of transportation services to determine whether they are shut down and to stay informed as they reopen. However, understanding how travelers seek information during extreme events can be challenging due to a lack of readily available data. For example, surveys are difficult to administer during unplanned events, and they can be expensive to collect retroactively. Therefore, this research aims to use a new data source to understand how travelers seek information during an extreme weather event. Specifically, the records from an urban transportation information smartphone application (or "app") are examined to provide insight into how travelers utilize new information and communication technologies during severe storms.

[1]Ecole Nationale des Travaux Publics de l'Etat, Vaulx-en-Velin Cedex, France
[2]University of Tennessee, Knoxville, TN
[3]City College of New York, New York, NY
[4]College of Staten Island, Staten Island, NY

Corresponding Author:
Address correspondence to Candace Brakewood: cbrakewo@utk.edu

Prior Research and Objectives

There is a growing body of literature that examines the transportation system impacts of extreme weather events (*2–4*) and large special events such as sports games (*5*, *6*). This brief review of prior research focuses specifically on studies that use new data sources to understand the impacts of extreme weather on the transportation system in the New York City region.

One noteworthy study measured the resilience of the New York City transportation system after Hurricane Sandy using data from taxi GPS traces (*7*). The findings suggest that Hurricane Sandy affected roadway traffic conditions for more than 5 days, and that significant traffic delays were encountered during the post-disaster response period. A second noteworthy study used two "big data" sources to assess the consequences of Hurricane Sandy and Hurricane Irene in New York City (*8*). Post-hurricane recovery patterns for the roadway and subway systems were analyzed using taxi trip data and subway turnstile data, respectively. The method involved calculating the rate of recovery, which was found to be lower for Hurricane Sandy compared with Hurricane Irene. A third study developed a method to detect demand fluctuations and correlate them with disruptive events (such as extreme weather conditions) using New York City taxi trip data (*9*).

In summary, there are several emerging data sources that can help to quantify the impacts of extreme events on transportation system performance. However, to the best of the authors' knowledge, prior research has not considered the relationship between extreme events and transportation app utilization. Therefore, the objective of this study is to explore the use of transportation information provided via smartphone applications during an extreme weather event. The specific dataset used in this analysis is from a smartphone application called Transit, which is available in numerous cities (*10–12*). Because of the multi-city nature of the Transit app, the dataset allows for a comparison between three cities that were affected by the snowstorm of January 2016: New York City, Philadelphia, and Washington, D.C.

Background on the Winter Storm

Winter Storm Jonas paralyzed the eastern United States from Friday, January 22, 2016 to Sunday, January 24, 2016. Between 20 in. and 30 in. of snow accumulated in New York City, Philadelphia, and Washington, D.C. (*13*). On Friday, January 22, government officials in Washington, D.C. declared a snow emergency (*14*), and the local government in Philadelphia did the same. The following day (January 23), officials in New York City declared a travel ban. In each of these three cities, transit services were gradually shut down, beginning before the storm in some areas and resulting in a complete shutdown during the worst periods of the snowstorm. Additionally, Uber service was officially shutdown in New York City (*15*).

The hour-by-hour status of public transit and shared mobility services (specifically, Uber) before, during, and after the snowstorm are shown in Figure 1. This information was compiled from transportation authority websites, local media sources, and social media (*16–29*).

Dataset

This section contains a general description of the smartphone app data used in this analysis. The first part gives a brief overview of the smartphone app, the second part defines the geographic area of analysis, and the last section includes a brief description of the data files.

Overview of the Smartphone Application

Transit is a smartphone application available on Android and iPhone that provides information about public transportation and shared mobility services such as bikeshare, carshare, and ridehailing (*10*). Transit covers New York City, Philadelphia, Washington, D.C., and approximately 125 other cities. The app includes many features, such as real-time transit vehicle location information (see left screenshot in Figure 2), trip planning (see right screenshot in Figure 2), transit service alerts, and multimodal support including Uber. To request an Uber during the period of analysis, a user would click "request an Uber" in the Transit app, and s/he would then be directed to Uber's app to complete the order. More recently, this Uber feature has been changed so that booking and payment can be completed directly within the Transit app.

Time Period and Geographic Area of Analysis

The time period for this study is Friday, January 8, 2016 to Friday, January 29, 2016: two weeks before the snowstorm and one week after. The geographic areas are the regions of New York City, Philadelphia, and Washington, D.C. The boundaries of the three regions were selected based on the Metropolitan Planning Organization areas (*30–34*). Specifically, three bounding boxes were drawn to define the maximum and minimum latitude and longitude for each area. The New York City area includes the region served by the New York Metropolitan Transportation Council (NYMTC) (*30*) and the North Jersey Transportation Planning Authority (NJTPA) (*31*), which is shown in Figure 3 in yellow. The Philadelphia area includes the regions served by the Delaware Valley Regional Planning Commission

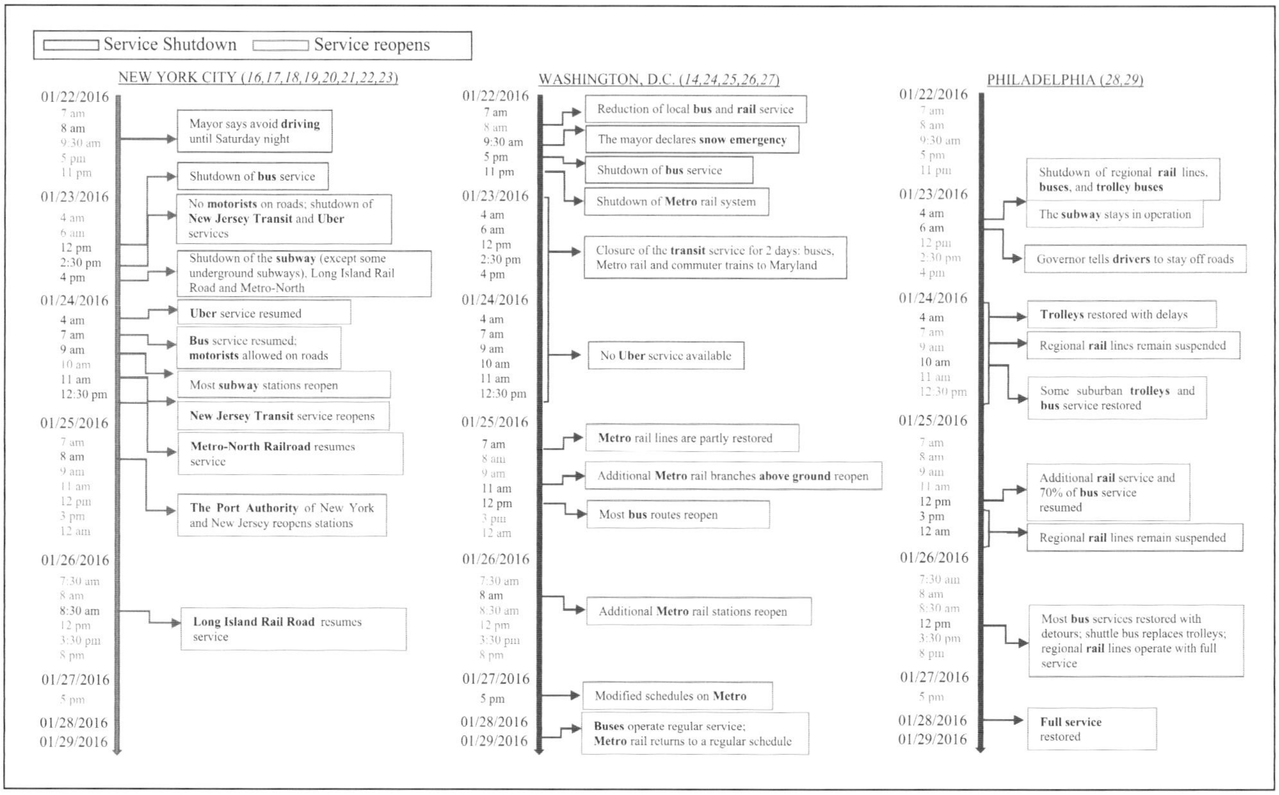

Figure 1. Timeline of the snowstorm.

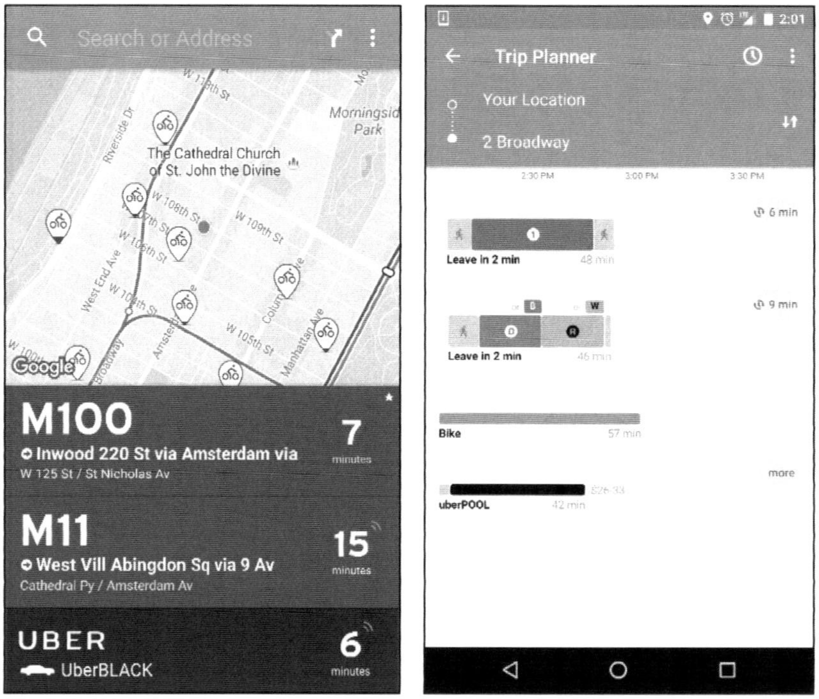

Figure 2. Transit app screenshots.

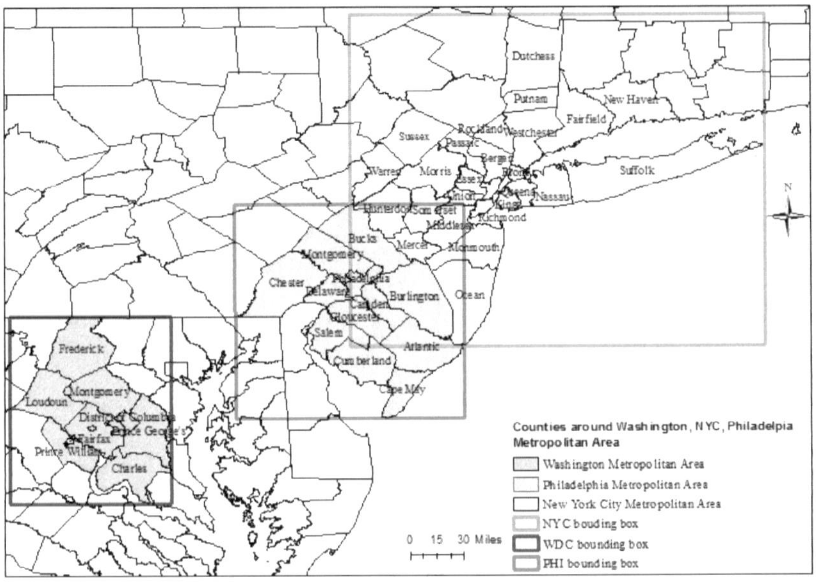

Figure 3. Bounding boxes around the New York City, Philadelphia, and Washington, D.C. Metropolitan areas.

(DVRPC) (*33*) and the South Jersey Transportation Planning Organization (SJTPO) (*34*), which is shown in green in Figure 3. Last, the Washington, D.C. area was selected using the region served by the Capital Region Transportation Planning Board (TPB) (*32*) shown in blue in Figure 3.

Data Description

The data used in this analysis was provided by Transit to the research team. The dataset contains user interactions with the app. Each time a user opens the app, a unique session identifier (ID) is created, the time is recorded, and the location of the user (latitude/longitude) is stored. Between January 8 and January 29, 2016, there were millions of Transit app sessions in the areas of analysis.

The following analysis also uses a subset of the data that includes all Uber requests made through the Transit app. Each time a user clicks to request an Uber in Transit, the user's location (latitude/longitude) and the time of the request are recorded. Between January 8 and January 29, 2016, there were thousands of Uber requests in the areas of analysis.

Analysis

An analysis of Transit app utilization during the January 2016 snowstorm was conducted in four parts. First, overall Transit app usage patterns were visually explored, and then the rate of use was calculated by comparing app use during the snowstorm with a regular period. Next, a regression model of total hourly app use was created. Last, the use of the Uber feature in the Transit app was investigated in a visual analysis. The results of this analysis are described in the following sections.

Part 1: Visualization of Transit App Usage

First, overall Transit app usage patterns were visually explored. During the 3-week period of analysis, the total Transit app usage per hour in each region was calculated. This was then compared with the mean app usage per hour during the two first weeks before the snowstorm when there were normal weather conditions (January 8 to January 15).

Figure 4 shows hourly Transit app usage (the black line) compared with the mean hourly usage (the red line) for each of the three regions. Utilization of the app has a strong periodic weekly pattern. There is a peak during the morning rush hour and another during the evening rush hour for weekdays, and app usage decreases during the weekend. However, this typical pattern was interrupted during the snowstorm period, as can be seen in Figure 4. The vertical line in yellow shows the period when the snowstorm hit each of the three cities, and the vertical line in blue displays when the transit systems were closed. During this period, the amount of overall app usage per hour in each city was less than the mean. In the New York City area and in the Philadelphia area, this decrease is more pronounced on Saturday, January 23 during the afternoon. For the Washington, D.C. area, the decrease in app usage began earlier (on Friday,

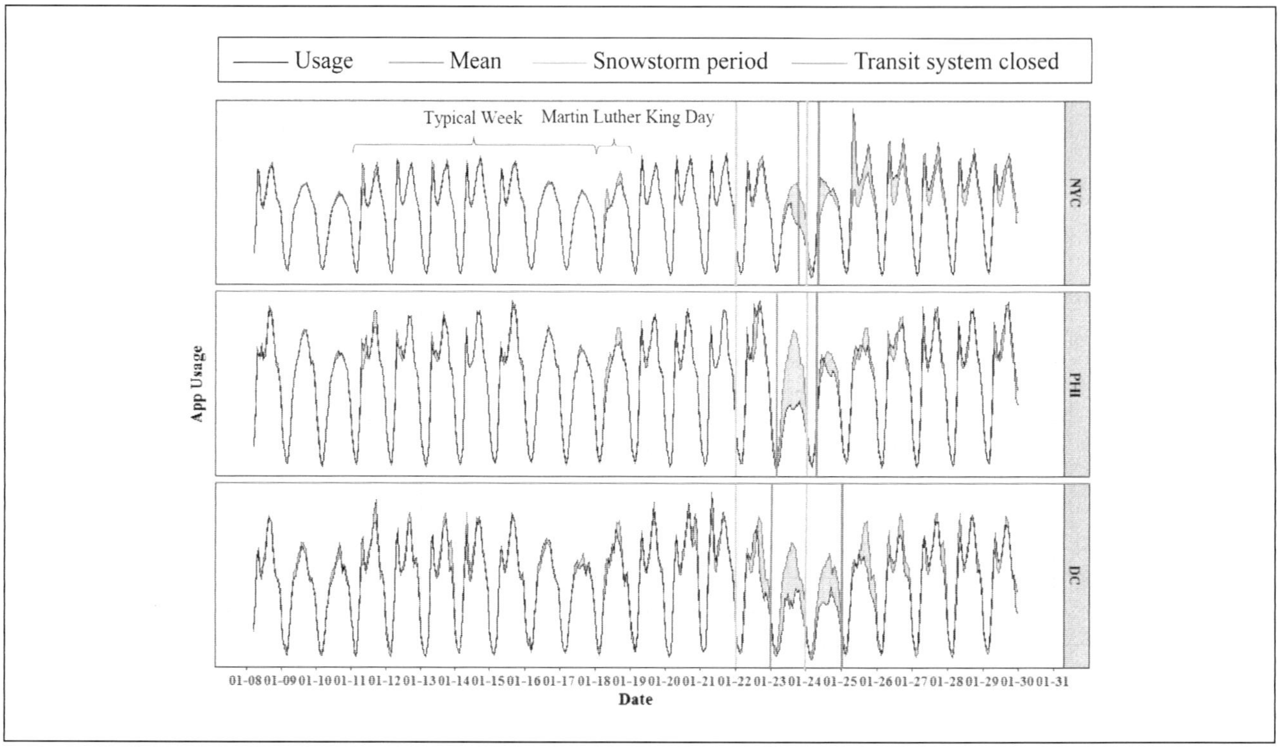

Figure 4. Transit app usage and mean usage in New York City Philadelphia, and Washington, D.C.

January 22) and stayed below the mean for longer than the two other areas.

Part 2: Rate of Transit App Use

Next, the rate of Transit app use was calculated in a similar manner to a previous paper on extreme events (*8*) by comparing mean hourly usage prior to the snowstorm with hourly app usage during the snowstorm for each city. For every hour t, the rate of use $R(t)$ was calculated by dividing the hourly app usage $U(t)$ by the mean hourly usage $Q(t)$ during the reference period, which was the two weeks before the snowstorm. Equation 1 shows the rate of use formula:

$$R(t) = \frac{U(t)}{Q(t)} \tag{1}$$

where

$R(t)$ = rate of use
$U(t)$ = hourly app use
$Q(t)$ = mean hourly app use

Figure 5 displays the results of the rate of use calculation for each of the three cities as a black line. The red line in Figure 5 shows 100%, which is the typical usage during the reference period. As can be seen in Figure 5, the three cities generally exhibit similar patterns. The rate of use is fairly consistent during the 2 weeks prior to the storm, and all three cities have a dip in use on the Martin Luther King, Jr. holiday, which was Monday, January 18, 2016. During the snowstorm, all three cities initially exhibited an increase in the rate of use, but this quickly decreased below the 100% level as the snowstorm continued.

In New York City, the rate of app usage dropped below 100% on Saturday, January 23 before noon and remained low until the following day (Sunday, January 24). On Saturday, most transit service in New York City was shut down during the snowstorm (with the exception of some underground areas of the subway), and it was during this time that the rate of use reached its lowest point (approximately 65% of mean usage). After the snowstorm finished and transit service was largely restored on Monday, January 25, a peak appeared during the morning rush hour, and the rate of use reached about 165%. This may be because Transit app users wanted more information as they went back to work to assure them that transit service was fully restored.

In the Philadelphia area, the rate of Transit app use initially increased on Friday, January 22 until noon; at this time, a snow emergency had been declared (*35*) but the transit system had not yet shut down. This spike in use is likely to be because Transit app users were searching for information to get home before the snowstorm. Shortly thereafter, during the nighttime on Friday, January 22, the rate of use dropped substantially below

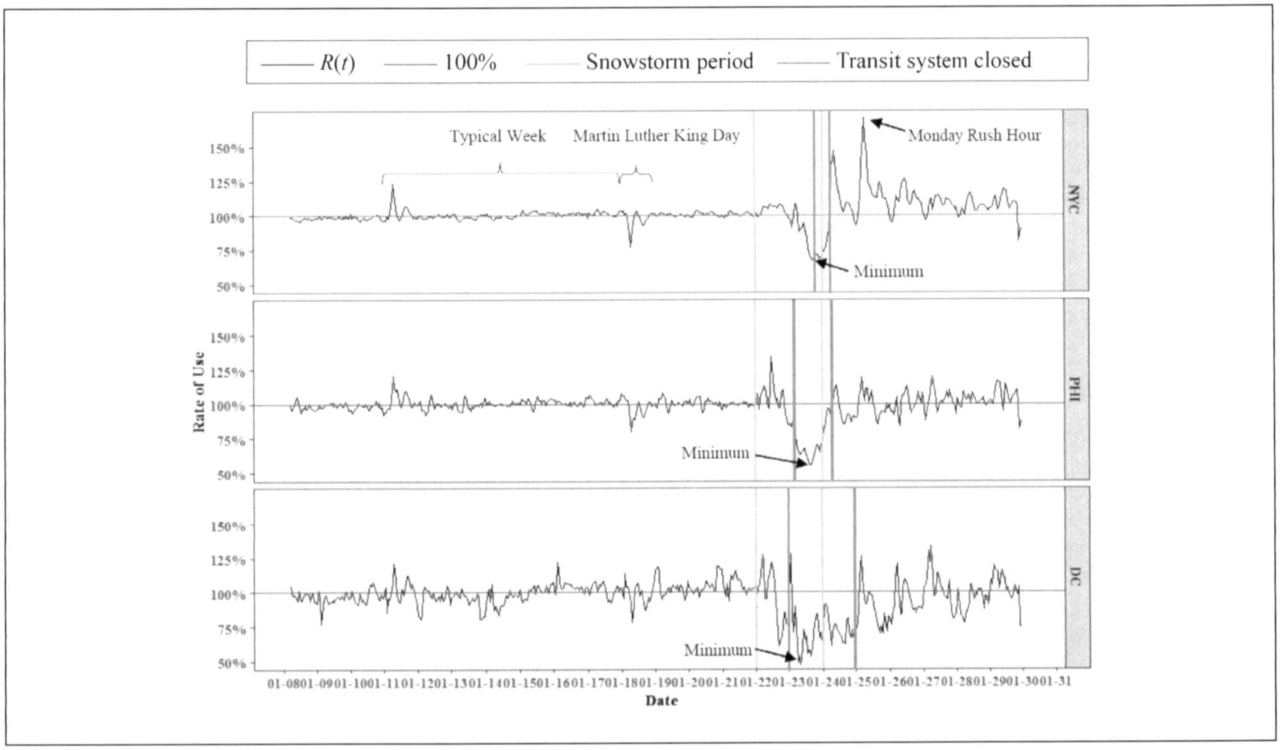

Figure 5. Rate of transit app use in New York City, Philadelphia, and Washington, D.C.

100% and stayed low until the morning of January 24, 2016. On Saturday 23, when Philadelphia transit authorities suspended all their services (Figure 1), the rate of use reached about 65% of the mean usage, which was the lowest point. After the snowstorm finished and transit service was largely restored on Monday, January 25, a small increase in the rate of use appeared during the morning rush hour.

In the Washington, D.C. area, the rate of Transit app use dropped below 100% for most hours from January 22 at noon until January 25 at midnight. After the storm, the rate of use took longer to stabilize compared with the other metropolitan areas, which may have been because the recovery time of the transit systems in this region was longer than the other areas (Figure 1). For example, it was not until Tuesday, January 26 that most of the Washington, D.C. metro system was reopened (with 86 of 91 Metrorail stations opened by Tuesday, January 26). Consequently, app users in Washington, D.C. seemed to be more affected by the snowstorm than those in New York City or Philadelphia.

Finally, it should be noted that some peaks in Figure 5 are not meaningful due to data limitations. For example, a peak in the rate of use appeared in each of the three cities on Monday, January 11. This is because the other Monday used to calculate the mean (reference) hourly usage was Martin Luther King Day. Future analyses should expand the period of analysis to accommodate this.

Part 3: Regression Model of Transit App Use

Next, three ordinary least squares (OLS) regression models were estimated: one for New York City, one for Philadelphia, and one for Washington, D.C. In each regression model, the dependent variable is the number of Transit app records per hour in the corresponding city. For independent variables, eight binary variables were defined to account for the effects of Martin Luther King Day (on Monday, January 18), 3 days of the snowstorm (Friday, January 22 to Sunday, January 24), and 4 days after the snowstorm (from Monday, January 25 to Thursday, January 28). Also, to control for the typical weekly app usage pattern, 167 binary variables were included in each model to account for the 168 hr in a week. Equation 2 below shows the regression model formulation, which is the same for each city:

$$\begin{aligned} \mathrm{Y_t} = \alpha + \sum_{i=1}^{167} \beta_i \mathrm{h_i} + \beta_{168}\mathrm{MLKD} \\ + \beta_{169}\mathrm{SD1} + \beta_{170}\mathrm{SD2} + \beta_{171}\mathrm{SD3} \\ + \beta_{172}\mathrm{RD1} + \beta_{173}\mathrm{RD2} + \beta_{174}\mathrm{RD3} + \beta_{175}\mathrm{RD4} \end{aligned} \quad (2)$$

where

Y_t = number of Transit app records per hour t

h_1 to h_{167} = binary variables representing each hour of the week (i.e., h_1 represents Friday at 12:00 a.m., h_{25}

Table 1. Regression Results for New York City, Philadelphia, and Washington, D.C.

	New York City	Philadelphia	Washington, D.C.
Intercept	79,173.600***	8,568.542***	10,523.240***
	(2455.90)	(271.23)	(496.43)
Mon Jan 18 (Martin Luther King Day)	−3,843.750***	−576.625***	−547.333**
	(1,215.36)	(134.23)	(245.67)
Fri Jan 22 (1st day of snowstorm)	2,773.521***	381.437***	−528.813**
	(1052.53)	(116.24)	(212.75)
Sat Jan 23 (2nd day of snowstorm)	−8,925.937***	2,200.625***	−2,202.167***
	(1052.53)	(116.24)	(212.75)
Sun Jan 24 (3rd day of snowstorm)	4,110.062***	−389.958***	−1,802.104***
	(1052.53)	(116.24)	(212.75)
Mon Jan 25 (1st day after snowstorm)	9,709.958***	−373.292***	−1,377.792***
	(1215.36)	(134.23)	(245.67)
Tue Jan 26 (2nd day after snowstorm)	7,013.187***	43.375	−525.708**
	(1052.53)	(116.24)	(212.75)
Wed Jan 27 (3rd day after snowstorm)	4,903.354***	250.750**	−48.19
	(1052.53)	(116.24)	(212.75)
Thu Jan 28 (4th day after snowstorm)	3,544.271***	173.25	−411.500*
	(1052.53)	(116.24)	(212.75)
Observations	504	504	504
R^2	0.97	0.977	0.949
Adjusted R^2	0.955	0.965	0.922
Residual standard error (df = 328)	4,210.11	464.97	851.014
F statistic (df = 175; 328)	61.650***	79.951***	35.105***

Note: Standard error shown in parentheses. One hundred sixty-seven hourly dummy variables not presented in the table; available upon request.
*$p < 0.1$. **$p < 0.05$. ***$p < 0.01$.

represents Saturday at 12:00 a.m., and h_{167} represents Thursday at 11:00 p.m.)

MLKD = binary variable representing Martin Luther King Day

SD1 to SD3 = binary variables representing the first, second, and third day of the snowstorm from Friday, January 22 to Sunday, January 24, respectively

RD1 to RD4 = binary variables representing the first to fourth days after the snowstorm from Monday, January 25 to Thursday, January 28

The results of the regression models for each city are presented in Table 1. All three models have high *R*-squared values, and more than 90% of variation in the data can be explained by the model for each city. The *F* statistic is significant for each model, which confirms the high joint significance of the variables.

For the New York City regression model, all eight binary variables representing Martin Luther King Day, the snowstorm days, and recovery days were significant (*p*-value < 0.01); this means that holidays and extreme weather strongly affect Transit app utilization. The coefficient for Martin Luther King Day was negative (−3,843.750), suggesting that app usage decreases on holidays. On Friday, January 22, when the snowstorm first hit New York City, the number of Transit app records increased (coefficient of 2,773.521); this increase may be due to a perceived disruption in the transit system. However, on the second day of the storm when the transit system was shut down and a travel ban for motorists was imposed, Transit app usage decreased significantly (coefficient of −8,925.937). On the last day of the snowstorm, when motorists were allowed on the roads and the transit system began to reopen, the number of Transit app records increased (coefficient of 4,110.062). The greatest change in Transit app usage was observed on the first day after the snowstorm on Monday, January 25, which had a large positive coefficient (9,709.958). On this day, users were most likely searching for transit information for their commute trip. On the subsequent days after the snowstorm, Transit app usage was larger than usage on normal days, which is indicated by the positive significant coefficients on the second, third, and fourth recovery days (7,013.187, 4,903.354, and 3,544.271, respectively).

In Philadelphia, the binary variables representing Martin Luther King Day, the 3 days of the snowstorm, and the first day after the snowstorm were strongly significant (*p*-value < 0.01). The effect on Martin Luther King Day is similar to that observed in New York City; Transit app usage was less than normal days (coefficient of −576.625) on the holiday. On the first and second days of the snowstorm, app usage was higher than usual (coefficient of 381.437 and 2,200.625, respectively). On the last day of the snowstorm (Sunday, January 24), the

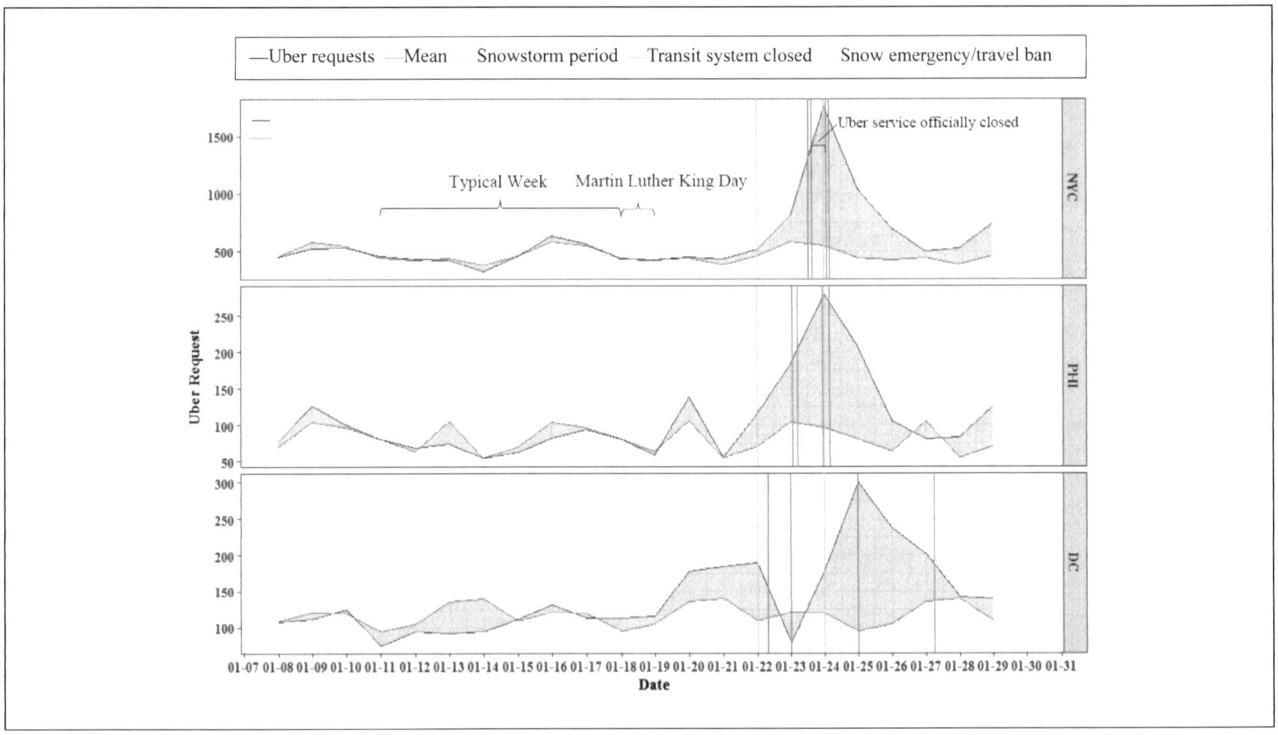

Figure 6. Daily Uber requests and mean Uber requests in New York City, Philadelphia, and Washington, D.C.

number of Transit app records decreased, which is expected due to the transit system shutdown over the weekend (coefficient of −389.958). The decrease in Transit app usage continued on the first day after the snowstorm (coefficient of −373.292) even though the transit system resumed service and it was a weekday (Monday, January 25). Two days after the snowstorm, the number of Transit app records increased, and this increase continued during the third day and fourth days of recovery after the storm.

In Washington, D.C., the coefficients were significant (p-value < 0.05) for the binary variables representing Martin Luther King Day, the 3 days of the snowstorm, and 1 day after the snowstorm. All of the coefficients for the storm days and the recovery days in this model were negative, which means that Transit app records decreased during the storm and continued to be lower than normal after the snowstorm. The transit system in Washington, D.C. resumed service on the first day after the snowstorm (Monday, January 25) but took longer to get back to normal. This model suggests that Washington, D.C. had the slowest recovery in Transit app usage of the three cities.

Part 4: Visualization of Uber Requests

In addition to overall Transit app usage, Uber requests made through the app were also considered. The dataset contains all sessions in which a user opened the app and clicked on "Request an Uber." Notably, during the period of analysis, users could not book an Uber directly within the Transit app; instead, they would be directed to Uber's app for fulfillment. Consequently, the Transit app dataset only highlights the users' intention of taking an Uber and does not reflect actual Uber trips.

Figure 6 shows daily Uber requests through the Transit app and a comparison with the mean of the previous 2 weeks. The amount of data was very small compared with overall Transit app use; thus, the analysis was conducted on a daily basis instead of hourly. As can be seen in Figure 6, the yellow line shows when the snowstorm hit the three cities, the line in blue is when the transit systems were closed, and the line in green represents the travel ban or snow emergency period. It should be noted that in New York City, Uber service was officially suspended from Saturday, January 23 at 2:30 p.m. until Sunday, January 24 at 4:00 a.m. (Figure 1). In Philadelphia and in Washington, D.C., a reference stating that Uber service was suspended could not be found; however, the authorities asked citizens to stay off of the roads in both cities (*34*, *36*, *37*), so it was assumed that Uber was not operating during this time.

In New York City, the total number of Uber requests via the Transit app was between 300 and 600 per day during the reference period prior to the storm. Interestingly, the number of Uber requests was higher during the snowstorm period. On Sunday, January 24, the number of Uber requests reached a peak of 1,750 in New York City.

In Philadelphia, the number was between 50 and 150 per day prior to the storm and reached 280 requests during the storm. For Washington, D.C., the number of Uber requests was between 75 and 180 per day prior to the storm, and the highest peak represented more than 300 requests on Monday, January 25. Unlike the two others cities, the lowest point in Washington, D.C. occurred on Saturday, January 23 when no vehicles were allowed on roads and the transit system was shut down.

The results imply that despite travel bans, the number of Uber requests was greater than usual. During the snowstorm, Transit app users were probably looking for other modes of travel when the transit systems were shut down. When the transit systems began to reopen service (on Sunday in New York City and in Philadelphia and on Monday in Washington, D.C.), the number of Uber requests via the Transit app reached the highest point. Last, it is important to emphasize that these are requests for Uber, which do not necessarily involve fulfilled trips; at the time of analysis, Transit app users were directed to Uber's app for fulfillment.

Conclusions and Future Research

This study assessed information utilization during an extreme weather event by analyzing data from a smartphone app called Transit, which provides real-time transit and shared mobility information in many North American cities. The analysis focused on a snowstorm that hit the northeastern United States in January 2016 and severely disrupted transit and shared mobility services. A four-part analysis was conducted for three cities: New York City, Philadelphia, and Washington, D.C. First, hourly Transit app utilization during the snowstorm was compared to the mean hourly app utilization prior to the storm. Second, the rate of app usage was calculated by dividing hourly utilization during the storm by the mean hourly volume before the storm. Third, an OLS regression model of hourly app usage was estimated for each city. Last, a feature within the app used to request Uber vehicles was examined.

The results of the first three analyses revealed that overall app usage decreased during the snowstorm in all three cities. After the storm, New York City experienced a significant increase in overall app use during the first Monday commuting period, and Washington, D.C. had the slowest recovery in app usage. The analysis of Uber data revealed that Transit app users continued to search for ridehailing services during the snowstorm, despite travel bans. Although some of the findings of this study are to be expected – such as the decrease in overall Transit app usage during the storm and an increase immediately afterward – the number of Transit app searches during the snowstorm was surprisingly large, particularly during the period when the transit systems were completely shut down (i.e., a decrease to only about 65% of mean use).

These findings are important for transportation operators and app developers to understand how travelers use information from smartphone apps during extreme weather events, which would be difficult to do using traditional data sources such as surveys. This new type of technology – smartphone apps – largely did not exist a decade ago and has rapidly changed how transit agencies communicate with travelers. However, traveler utilization of this new technology during unusual or extreme events has not been well documented. If usage patterns are better understood, it could help transit agencies to improve communication with travelers during extreme events, potentially helping transit systems to recover faster. This analysis shows that travelers want to be informed of changes in transportation service, as evidenced by large spikes in Transit app usage after the snowstorm, and therefore, transportation providers should anticipate increased communication levels immediately after extreme weather events. Transportation operators should also consider multiple means of communication (beyond just apps) to ensure that they reach all of the traveling public.

Finally, many interesting areas for future research emerged from this analysis. First, the analysis of Winter Storm Jonas could be expanded to other regions, including areas that were not as heavily affected by the snow as New York City, Philadelphia, and Washington, D.C. Another interesting area for future research would be to examine Transit app usage patterns at the individual level, instead of at the metropolitan level. Additionally, the continuous nature of this dataset enables other anomalous events to be studied, including both planned events, such as concerts or sporting events, and unplanned events, such as hurricanes. Finally, it would be interesting to compare utilization of the Transit app to use of other transportation-related smartphone applications; however, in order for comparative analyses to be possible, the developers of other applications (which are often private companies) must be willing to share usage data. Hopefully there will be increased willingness to share data in the future as utilization of smartphone apps continues to grow.

Acknowledgments

The authors acknowledge the Transit app for sharing data, particularly Jake Sion. This research was supported in part by a 2015 City University of New York Collaborative Incentive Research grant (CIRG Project #2218) and a 2016 University Transportation Research Center faculty-initiated grant (UTRC Grant #49198-44-28).

Author Contributions

The authors confirm contribution to the paper as follows: study conception and design: CR, CB, NG, EJK, and JP; data collection and cleaning: CR, NG, and EJK; analysis and interpretation of results: CR, CB, NG, and JP; draft manuscript preparation: CR, CB, NG, EJK, and JP. All authors reviewed the results and approved the final version of the manuscript.

References

1. The Weather Channel. *Winter Storm Jonas: Where Does It Rank Historically?* https://weather.com/storms/winter/news/winter-storm-jonas-rank-in-history. Accessed August 1, 2017.
2. Chang, S. E., and N. Nojima. Measuring Post-Disaster Transportation System Performance: The 1995 Kobe Earthquake in Comparative Perspective. *Transportation Research Part A: Policy and Practice*, Vol. 35, 2001, pp. 475–494.
3. Knapp, K. K., L. D. Smithson, and A. J. Khattak. The Mobility and Safety Impacts of Winter Storm Events in a Freeway Environment. *Proc., Mid-Continent Transportation Symposium*, Ames, IA, 2000, pp. 67–71.
4. He, X., and H. X. Liu. Modeling the Day-to-day Traffic Evolution Process after an Unexpected Network Disruption. *Transportation Research Part B*, Vol. 46, 2012, pp. 50–71.
5. Cottrill, C., P. Gault, G. Yeboah, J. D. Nelson, J. Anable, and T. Budd. Tweeting Transit: An Examination of Social Media Strategies for Transport Information Management during a Large Event. *Transportation Research Part C: Emerging Technologies*, Vol. 77, 2017, pp. 421–432.
6. Pereira, F. C., F. Rodrigues, and M. Ben-Akiva. Using Data from the Web to Predict Public Transport Arrivals under Special Events Scenarios. *Journal of Intelligent Transportation Systems*, Vol. 19, No. 3, 2015, pp. 273–288.
7. Donovan, B., and D. B. Work. Empirically Quantifying City-scale Transportation System Resilience to Extreme Events. *Transportation Research Part C*, Vol. 79, 2017, pp. 333–346.
8. Zhu, Y., K. Ozbay, K. Xie, and H. Yang. Using Big Data to Study Resilience of Taxi and Subway Trips for Hurricanes Sandy and Irene. *Transportation Research Record: Journal of the Transportation Research Board*, 2016. 2599: 70–80.
9. Markou, I., F. Rodrigues, and F. C. Pereira. Demand Pattern Analysis of Taxi Trip Data for Anomalies Detection and Explanation. Presented at 96th Annual Meeting of the Transportation Research Board, Washington, D.C., 2017.
10. Ghahramani, N., and C. Brakewood. Trends in Mobile Transit Information Utilization: An Exploratory Analysis of Transit App in New York City. *Journal of Public Transportation*, Vol. 19, No. 3, 2016, pp. 139–160.
11. Davidson, A., J. Peters, and C. Brakewood. Interactive Travel Modes: Uber, Transit and Mobility in New York City. Presented at 96th Annual Meeting of the Transportation Research Board, Washington, D.C., 2017.
12. Ghahramani, N., C. Brakewood, and J. Peters. An Exploratory Analysis of Intercity Travel Patterns using Backend Data from a Transit Smartphone Application. Presented at 96th Annual Meeting of the Transportation Research Board, Washington, D.C., 2017.
13. National Oceanic and Atmospheric Administration. *The Historic Nor'easter of January 2016*. https://www.weather.gov/media/publications/assessments/16Northeast_Blizzard.pdf. Accessed July 29, 2017.
14. Office of the Mayor. *Executive Office of the Mayor*. https://mayor.dc.gov/release/mayor-bowser-declares-public-emergency-district-columbia-ahead-winter-storm. Accessed July 25, 2017.
15. Forbes. *Winter Storm Jonas Knocks Out Uber In New York, New Jersey*. https://www.forbes.com/sites/briansolomon/2016/01/23/winter-storm-jonas-knocks-out-uber-in-new-york-new-jersey/#12b164934cb2. Accessed July 25, 2017.
16. New York State. https://www.governor.ny.gov/news/governor-cuomo-issues-update-mta-service-suspensions. Accessed July 24, 2017.
17. New York City Metropolitan Transportation Authority. *MTA Archive Alert*. https://www.mymtaalerts.com/messagearchive.aspx. Accessed July 24, 2017.
18. New York City Metropolitan Transportation Authority. *MTA Service Advisory*. http://web.archive.org/web/20160123182337/http:/alert.mta.info/. Accessed July 24, 2017.
19. New York City Metropolitan Transportation Authority. *MTA Info*. http://www.mta.info/news-blizzard/2016/01/24/approximately-80-lirr-customers-can-use-rail-service-tomorrow. Accessed July 24, 2017.
20. CBS New York. *PATH Service Restored Between Newark And Journal Square*. http://newyork.cbslocal.com/2016/01/26/path-service-restored-newark-journal-square/. Accessed July 24, 2017.
21. NJ.com. *NJ Transit to Resume Service Sunday Following Blizzard*. http://www.nj.com/news/index.ssf/2016/01/nj_transit_back_up_and_running_after_blizzard.html. Accessed July 31, 2017.
22. Twitter. *Uber NYC Account*. https://twitter.com/Uber_NYC/status/691235011488522240. Accessed July 24, 2017.
23. NYC Housing Preservation & Development. *Public Service Announcement*. http://www1.nyc.gov/site/hpd/about/public-service-announcements/1-22-16.page. Accessed July 25, 2017.
24. Washington Metropolitan Area Transit Authority. *Daily Service Report Archives*. https://www.wmata.com/service/daily-report/archives.cfm?year=2016. Accessed July 24, 2017.
25. Maryland Department of Transportation. *Maryland Transit Administration*. https://mta.maryland.gov/mta-temporarily-suspend-weekend-transit-services. Accessed July 24, 2017.
26. CBS Maryland. *Limited Service Continues For MTA Maryland*. http://baltimore.cbslocal.com/2016/01/26/limited-service-continues-for-mta-maryland/. Accessed July 24, 2017.
27. CBS Maryland. *MTA Offering Free Rides Monday Due To Snow*. http://baltimore.cbslocal.com/2016/01/24/mta-offering-free-rides-monday-due-to-snow/. Accessed July 24, 2017.

28. Southeastern Pennsylvania Transportation Authority. *Press Releases*. http://www.septa.org/media/releases/. Accessed July 24, 2017.
29. Philly.com. *Winter Storm Recovery Updates: Jan. 28, 2016*. http://www.philly.com/philly/blogs/real-time/Winter-storm-recovery-updates-Jan-25-2016.html. Accessed July 24, 2017.
30. New York Metropolitan Transportation Council. https://www.nymtc.org/. Accessed July 24, 2017.
31. New Jersey Transportation Planning Authority. http://njtpa.org/home. Accessed July 24, 2017.
32. Metropolitan Washington Council of Governments. https://www.mwcog.org/tpb/. Accessed July 24, 2017.
33. Delaware Valley Regional Planning Commission. http://www.dvrpc.org/. Accessed July 24, 2017.
34. South Jersey Transportation Planning Authorization. http://sjtpo.org/. Accessed July 24, 2016.
35. CBS Philadelphia. *Snow Emergency Begins in Philadelphia 9pm Friday*. http://philadelphia.cbslocal.com/2016/01/22/snow-emergency-begins-in-philadelphia-9-pm-friday/. Accessed August 1, 2017.
36. Office of the Mayor. *Executive Office of the Mayor*. https://mayor.dc.gov/release/mayor-bowser-declares-public-emergency-district-columbia-ahead-winter-storm. Accessed July 25, 2017.
37. Governor of Pennsylvania. *Press Releases*. https://www.governor.pa.gov/governor-wolf-gives-winter-storm-update-strongly-urges-public-not-to-travel/. Accessed July 25, 2017.

The Standing Committee on the Logistics of Disaster Response and Business Continuity (ABR20) peer-reviewed this paper (18-00302).

Article

TRR
JOURNAL OF THE TRANSPORTATION RESEARCH BOARD

Transportation Research Record
2018, Vol. 2672(1) 101–112

DOI: 10.1177/0361198118774239
journals.sagepub.com/home/trr
SAGE

Developing Priority Index for Managing Utility Disruptions in Urban Areas with Focus on Cascading and Interdependent Effects

Srijith Balakrishnan[1] and Zhanmin Zhang[1]

Abstract

Unanticipated events such as natural disasters, terrorist attacks, cyber-attacks, and so forth, could cause prolonged disruptions in major utility service networks including, for example, water and electricity, in urban areas. Owing to the presence of complex interdependencies among infrastructure systems in an urban network, the disruption of one system may trigger a chain of events that degrades the proper functioning of several other dependent systems. Consequently, many parts of the city may not have access to multiple utility services and amenities. Identifying the most vulnerable communities exposed to such utility disruptions is key to performing immediate relief operations. In this paper, the concept of Priority Index, introduced as a measure of the susceptibility of communities to the event, is presented to rank urban regions based on the extent of the impact of disruptions (both cascading and interdependent impacts) caused by an event, as well as the social vulnerability of communities. Agent-based models are employed to simulate the consequences of a disruptive event on a semi-realistic urban infrastructure network. Later, the extent of impact on communities is evaluated using the simulation results and the American Community Survey data. The proposed Priority Index could help city administrations and utility service agencies identify the regions in a city that require immediate attention after a disruptive event occurs in the infrastructure network. A case study based on a semi-realistic infrastructure network in Austin, Texas is presented to demonstrate the implementation of the concept of Priority Index and the methodological framework.

Human civilizations gave shape to cities, and technology made cities better. Today, cities constitute the most critical element of the social and economic development of nations. The two most important components of cities are its infrastructure network and its communities. The mutual dependence of these two components has resulted in efficient and constantly evolving cities. However, this integral nexus, while being an opportunity, is also a challenge for urban planners, because a disturbance on either of these two components by any external incident will have far-reaching consequences in the other, and has the potential to bring a city to halt. Owing to the networked structure of urban infrastructure systems, failure of a system could further trigger failure of dependent systems in the network, and consequently cause widespread disruptions in public utility services. This has been evident in many past incidents such as the World Trade Center attack in 2001, the Northeast blackout in 2003, the Indian Ocean earthquake in 2004, and so forth. These events were unforeseen, and their consequences on infrastructure systems were aggravated by the complex and interdependent structure of urban infrastructure network. Thus, during such disasters, urban communities are susceptible not only to the direct impacts of the disaster, but also to prolonged utility disruptions resulting from the inability of infrastructure systems to function at satisfactory performance levels. At present, there are vulnerability assessment tools to predict the direct impact of disasters on communities. However, there is a lack of models to evaluate the indirect impacts, such as large-scale utility disruptions. In this paper, the authors present a framework and modified measure called Priority Index (PI) to prioritize relief operations during

[1]Department of Civil, Architectural and Environmental Engineering, The University of Texas at Austin, TX

Corresponding Author:
Address correspondence to Srijith Balakrishnan: srijith@utexas.edu

emergencies that would simultaneously augment the infrastructure and community resilience of a city.

The rest of the paper is organized as follows: the Literature Review section presents a review of existing literature pertaining to this study; the Methodology section elaborates on the framework adopted for developing PI; the Case Study section demonstrates the implementation of the methodological framework; and the Conclusion section lists the findings.

Literature Review

The characteristics of a city are largely dependent on its infrastructure systems, such as utility services, economic institutions, and transportation infrastructure. Infrastructure systems are critical for ensuring an adequate supply of resources and services to communities, economic sectors, and other social institutions in a city. In addition, several studies in the past have validated the constructive role of urban infrastructure in stimulating the social and economic development of cities and nations. Some of the areas in which infrastructure development made significant improvements are economic development (*1*), poverty alleviation (*2*), social equity (*3*), and agricultural and regional development (*4*).

A modern urban infrastructure network can be considered as a complex and dynamic system of interconnected and interdependent infrastructures. Rinaldi et al. classified the interdependencies existing among infrastructure systems into four categories, namely, physical, geographic, cyber, and logical interdependencies (*5*). Physical interdependency is linked to material flows, whereas cyber interdependency pertains to information flows. Geographic interdependency relates to physical proximity, and finally, logical interdependency encompasses all other types of interdependencies. In an interconnected network, the performance of an infrastructure system is influenced not only by its functional capability, but also by the performance of its dependee systems. Though the interdependent nature of urban infrastructure ensures operational efficiency of component systems, it can also increase system vulnerability (*6*). A disruptive event that affects an infrastructure system could trigger cascading and interdependent effects on its dependent systems, degrading their functional efficiency. The ability of communities to endure such disruptions depends on their socio-economic characteristics. However, major social and economic disparities among urban communities are commonplace in cities. Therefore, from an emergency management perspective, the disaster risks arising from such interdependencies on infrastructure systems and communities need to be assessed using an integrated approach. The first step in the process of evaluating the vulnerability in an infrastructure network is to quantify the consequences arising from interdependencies.

Methods for Quantifying Interdependent Effects of Hazards

Ouyang identified and classified various methods for quantifying the interdependent effects in networked infrastructures into five categories, namely, empirical-, agent-, system dynamics-, economic theory-, and network-based approaches (*7*). In the empirical approach, the interdependencies are quantified based on historical failure patterns and expert experience (*8*). Agent-based models consider each component of a complex infrastructure network as an autonomous agent whose functions are decided by a set of well-described rules. This is a bottom-up approach which can be used to simulate the interdependent effects in large-scale urban systems (*9*). Studies like the one by Brown et al. used principles of system dynamics to evaluate the potential risks to critical systems arising from the interdependencies in infrastructure networks (*10*). Another important approach is the input–output inoperability model (IIM) derived from the Leontief Input–Output economic model used for assessing the stability of economic systems (*11*). This approach has been used in several hybrid models for evaluation of infrastructure interdependencies. Recent studies have focused more on models based on network theory (*12*). An infrastructure system can be treated as a graph, where nodes represent infrastructure components and links represent interdependent relationships.

Methods for Evaluating Social Vulnerability to Hazards

The approaches summarized in the previous section have been used for evaluating the effect of interdependencies in aggravating the risks on infrastructure networks arising from unanticipated events. The authors also conducted another phase of the literature review to understand how hazard risks on urban communities are evaluated, and whether interdependent effects are considered in the evaluation.

Social vulnerability indicates how sensitive communities are to hazards, and their ability to respond during such events. The metrics to evaluate the vulnerability of communities to hazards can be broadly classified into two categories: generic vulnerability measures and event-specific vulnerability measures. The generic vulnerability measures use surrogate socio-economic, built-environment, and demographic variables to identify the most vulnerable communities. Cutter et al. constructed a Social Vulnerability Index (SoVI) for evaluating the social vulnerability of communities to environmental hazards (*13*). The researchers identified 42 relevant variables from US Census data to develop SoVI for the counties in the United States. In a similar effort, Flanagan et al. developed another generic social vulnerability index (SVI) for the Agency for Toxic Substances and Disease Registry using fifteen census variables pertaining to socio-economic status, household characteristics, demographic characteristics, and housing characteristics (*14*). SVI is an ordinal measure which indicates the percentile rank of census tracts based on the variables. Though the index provides the vulnerability ranks of census tracts, relative comparisons about the magnitude of vulnerability are not possible. Huang and London (*15*) developed a census block-level SVI as a measure of the health challenges posed by hazards, based on six variables, namely, proximity to health care facilities, poverty rate, education, linguistic isolation, race/ethnicity, and age. The advantages of this index are that it is a normalized value between 0 and 1, and that it can be used for relative vulnerability comparisons.

Event- or hazard-specific social vulnerability measures, along with generic vulnerability factors, also account for the factors or policies that can augment the community's response to that hazard. For example, Chakraborty et al. combined generic vulnerability index and geophysical risk index (an index based on vulnerability to floods) to prioritize evacuation assistance needs in Florida (*16*). Schmidtlein et al. simulated the potential earthquake losses in South Carolina using the HAZUS-MH software (developed by the Federal Emergency Management Agency) and assessed the correlations with SoVI values (*17*). The study also identified the most significant census variables that are correlated with losses to earthquakes. Rygel et al. developed a social vulnerability indicator for assessing the impact of storm surges resulting from hurricanes (*18*). The researchers performed principal component analysis to identify the most relevant set of factors affecting vulnerability to the hazard under consideration.

The review revealed that literature is scarce when it comes to evaluating the vulnerability of communities to widespread utility service outages resulting from various types of hazards. The direct impacts of disasters on communities, such as health hazards and physical harm, may be limited to the location of occurrence. However, the indirect impacts, such as prolonged utility service disruptions arising from infrastructure system failures during a disaster, are more likely to affect communities in other regions of cities, as well. This warrants the need for a framework to assess and prioritize communities who are most vulnerable to the indirect impacts of disasters resulting from the interdependent nature of infrastructure systems.

Methodology

Figure 1 presents the methodological framework adopted for developing the PI. The index is designed to rank census tracts based on their vulnerability to utility disruptions resulting from an unanticipated event. The framework consists of two independent components. The first component is the quantification of impacts of the event on the performance of various utility systems in the infrastructure network (from which a community's exposure to such disruptions can be quantified) using agent-based modeling (ABM) approach. The exposure would depend on the structure of the infrastructure network, and the interdependent relationships existing among its various components. The second component is the evaluation of the social vulnerability of communities residing in the affected regions from publicly available American Community Survey (ACS) data. Social vulnerability variables could be used as surrogate measures to evaluate the capability of communities in a census tract to endure prolonged utility disruptions. Once the exposure and social vulnerability are quantified, they are combined to develop the PI. In the rest of the section, the various stages in the development of PI are elaborated.

Development of the Infrastructure Network and Defining Interdependencies

Every urban infrastructure network consists of numerous interconnected and interdependent infrastructure systems (or utility services) such as electricity grid, water supply system, sewage disposal system, and so on. Though each of these infrastructure systems is established independently, they depend on other infrastructure systems for their proper functioning. The degree of dependency of an infrastructure system on another is affected by various factors, including the resources and services required by the former, the resources and services produced by latter, geographical proximity, and so forth. Any degree of performance drop in one infrastructure node in the network could trigger further performance drops in its dependent nodes in the presence of such dependencies and/or interdependencies. This may disrupt the normal supply of multiple utility services in the affected regions.

As mentioned before, the first stage in the process of developing PI is to analyze and estimate the network level performance drop in various utilities which might occur as a result of the disruptive event. As the urban infrastructure network consists of several infrastructure systems which are composed of individual infrastructure units, it could be modeled as a graph over a geographical information system (GIS) framework. An infrastructure system can be regarded as a subgraph and individual infrastructure units as nodes. The dependencies could be modeled as unidirectional links in the graph, each of

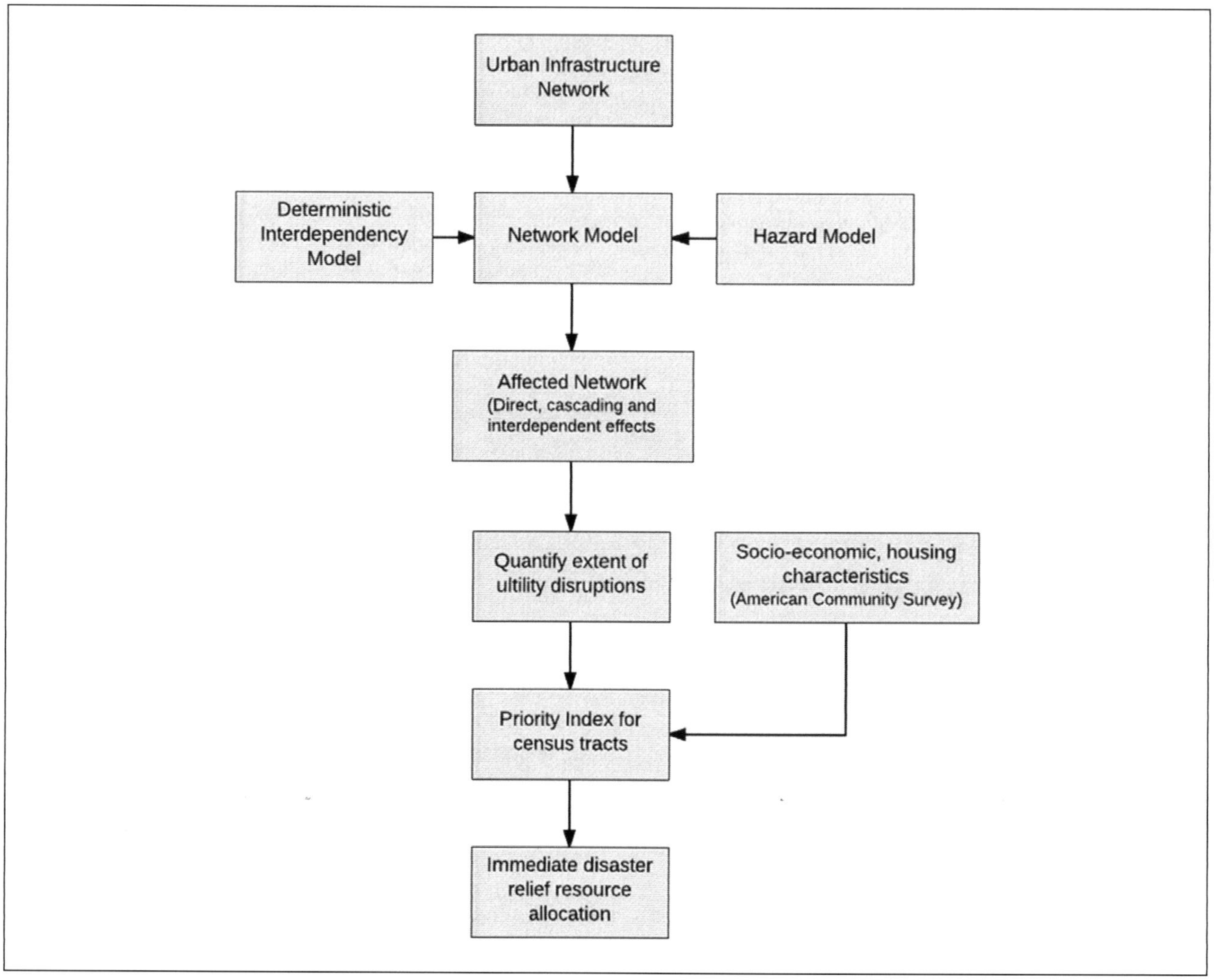

Figure 1. Methodological framework for developing priority index.

them carrying a weight according to the degree of interdependency. Mathematically, the network model can be represented as follows Equation 1:

$$G_\Omega = (N_\Omega, D_\Omega) \tag{1}$$

where G_Ω is the infrastructure network model on a GIS framework Ω, N_Ω is the set of infrastructure nodes, and D_Ω is the set of dependencies among the infrastructure nodes. Each of the nodes, $i \in N_\Omega$, belongs to an infrastructure type, K, from a set $\kappa : \kappa(i) = K$. If a node i is dependent on another node j, the dependency between the two nodes is represented as follows Equation 2:

$$d_{ij} \in D_\Omega \Leftrightarrow \exists d_{ij} \tag{2}$$

Whether i is connected to j is dependent on their respective infrastructure types and the service area of j. If there exists both d_{ij} and d_{ji}, it is called an interdependency. Equations 1 and 2 define the structure of the infrastructure network model. However, the degree of dependencies among various nodes vary depending on the infrastructure types. The interdependency model W is a set of dependency values in the range of [0, 1] mapped against every d_{ij} that exists in the network. For the present study, a deterministic interdependency model is adopted. Mathematically:

$$w_{ij} \in W = w_{\kappa(i)\kappa(j)} \Leftrightarrow \exists d_{ij} \tag{3}$$

where w_{ij} is the dependency value of i on j, $w_{\kappa(i)\kappa(j)}$, is the dependency value for infrastructure type $\kappa(i)$ dependent on $\kappa(j)$.

Simulation of Disruptive Event and Evaluation of Post-Event Performance of Infrastructure Network

Consider a disruptive event, H, occurring at a point $p(x, y)$, where x and y are the longitude and the latitude of the point p, respectively. The disruptive event affects a certain geographical region, $\omega \subset \Omega$. The event could be modeled as a function of initial intensity at p, ι_0, time of occurrence, T, and the speed of propagation of impact, ν, as follows:

$$H = f(\iota_0, T, \nu) \tag{4}$$

The impact of H on node i is dependent on initial intensity and distance of the node from the point of occurrence, and is given by Equation 5:

$$\iota_H^i = f(\iota_0, \ell ip) : 0 \le \iota_H^i \le 1 \tag{5}$$

where ℓ_{ip} is the distance between i and p. If initial expected performance of i at time T is represented by $P_i(0) : 0 \le P_i(0) \le 1$, then the expected performance of i after the occurrence of the event can be modeled using Equation 6:

$$P_i(T+\tau) = \begin{cases} P_i(0) & \text{if } \tau < \dfrac{\ell ip}{\nu} \\ P_i(0) - \iota_i^H & \text{if } \tau \ge \dfrac{\ell ip}{\nu} \end{cases} \tag{6}$$

where τ is time elapsed after T until the instance under consideration. Along with physical impact of an event on a node, the expected performance can also degrade because of the shortage of the supply of adequate resources on time. The latter becomes obvious if infrastructure systems in the network are interdependent. The reduction in expected performance of node i will reflect in the expected performance of those nodes that are dependent on i for their functioning. However, Equation 6 does not account for such interdependent and cascading effects of the event on the network. Accounting for this factor in the performance of nodes, the expected performance at any time t can be simulated using the following generalized expression:

$$P_i(t) = \max\left(0, P_i(0) - \left[\sum_j (1 - P_j(t - \Delta t)) w_{\mathrm{ij}}\right] - \rho_i \iota_i^H\right) : 0 \le P_i(t) \le 1 \tag{7}$$

where Δt is the simulation step size, J is set of all dependee nodes of i, and ρ is an indicator variable with value 1 if $t - T > \frac{\ell_{\mathrm{ip}}}{\nu}$, and 0 otherwise. The expected performance loss in any node i, $1 - P_i(t)$ is the exposure of i to the event H.

Spatial Analysis of Hazard Exposure and Social Vulnerability of Communities

Once the post-event expected performance of utility services is estimated, the analysis must identify how the disruptions affect the communities. For this purpose, the following steps are adopted:

1. Create an adequate sample of buildings that are dependent on the disrupted utility services based on the actual spatial distribution of building footprint in the study area.
2. Identify and map the dependent utility service nodes to each building in the sample and their corresponding post-event performance levels.
3. Identify and map the buildings to the corresponding census tract to which they belong.
4. Evaluate the distribution of utility service disruptions in each census tract from the building data. This provides information about the exposure of each census tract to the utility service disruptions resulting from the reduced performance levels of the infrastructure nodes.

Next, the generic social vulnerability characteristics of communities are analyzed. The underlying assumption is that those populations with higher social vulnerability to disasters are more likely to be affected than their counterparts with lower social vulnerability during a disaster or disruptive event. Data from censuses and related updates can be used to obtain relevant information about the demography, infrastructure, and socio-economic characteristics of communities residing in the city. In the United States, the Census Bureau conducts the ACS to collect data regarding communities at various geographic levels every year. Based on the ACS data, the Agency for Toxic Substances and Disease Registry (*14*) has developed a SVI for each of the census tracts using fifteen social factors that describe a community's social vulnerability (Table 1). However, the limitation of SVI is that it is an ordinal measure, and therefore does not convey the magnitude of relative differences in vulnerability across regions. Thus, a modified Social Vulnerability Index (mSVI) is developed based on the same set of variables, which also accounts for the relative difference in social vulnerabilities. An approach like the one suggested by Huang and London (*15*) is adopted for developing mSVI.

Consider a census tract, m, with the social factors listed in Table 1 denoted by a set, S. The corresponding mSVI is given by Equation 8:

$$mSVI_m = \frac{\sum_{s \in S} s_m}{15 \times 100} \tag{8}$$

In order to convert $mSVI_m$ to its normalized form, Equation 9 is employed:

$$mSVI_{\mathrm{m,\,norm}} = \frac{mSVI_m}{\max(mSVI_m)} : 0 \le mSVI_{\mathrm{m,\,norm}} \le 1 \tag{9}$$

Table 1. Social Factors Considered in Developing SVI (Agency for Toxic Substances and Disease Registry)

Category	Variables
Socio-economic status	• % of population below poverty • % population unemployed • per capita income[a] • % population without high school diploma
Household composition & disability	• % population older than 65 years of age • % population younger than 17 years of age • % civilians with disability • % single parent household
Minority status and language	• % population belonging to minority communities • % population speaking English "less than well"
Housing and transportation	• % multi-units structures • % mobile homes • % housing units with more people than rooms • % households with no vehicle • % persons in institutionalized group quarters

[a]During the calculation of mSVI, per capita income (pci) is converted into a percentage variable which is equivalent to the ratio of the difference between the highest tract-level pci in the region and that in the current census tract to the same highest tract-level pci.

Development of Priority Indices

The final step is to combine the exposure and social vulnerability of census tracts to develop the PI. If the expected performance of utility service K in census tract m is denoted by P_K^m, then the weighted exposure in census tract, E_m is given by Equation 10:

$$E_m = \sum_K \left(1 - P_K^m\right) \times \omega_K^m : 0 \leq E_m \leq 1 \quad (10)$$

where ω_K^m is the weight for utility service K in census tract m. PI of a census tract for a given disruptive event, PI_m, is defined as the product of weighted exposure (E_m), and normalized social vulnerability index ($mSVI_{\mathrm{m,\,norm}}$) corresponding to m:

$$PI_m = E_m \times mSVI_{m,\,norm} : 0 \leq PI_m \leq 1 \quad (11)$$

If a census tract has high social vulnerability and high exposure to the event, the PI will be close to 1. Conversely, if the census tract has low social vulnerability and is subjected to low exposure, the PI will be close to 0. Thus, the PI could be used to identify those census tracts that require immediate attention after a hazard occurs in the infrastructure network. This would help in making decisions that ensure rational utilization of available resources during emergencies.

Case Study

Description of Infrastructure Network

The City of Austin, Texas is chosen for the case study. The primary goal of the paper is to present a methodology to prioritize urban regions for managing utility service disruptions during a hazard or disruptive event, rather than the accurate quantification of interdependent risks. Therefore, the authors created a semi-realistic network by limiting the number of utility services and adopting hypothetical values for interdependencies. The infrastructure systems chosen for the study are power plants, substations, electricity maintenance service, water treatment plants, and sewage treatment plants. The location details of the infrastructure nodes are obtained from public data and reports published by the City of Austin (*19*, *20*). For simplicity of the network, it is assumed that each of the infrastructure units has a unique service area, which do not overlap with each other. The details regarding the service areas of water treatment plants, sewage treatment plants, and electrical maintenance service stations are obtained from respective utility websites and official reports published by City of Austin. For those infrastructure systems such as substations and power plants (for which the service area details are not available), the service area is decided by the distance criterion, that is, all other utilities and buildings under study depend on the nearest substation and every substation depends on nearest power plant for its functioning. In addition to the locations of infrastructure nodes and their respective service areas, information regarding dependencies is also required for simulating the indirect impacts of the disruptive event on the network. Since this information was not available, appropriate values are assumed and are presented in Figure 2. The values represent the reduction in performance at the consumer node if the producer node, which it is dependent on, is completely failed.

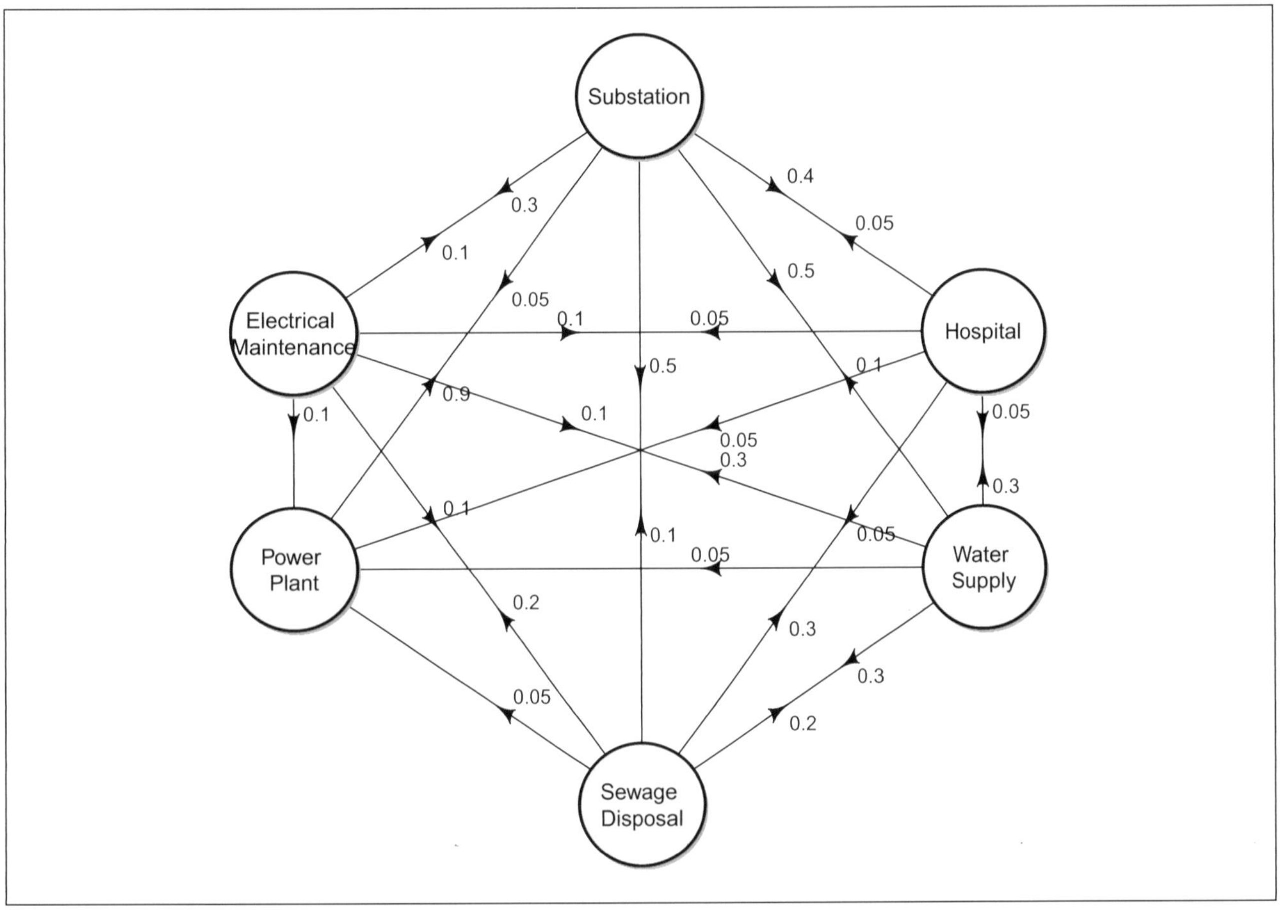

Figure 2. Assumed dependencies among infrastructure systems.
Note: Every edge represents dependency, and the direction of the arrows represents the direction of dependency (flow of resource or service). The number adjacent to each arrow represents the degree of dependency. For example, a dependency of 0.3 from the water supply node to the sewage disposal node suggests that if the water supply completely fails, the percentage of performance reduction in the sewage disposal system is 30%. An edge with arrows in both directions represents interdependency.

Simulation Results

As discussed earlier in the paper, an ABM approach is used to simulate the propagation of cascading and interdependent effect on the infrastructure network. To demonstrate this, an artificial disruptive event is generated which caused a water treatment plant to stop functioning entirely as shown in Figure 3 (denoted by red rectangle). It is assumed that all the infrastructure nodes in the network function at maximum performance level prior to the event. The Anylogic® software package (*21*) is used to construct the infrastructure network model and simulate the impact of the event on the network. A step size of 1 minute is used for the simulation.

Figure 4 presents the simulated progression of expected system performance of various utility services after the disruptive event is generated. The failure of the water treatment as a result of the initial event is reflected in the expected performance of utility services such as health care, electricity, and sewage disposal, as a result of the existing interdependencies in the network. As can be seen in the figure, the initial reductions in expected performance (lower order) in all utility systems are abrupt compared with the subsequent reductions. The initial reductions result from the direct and cascading impacts caused by dependencies. The comparatively smaller, and higher order reductions in expected performance in the later stages could be attributed to indirect and interdependent effects. Once the higher order effects become negligible, the affected utility systems stabilize and attain new equilibrium, but with reduced performance levels.

Figure 5 presents the distribution of the expected performance levels of utility services in various census tracts. The expected performance level of electricity supply in various census tracts after the event ranges from 35 percent to 92 percent with an average of 70 percent (Figure 5*a*).

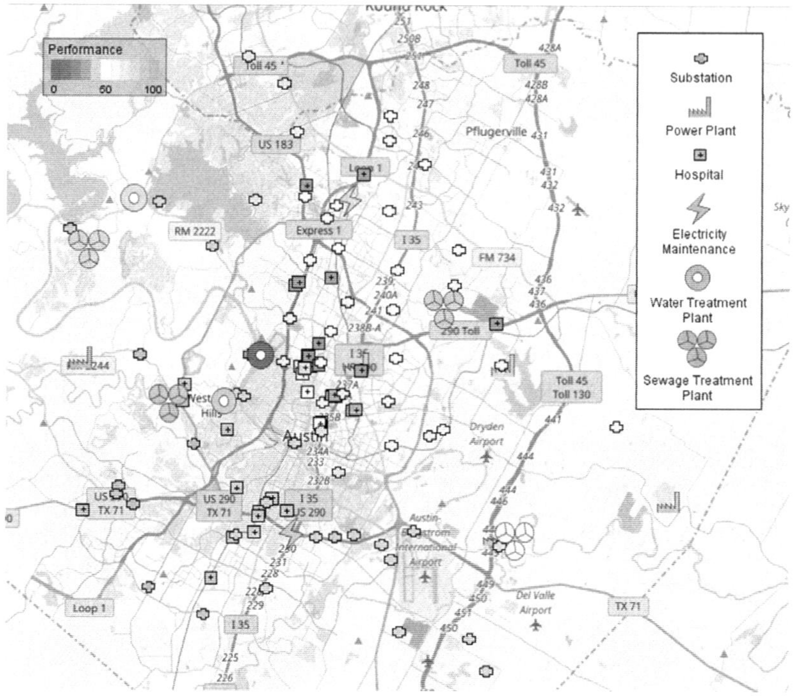

Figure 3. Semi-realistic infrastructure network and impacts of the failure event.

Note: The initial failure occurs in the water treatment plant (marked in red line). Because of the interdependencies, the effect propagates to other infrastructure systems. The color of the infrastructure nodes represents their performance level after the event occurs (red indicates that the node failed, and green indicates that the node is still functioning).

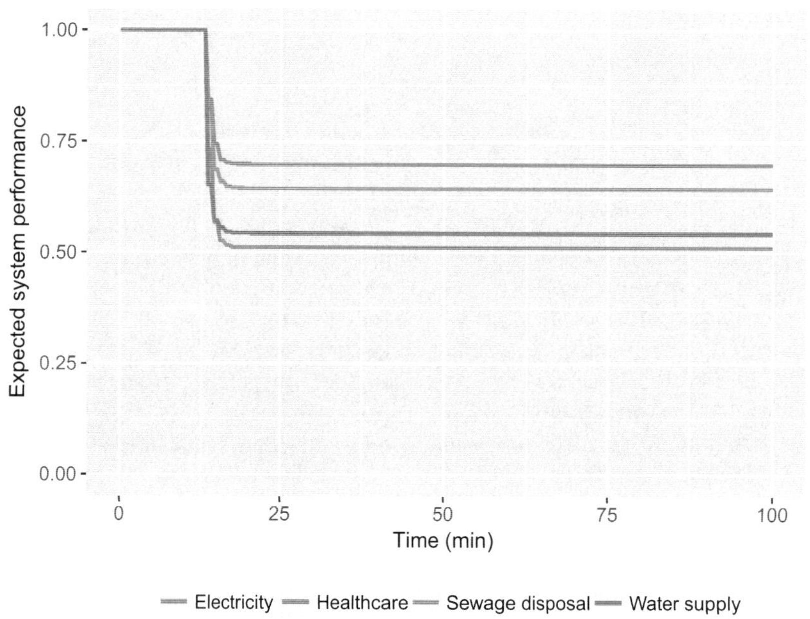

Figure 4. Simulated impact of the disruptive event on the expected performance of infrastructure network (cascading and interdependent effects combined).

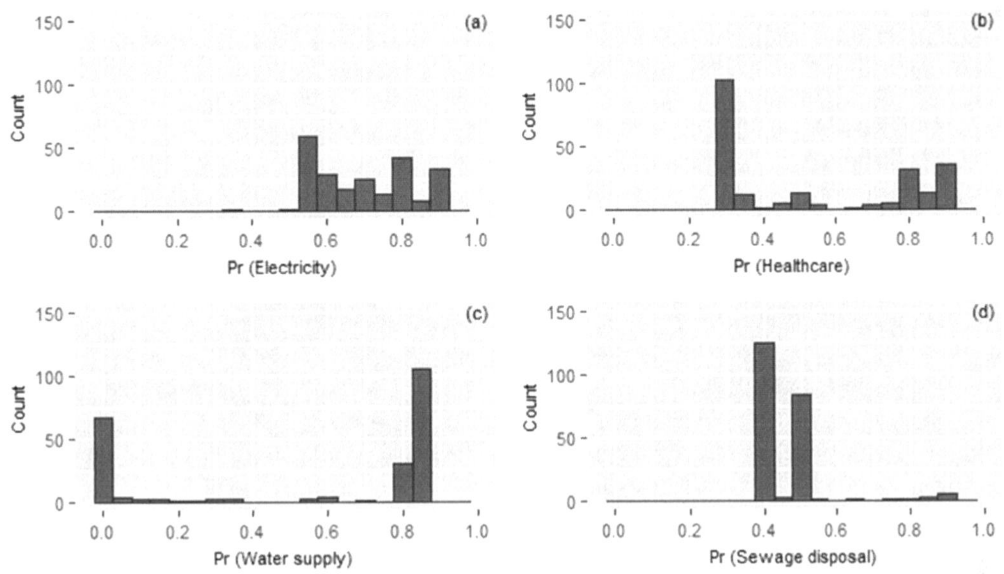

Figure 5. Distribution of expected utility service levels in census tracts after the event: (*a*) electricity; (*b*) health care; (*c*) water supply; and (*d*) sewage disposal.

Similarly, the expected service level of hospitals ranges from 28 percent to 87 percent with an average of 52 percent (Figure 5*b*). The expected performance of water supply service ranges between 0 percent and 84 percent with average of 54 percent (Figure 5*c*), and that of sewage treatment service ranges between 39 percent and 89 percent with an average of 45 percent (Figure 5*d*).

Though Figures 4 and 5 provide an overall idea about the post-event expected network performance, this is of insignificant use from a decision maker's perspective unless the spatial distributions of the utility services disruption levels are understood. For this purpose, a random sample of 10,925 buildings (approximately 2% of total building footprint) distributed in the study area is selected. The buildings are then classified according to the respective utility nodes they depend on for their functioning (electricity, health care, water supply, and sewage treatment), and corresponding performance levels are identified. The buildings are then mapped to the census tracts to which they belong. In this way, the distribution of utility disruptions in all the 222 census tracts under study are obtained. The expected levels of utility disruptions in each of the census tract are presented in Figure 6.

Figure 6*a* shows the levels of electricity supply in the study area after the occurrence of the event. It can be observed that electricity supply disruptions are more likely to occur in census tracts in the northeastern and southeastern parts as shown in Figure 3. Similarly, Figure 6*b* shows that the functioning of hospitals in the north and northeastern parts are more likely to be disrupted. This suggests that a significant fraction of the communities will have access to no or less efficient health care. In addition, Figure 6, *c* and *d* illustrate how the failure of the water treatment plant could affect the water supply and sewage treatment services in different parts of the city, respectively.

Estimation of Weighted Exposure and Social Vulnerability

Analyzing the spatial distribution of the performance of utility services, it is evident that the degree of utility disruptions differs significantly across census tracts. For example, in the northeastern parts of the city, the simulation results predict that there will be almost no water supply and considerable reduction in the performance of health care and sewage disposal services. However, in the census tracts in the southwest, the probability of disruptions in sewage disposal service is high, whereas other utility services are expected to function satisfactorily. To identify the census tracts most affected by the disruption in utility services, a weighted performance measure is calculated. In reality, the dependency of buildings and communities on utility services differ from one region to another, depending on socio-economic factors. In such scenarios, it is possible for the decision maker to provide actual weights for the utilities to obtain the weighted exposure on the urban region. For this case study, the weighted exposure is obtained by assuming a weight

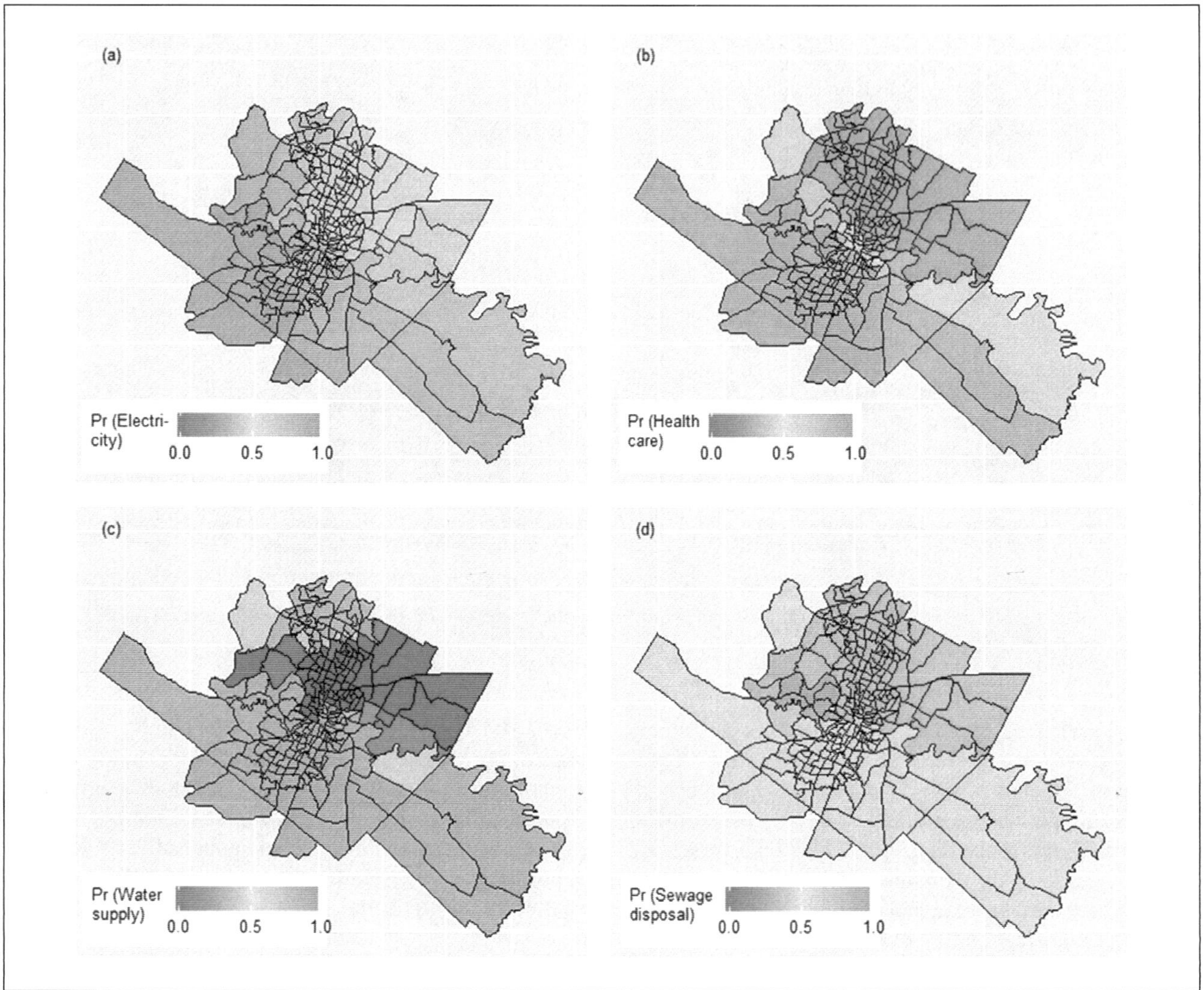

Figure 6. Spatial distribution of utility disruptions based on census tracts: (*a*) electricity; (*b*) health care; (*c*) water supply; and (*d*) sewage disposal.

factor of 0.25 for all the four utility services. The weighted exposure values of census tracts are presented in Figure 7*a*.

In addition to the geographical distribution of exposure to the disruptive event, it is also important to analyze the generic socio-economic, demographic, and housing characteristics of the census tracts. To estimate the vulnerability of the communities, normalized social vulnerability indices ($mSVI_{norm}$) of census tracts were computed from the 2014 ACS data, based on the methodology explained in the methodology section. The $mSVI_{norm}$ values are presented in Figure 7*b*. The results show that the communities residing in the eastern part of the city are comparatively more vulnerable to disasters and service disruptions than those in the western part. This can be directly linked to the socio-economic disparities existing between the communities residing in the eastern and western sides of Interstate-35, which runs north–south of Austin.

Calculation of Priority Index

The PIs for the census tracts are calculated using Equation 11. The results are presented in Figure 8. The results show that the northeastern regions of the city have high index values, indicating that those regions are more likely to be affected by the given water treatment plant failure. Communities in the southwestern region are the least likely to be affected by the same hazard. More importantly, it is possible from the figure to easily identify the census tracts that require immediate attention, and those do not. This property of PI makes it an easy and efficient tool for managing utility disruptions.

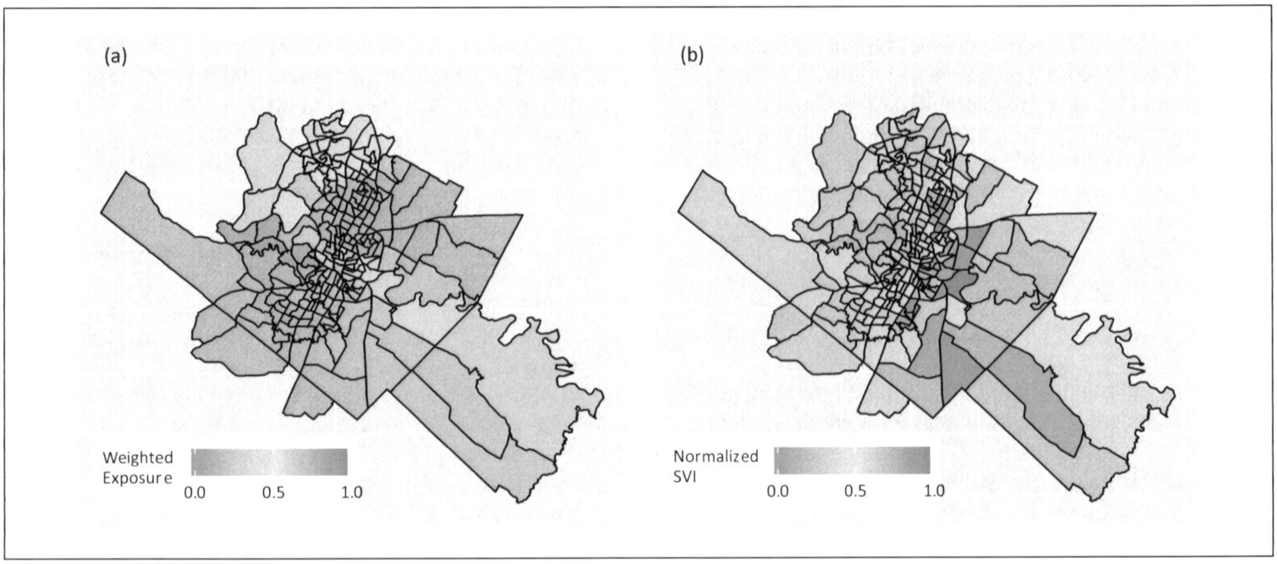

Figure 7. (*a*) Normalized mSVI values of census tracts and (*b*) weighted exposure of the combined effect of the disruptive event on communities.

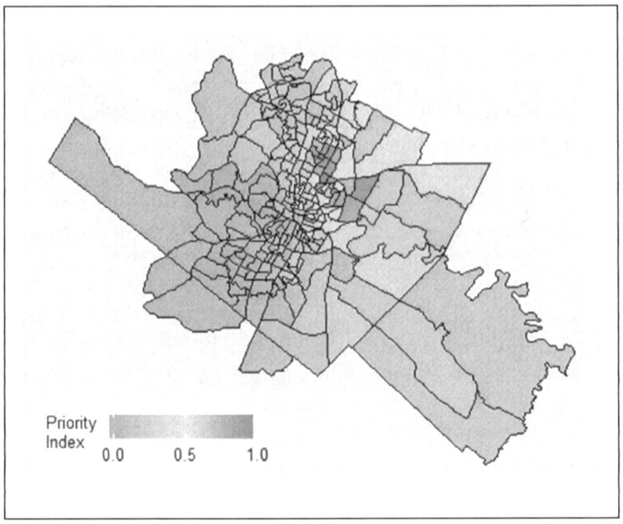

Figure 8. Priority index values of census tracts under study for the generated disruptive event.

Note: Priority index of 1 = highly susceptible; priority index of 0 = not susceptible

Conclusion

In the present study, a ratio-scale measure, PI, is proposed to evaluate the susceptibility of communities to unanticipated events and the resultant utility disruptions. The advantage of PI is that it is equally dependent on the generic social vulnerability of communities, as well as the degree of exposure to a given disruption on the network, enabling it to reflect the real conditions of communities in various parts of the urban region during utility disruptions and hazards. In addition, it enables comparison of susceptibility of two census tracts using a linear scale. The performance drop in the infrastructure network is evaluated by giving due consideration to both the direct and indirect impacts of hazards arising from its interdependent structure.

The framework could be employed for emergency planning and disaster risk assessment, as well as for managing immediate relief operations, such as distribution of food and water, during a disaster. The framework could find potential applications in cities where backup mechanisms to withstand prolonged and uncertain utility service disruptions are unreliable or absent. The results from the study underscore the importance of proper management of interdependent infrastructure systems to ensure the well-being of urban communities.

A key limitation of the present study is that it has not accounted for either the redundancy of the infrastructure network or the ability of individual infrastructure nodes to negate the impacts of utility service failures through backup systems. These factors warrant due consideration while implementing the framework.

Author Contributions

The authors confirm contribution to the paper as follows: study conception and design: Srijith Balakrishnan, Zhanmin Zhang; data collection: Srijith Balakrishnan; analysis and interpretation of results: Srijith Balakrishnan, Zhanmin Zhang; draft manuscript preparation: Srijith Balakrishnan. All authors reviewed the results and approved the final version of the manuscript.

References

1. Kumari, A., and A. Sharma. Physical & Social Infrastructure in India and Its Relationship with Economic Development. *World Development Perspectives*, Vol. 5, 2017, pp. 30–33.

2. Ogun, T. P. Infrastructure and Poverty Reduction: Implications for Urban Development in Nigeria. *Urban Forum*, Vol. 21, No. 3, 2010, pp. 249–266.
3. Calderon, C., and L. Serven, Infrastructure and Economic Development in Sub-Saharan Africa. *Journal of African Economies*, Vol. 19, 2010, pp. i13–i87.
4. Pinstrup-Andersen, P., and S. Shimokawa. Rural Infrastructure and Agricultural Development. In *Annual World Bank Conference on Development Economics Global: Rethinking Infrastructure for Development* (F. Bourguignon, and B. Pleskovic, eds.), The World Bank, Washington, D.C., 2007, pp. 175–203.
5. Rinaldi, S. M., J. P. Peerenboom, and T. K. Kelly. Identifying, Understanding, and Analyzing Critical Infrastructure Interdependencies. *IEEE Control Systems*, Vol. 21, No. 6, 2001, pp. 11–25.
6. Ouyang, M., and Z. Wang. Resilience Assessment of Interdependent Infrastructure Systems: With a Focus on Joint Restoration Modeling and Analysis. *Reliability Engineering & System Safety*, Vol. 141, 2015, pp. 74–82.
7. Ouyang, M. Review on Modeling and Simulation of Interdependent Critical Infrastructure Systems. *Reliability Engineering & System Safety*, Vol. 121, 2014, pp. 43–60.
8. Chang, S. E., T. L. McDaniels, J. Mikawoz, and K. Peterson. Infrastructure Failure Interdependencies in Extreme Events: Power Outage Consequences in the 1998 Ice Storm. *Natural Hazards*, Vol. 41, No. 2, 2007, pp. 337–358.
9. Dudenhoeffer, D. D., M. R. Permann, and M. Manic. CIMS: A Framework for Infrastructure Interdependency Modeling and Analysis. *Proc., 38th Conference on Winter Simulation*, Monterey, CA, ACM Digital Library, 2006, pp. 478–485.
10. Brown, T., W. Beyeler, and D. Barton. Assessing Infrastructure Interdependencies: The Challenge of Risk Analysis for Complex Adaptive Systems. *International Journal of Critical Infrastructures*, Vol. 1, No. 1, 2004, pp. 108–117.
11. Haimes, Y., and P. Jiang. Leontief-Based Model of Risk in Complex Interconnected Infrastructures. *Journal of Infrastructure Systems*, Vol. 7, No. 1, 2001, pp. 1–12.
12. Svendsen, N., and S. Wolthusen. Graph Models of Critical Infrastructure Interdependencies. In *Inter-Domain Management*, Springer, Berlin, Germany, 2007, pp. 208–211.
13. Cutter, S. L., B. J. Boruff, and W. L. Shirley. Social Vulnerability to Environmental Hazards. *Social Science Quarterly*, Vol. 84, No. 2, 2003, pp. 242–261.
14. Flanagan, B. E., E. W. Gregory, E. J. Hallisey, J. L. Heitgerd, and B. Lewis. A Social Vulnerability Index for Disaster Management. *Journal of Homeland Security and Emergency Management*, Vol. 8, No. 1, 2011.
15. Huang, G., and J. K. London. Cumulative Environmental Vulnerability and Environmental Justice in California's San Joaquin Valley. *International Journal of Environmental Research and Public Health*, Vol. 9, No. 5, 2012, pp. 1593–1608.
16. Chakraborty, J., G. A. Tobin, and B. E. Montz. Population Evacuation: Assessing Spatial Variability in Geophysical Risk and Social Vulnerability to Natural Hazards. *Natural Hazards Review*, Vol. 6, No. 1, 2005, pp. 23–33.
17. Schmidtlein, M. C., J. M. Shafer, M. Berry, and S. L. Cutter, Modeled Earthquake Losses and Social Vulnerability in Charleston, SC. *Applied Geography*, Vol. 31, No. 1, 2011, pp. 269–281.
18. Rygel, L., D. O'Sullivan, and ,B. Yarnal, A Method for Constructing a Social Vulnerability Index: An Application to Hurricane Storm Surges in a Developed Country. *Mitigation and Adaptation Strategies for Global Change*, Vol. 11, No. 3, 2006, pp. 741–764.
19. City of Austin. *Community Inventory Report Draft: Public Utilities*. Technical Report. City of Austin, 2009. ftp://ftp.ci.austin.tx.us/GIS-Data/planning/compplan/8-community_inventory_PublicUtilities_v3.pdf. Accessed July 21, 2017.
20. City of Austin. *Open Data Portal*, 2017. https://data.austintexas.gov. Accessed July 21, 2017.
21. The Anylogic Company. *Anylogic®*. Computer Software, Chicago, IL, 2017. https://www.anylogic.com.

The Standing Committee on the Logistics of Disaster Response and Business Continuity (ABR20) peer-reviewed this paper (18-04796).

Article

TRR
JOURNAL OF THE TRANSPORTATION RESEARCH BOARD

Transportation Research Record
2018, Vol. 2672(1) 113–124

DOI: 10.1177/0361198118787937
journals.sagepub.com/home/trr
SAGE

Telecom, Traffic Cones, and the Big One: Identifying Transportation and Communications Emergency Support Workforces and Calculating Their Exposure to Seismic Peak Ground Accelerations

Pierre M. Auza[1], Diana C. Lavery[2], R. Jayakrishnan[1], and Yuko J. Nakanishi[3]

Abstract

Successful post-disaster response and recovery depends on prompt restoration of infrastructure, including transportation or communications. However, disasters can have an impact on the workforce responsible for restoration, for example, by damaging their homes. This study has two goals: 1. Identify workers potentially participating in restoring transportation and communications infrastructure; 2. Calculate these workers' exposure to the peak ground accelerations (PGAs) of a 7.8 magnitude earthquake in a Southern California scenario, and compare it with the rest of the working population's exposure. Four steps are required. First, calculate the mean PGA for each affected public use microdata area (PUMA). Second, identify the infrastructure restoration workforce by specifying Standard Occupational Classification (SOC) and North American Industry Classification System (NAICS) codes. When specifying, use the Emergency Support Function (ESF) Annexes for Transportation (ESF#1) and Communications (ESF#2) to clarify workers' roles and responsibilities. This ESF-specific listing of codes is a novel contribution. Third, via frequency table, calculate the mean and standard deviation of transportation and communications workers' exposure to PGAs in their PUMAs of residence. Finally, test the difference in mean PGA exposures between two populations: (a) transportation or communications workers and (b) the rest of the working population. This study finds that, for this scenario, transportation workers are exposed to statistically significant higher PGAs than non-transportation workers, and communication workers to significantly lower PGAs. For practitioners, knowing which worker categories a disaster disproportionately affects could justify pre-event investments in workforce preparedness and recovery planning efforts.

After disasters like earthquakes or wildfires, successful response and recovery depends upon prompt restoration of infrastructure, including transportation and communications. Restoration requires public or private owners or operators of infrastructure to mobilize their workforce. However, disasters can affect this workforce (for example, by damaging their homes), thus impeding workers' ability to contribute to the restoration effort.

Transportation agencies' physical infrastructure includes roadways, bridges, and critical facilities like regional offices, maintenance yards, and transportation management centers. Worker categories like highway maintenance workers and bridge inspectors become critical to restoring functionality. Any methodology that can determine the worker categories disproportionately affected by a disaster could thus aid practitioners in planning for workforce preparedness, or response and recovery operations.

Aside from public agencies and jurisdictions, for several transportation modes, private entities own and operate the infrastructure. The Transportation Research Board maintains standing committees for eight modes (*1*); for example, private ownership is common for

[1]University of California, Irvine, CA
[2]Independent Consultant, Long Beach, CA
[3]Nakanishi Research & Consulting, LLC, Forest Hills, NY

Corresponding Author:
Address correspondence to Pierre M. Auza: pauza@uci.edu

aviation. The Department of Homeland Security recognizes seven modes as subsectors of the Transportation critical infrastructure sector (CIS); of these, private ownership is common for Maritime, Mass Transit & Passenger Rail, Pipeline Systems, Freight Rail, and Postal & Shipping (*2*).

The Department of Homeland Security also recognizes Transportation as Emergency Support Function (ESF) #1. The *National Response Framework* (*3*) and its associated ESF Annex (*4*) define the scope of ESF#1 responsibilities. This study employs the ESF#1 Annex to assist in identifying the worker and industry categories responsible for implementing ESF#1 across both public and private sectors. For worker and industry categories, this study uses the Bureau of Labor Statistics' Standard Occupational Classification (SOC) system and the North American Industry Classification System (NAICS), respectively.

Resilient transportation operations increasingly rely upon communications infrastructure. For example, the Volpe Center found over 50 different systems and applications that rely upon GPS positioning, navigation, and timing services (*5*). Another sign of interdependence is transportation infrastructure's vulnerability to cyber attacks. For example, the American Public Transportation Association publishes recommended practices on securing the control and communications systems of transit systems (*6*).

The rise of cloud or network computing means that, increasingly, "control system components and networks are now accessible from anywhere and are increasingly connected to enterprise data" and thereby to human safety (*7*). First-generation intelligent transportation systems (ITS) devices had fewer human safety ramifications (*7*). However, the near future will see increases in both connected vehicles and connected vehicle infrastructure. In 2016, the National Highway Traffic Safety Administration published its *Federal Automated Vehicles Policy* (*8*). In December 2013, the Intelligent Transportation Society of America published a market report which found that 54% of respondents to a survey of its members belonged to departments which had developed a regional or agency-specific ITS architecture; 63% of state governments had an ITS strategy in place, and 71% of survey respondents that have a state ITS plan said their agency implemented ITS to reduce congestion and traffic delays (*9*). Meanwhile, the 2013 National ITS Deployment Tracking survey suggests that one-half to three-fourths of surveyed agencies planned to expand current deployments or deploy new technologies (*10*).

Consequently, restoring transportation infrastructure could depend upon restoring communications infrastructure. The Department of Homeland Security recognizes Communications as ESF #2. The present study also employs the ESF#2 Annex (*11*) to help identify relevant worker and industry categories.

While Southern California is susceptible to other (and more frequent) natural hazards (e.g., wildfires), the region is due for a major San Andreas earthquake (*12*). The present study uses this region as its study area. A 2016 CoreLogic report found that an 8.3 magnitude San Andreas earthquake could result in 1.6 to 3.5 million damaged homes (potentially costing $289 billion) (*12*). The United States Geological Survey's (USGS's) landmark 2008 ShakeOut scenario report simulated a 7.8 magnitude San Andreas earthquake and found $213 billion in economic losses; the study demonstrated that this scenario could disrupt several highways in Southern California (e.g., I-10, I-15, SR-14) (*13*).

Research Goals

This study seeks to achieve the following goals:

1. Identify the workers likely to participate in restoring transportation and communications infrastructure after an earthquake in Southern California. That is, identify those workers whose occupational and industry categories correspond with the scopes and responsibilities described in the Transportation (ESF#1) and Communications (ESF#2) Emergency Support Function annexes (*4*, *11*). To the authors' knowledge, no such listing of worker and industry codes exists.
2. Calculate the exposure of these workers to the peak ground accelerations (PGAs) of the scenario earthquake, and compare these values to the exposure of the rest of the working population. Are these workers exposed to statistically significant higher or lower levels of PGAs compared with their counterparts in other occupations and industries?

This study designates PGA exposure as the proxy for a worker's vulnerability to earthquake hazards. It assumes that exposure to PGAs at workers' areas of residence is positively correlated to damage to their homes, in turn affecting their ability to return to work restoring infrastructure. For example, one study highlighted one example action that would assist in recovery: arrange onsite housing for critical staff, emergency responders, and relief workers (*14*).

This study employs the U.S. Census Bureau's American Community Survey Public Use Microdata Sample (PUMS), which uses PUMS areas (PUMAs) as its geography level. PUMAs are statistical geographic areas that aggregate census tracts containing a minimum of 100,000 people. Counts of ESF#1 and ESF#2 workers are estimated by their PUMA of residence.

Literature Review

In attempting to match SOC and NAICS codes to ESF roles and responsibilities, the keywords "Standard Occupational Classification" and "North American Industry Classification System" must accompany keywords relating to "emergency" and "disaster." Similarly, the keywords "emergency support function" and "peak ground acceleration" must accompany keywords like "workers," "workforce," or "labor force". However, a search on the Transportation Research International Documentation (TRID) database and the Transportation Research Board Annual Meeting Online compendium of papers currently returns few studies and publications.

Freight modeling studies are the publications that most often mention both disasters and the NAICS. These studies use NAICS to identify industry sectors. One freight commodity modeling study needs NAICS codes to interpret the U.S. Census Bureau's County Business Patterns date, where employment is categorized by industry (*15*). One goods movement estimation study uses the Commodity Flow Survey, wherein freight activities by business establishments are classified by NAICS code (*16*). Two studies – (*16*) and the aggregate freight generation modeling study (*17*) – mention the hazard that disasters pose in disrupting these modeled freight flows. However, only study (*15*) raises the possibility of modeling disaster scenarios in the national freight model.

Neither TRID nor the Annual Meeting Online compendium returned any studies which employed the SOC system in an emergency or disaster environment. Similarly, neither database returned a paper that analyzes ESFs in terms of workforce or labor force planning. Whenever papers or presentations mention ESFs and the keyword "worker" simultaneously, it is in the context of providing critical workers with waivers or ensuring that they have proper credentialing (*18*, *19*).

When used alone, the phrase "peak ground acceleration" returns engineering papers in TRID or the Annual Meeting Online compendium. However, narrowing the search to include the keywords "workforce," "worker," or "labor force" returns no papers – thus demonstrating a gap in the literature which employs PGAs to estimate workforce impacts.

Data

PGA data in this paper are taken from the USGS Southern California earthquake scenario catalog (*20*). To estimate counts of workers in ESFs #1 and #2, the authors use the PUMS of the 2011–2015 American Community Survey (U.S. Census Bureau).

Peak Ground Acceleration (PGA) Data: Ardent Sentry 2015

This paper uses PGAs that were simulated for the 2015 Ardent Sentry exercise. Ardent Sentry is an exercise program conducted jointly by North American Aerospace Defense and the U.S. Northern Command (USNORTHCOM) (*21*). In 2015, Ardent Sentry's scenario involved a magnitude 7.8 earthquake originating on the San Andreas fault near the Salton Sea (*22*). This scenario earthquake has the same magnitude and epicenter as the 2008 ShakeOut scenario (*23*). However, the shaking intensity maps (Figure 1) differ slightly; for ShakeOut 2008 (Figure 1*a*), the Violent and Extreme regions extend further away from the San Andreas fault than for the 2015 Ardent Sentry scenario (Figure 1*b*).

Working Population Data

This study employs the five-year ACS PUMS from 2011 to 2015. Using the five-year survey yields smaller margins of error than using the one-year survey. We thus use the most accurate estimates available, even though not the most current (five-year averages from 2011–2015, rather than 2015 estimates alone).

To isolate the worker populations of interest in the study area, this study first proposes sets of SOC and NAICS codes which the authors believe best correspond to ESF#1 (Transportation) and ESF#2 (Communications). Only after selecting these codes can one obtain counts and descriptive statistics at the PUMA level. This study uses the workers' PUMAs of residence.

For SOC and NAICS codes, this study uses the appropriate 2010 SOC or 2012 NAICS codes found within the 2011–2015 ACS PUMS Data Dictionary (*24*). While the NAICSP variable within the 2011–2015 ACS PUMS is based upon the 2012 NAICS codes, the NAICSP variable is subject to being abridged and thus losing the ability to distinguish between specialized industries or even industry groups. For example, all construction workers in PUMS are labeled as being in the NAICS sector "Construction" (NAICS = 23). This two-digit sector code is applied to all construction workers, from Highway, Street, and Bridge Construction workers (NAICS = 237310) to Industrial Building Construction workers (236210). It is thus not possible to isolate Highway, Street, and Bridge Construction workers. In contrast, the entire 2010 SOC structure is available within those PUMS variables relating to SOC codes (SOCP10 and SOCP12).

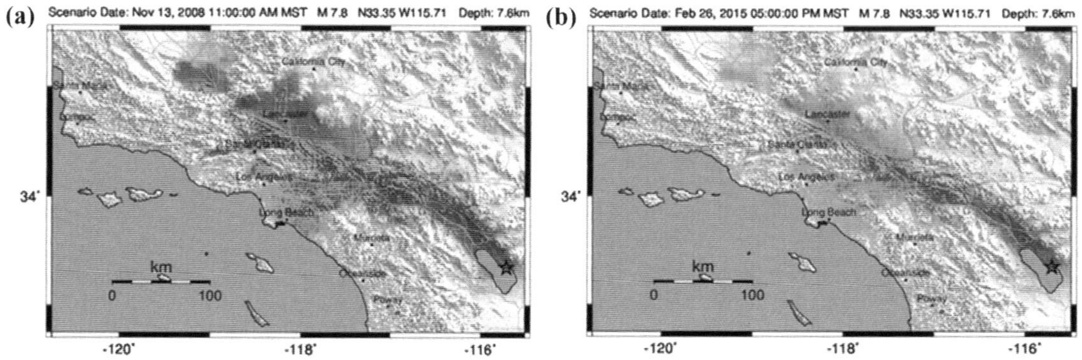

PERCEIVED SHAKING	Not felt	Weak	Light	Moderate	Strong	Very strong	Severe	Violent	Extreme
POTENTIAL DAMAGE	none	none	none	Very light	Light	Moderate	Mod./Heavy	Heavy	Very Heavy
PEAK ACC.(%g)	<0.05	0.3	2.8	6.2	12	22	40	75	>139
INSTRUMENTAL INTENSITY	I	II-III	IV	V	VI	VII	VIII	IX	X+

Scale based upon Worden et al. (2012)

Figure 1. Shaking intensity maps for (*a*) 2008 ShakeOut and (*b*) 2015 Ardent Sentry. Sources: (*22*, *23*), USGS.

Using PUMAs yields the smallest geographical resolution available with large enough sample sizes for both ESF#1 and ESF#2 workers. For all workers, this study uses the PUMA of residence variables (PUMA00 and PUMA10), not the Place of Work PUMA (POWPUMA). Rather than PUMA of employment, PUMA of residence is most of interest, since the authors seek to capture disruption to homes, personal assets, and family life that could prevent a worker from being able to report to work post-disaster. It is possible that a worker's PUMAs of residence and employment are the same, particularly for those living far from urban centers, or for those with very short commutes.

Methodology

Step 1: Determine Extent of PUMAs Affected by Scenario Earthquake and Calculate Mean PGA for Each PUMA

In ArcGIS 10.3.1, the authors first add two layers: (a) California state PUMA boundaries from the 2011–2015 ACS PUMS, and (b) polygon shapefiles for the Ardent Sentry scenario's PGAs. The Intersect geoprocessing tool isolates the PGA contours into discrete regions within each PUMA boundary. Each of these discrete regions corresponds to a PGA value. Thus, to estimate the whole PUMA's mean PGA, each region's PGA is weighted by its area.

Initially, this procedure yielded PGAs for 152 PUMAs, but did not capture four PUMAs in San Diego County: 0607318 San Diego, 0607320 Sweetwater/Chula Vista, 0607321 National City, and 0607322 San Diego. These were added manually by taking the mean PGA of the adjacent PUMAs. Ultimately, 156 PUMAs were affected by the Ardent Sentry scenario earthquake (Figure 2).

Step 2: Identify ESF#1 and ESF#2 Workers in Study Area and Calculate Descriptive Statistics

To identify ESF#1 and ESF#2 workers, the authors first produce a crosswalk of SOC and NAICS codes on the rows and ESF#1 and ESF#2 on the columns. This crosswalk is converted into a list of proposed SOC and NAICS codes for these ESFs (Table 1).

ESF#1 and ESF#2 workers are those workers: (a) whose SOC is one of the SOC codes proposed for that ESF, *and* (b) whose NAICS code is one of that ESF's proposed NAICS codes, *and*, for ESF #1 only, (c) specifically those working in the public sector for those in architecture and engineering occupations. Subsequently, using the proposed SOC and NAICS codes, the authors can flag ESF#1 and ESF#2 workers within the California PUMS data. For each PUMA, counts are obtained for ESF#1, ESF#2, non-ESF#1, and non-ESF#2 workers who reside in it. The authors also estimate distributions for housing type and household income (Figure 4 and Table 2). Figure 3 depicts the geographical distribution of ESF#1 and ESF#2 workers within the study area.

Both SOC and NAICS codes are necessary to identify both ESF#1 and ESF#2 workers within PUMS. For example, while Civil Engineers (SOC = 17-2051) could

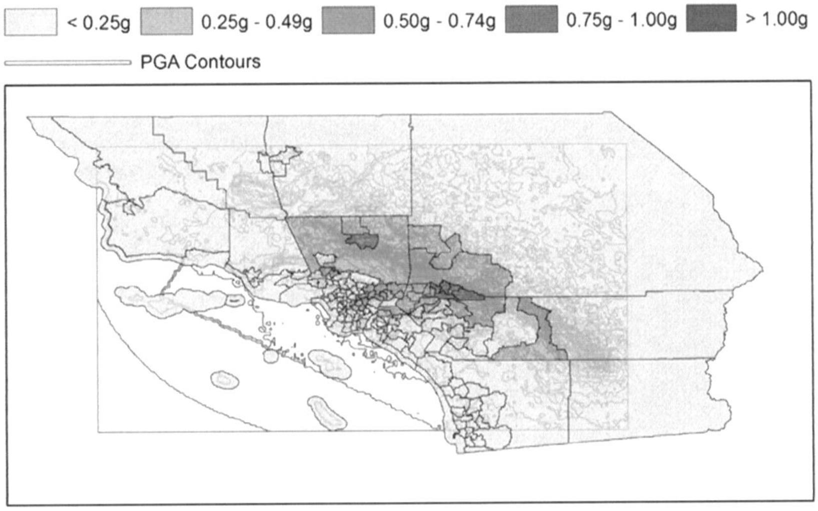

Figure 2. Mean PGA (in g) for each PUMA in study area.

be employed in many industry sectors both public and private, the ones most relevant to ESF#1 would be those in such industries as Highway, Street, and Bridge Construction (NAICS = 237310), the Regulation and Administration of Transportation Programs (NAICS = 926120), and others.

For most SOC and NAICS codes, the relationship with the corresponding ESF is self-evident. For example, the authors propose that Radio and Telecommunications Equipment Installers and Repairers (SOC = 49-2020) are workers who would be part of ESF#2.

For those SOC and NAICS codes where the relationship to an ESF was less clear, the authors would consult the ESF#1 (*4*) and ESF#2 (*11*) Annexes. For example, a reference to pipelines in the ESF#1 Annex (*4*) led to the crosswalk's inclusion of the Pipeline Transportation subsector (NAICS = 486XXX) in ESF#1.

Step 3: Construct Frequency Table to Calculate Mean and Standard Deviation of PGA Exposure for Average ESF#1 and ESF#2 Worker

If a frequency table is initialized with a PUMA's resident worker counts and mean PGA (calculated in Step 1), one can calculate the mean and standard deviation of PGA exposure for ESF#1, ESF#2, non-ESF#1, and non-ESF#2 workers.

All four sets of means and standard deviations are necessary to maintain statistical independence between workers in an ESF and workers not in that ESF. Ultimately, this study uses these four sets to conduct hypothesis testing on the difference between the two population means (see "Frequency Table Means and Standard Deviations" in Appendix).

Step 4: Conduct t-Test to Determine if PGA Exposure is Higher for ESF#1 and ESF#2 (Than for Rest of Working Population)

A *t*-test can determine if there is a statistically significant difference between the mean PGA exposures of ESF#1 and ESF#2 workers and those of non- ESF#1 and non-ESF#2 workers, respectively. With the means and standard deviations of PGA exposure for these worker categories, the authors can test the following null hypotheses (see "*t*-test Hypothesis Testing" in Appendix):

1. $H_0^{ESF\#1}$: $\mu_{ESF\#1} = \mu_{non-ESF\#1}$, no statistically significant difference between the mean PGA exposure of ESF#1 (Transportation) workers compared with non-ESF#1 workers.
2. $H_0^{ESF\#2}$: $\mu_{ESF\#2} = \mu_{non-ESF\#2}$, no statistically significant difference between the mean PGA exposure of ESF#2 (Communications) workers compared with non-ESF#2 workers.

Results

Result 1: Mean PGA for Each PUMA

Figure 2 presents the mean PGAs for each of the 156 PUMAs in the study area. For reference, the figure also includes the PGA contours for the Ardent Sentry scenario. The dense PGA contours are generally collinear with the San Andreas fault. The PUMAs with the

Table 1. SOC and NAICS Codes Proposed for Identifying ESF#1 and ESF#2 Workers

Standard Occupational Classification (SOC) codes	North American Industry Classification System (NAICS) codes
ESF#1: Transportation	
11-3071 Transportation, Storage, and Distribution Managers 11-9021 Construction Managers 11-9041 Architectural and Engineering Managers 17-1022 Surveyors[b] 17-2051 Civil Engineers[b] 17-2121 Marine Engineers and Naval Architects[b] 17-3011 Architectural and Civil Drafters[b] 17-3022 Civil Engineering Technicians[b] 17-3031 Surveying and Mapping Technicians[b] 33-9093 Transportation Security Screeners 47-4011 Construction and Building Inspectors 47-4051 Highway Maintenance Workers 47-4061 Rail-Track Equip. Operators 49-2091 Avionics Technicians 49-2093 Electrical & Electronics Installers and Repairers, Transportation Equipment 49-2096 Electronic Equipment Installers and Repairers, Motor Vehicles 49-30XX Vehicle and Mobile Equipment Mechanics, Installers, Repairers 53-1011 Aircraft Cargo Handling Supervisors 53-20XX Air Transportation Workers 53-3020 Bus Drivers 53-3032 Heavy and Tractor-Trailer 53-3033 Light Truck or Delivery Services 53-40XX Rail Transportation Workers 53-50XX Water Transportation Workers 53-6011 Bridge and Lock Tenders 53-6041 Traffic Technicians 53-6051 Transportation Inspectors	23XXXX Construction[a] *237310 Highway, Street, and Bridge Construction* 481XXX Air Transportation 482XXX Rail Transportation 483XXX Water Transportation 484XXX Truck Transportation 485M Bus & Urban Transit[a] *485XXX Transit & Ground Passenger Transportation* 486XXX Pipeline Transportation 488XXX Support Activities for Air Transportation 92M2 Administration of Economic Programs and Space Research[a] *926120 Regulation & Administration of Transportation Programs*
ESF#2: Communications	
11-3021 Computer and Information Systems Managers 15-11XX Computer Occupations 43-20XX Communications Equipment Operators 43-9011 Computer Operators 49-2020 Radio and Telecommunications Equipment Installers & Repairers 49-9052 Telecommunications Line Installers and Repairers	237130 Power & Communication Line & Related Structure Construction 5171 Wired Telecommunications Carriers 517Z Telecommunications, Except Wired Telecommunications Carriers 92M2 Administration of Economic Programs and Space Research[a] *926130 Regulation and Administration of Communications, Electric, Gas, & Other Utilities*

[a]The italicized entries below this entry are NAICS codes which are more specialized and relevant to this ESF, but which are not available in the 2011–2015 PUMS Data Dictionary (*24*).
[b]Only public-sector workers taken (class of worker = 3, 4, or 5).

Table 2. Distribution of Household Income (in Percentiles) for the General Working Population, ESF#1 Workers, and ESF#2 Workers in Southern California, 2011–2015

	Household income percentile (expressed in 2015 US$)				
Worker category	10%	25%	50% (median)	75%	90%
Working population	$23,985.60	$45,220.18	$82,103.65	$136,092.19	$209,091.83
ESF#1 (transportation)	$32,328.41	$57,748.26	$95,607.70	$144,910.61	$205,043.00
ESF#2 (communications)	$39,778.34	$68,664.90	$105,147.97	$155,832.68	$217,774.92

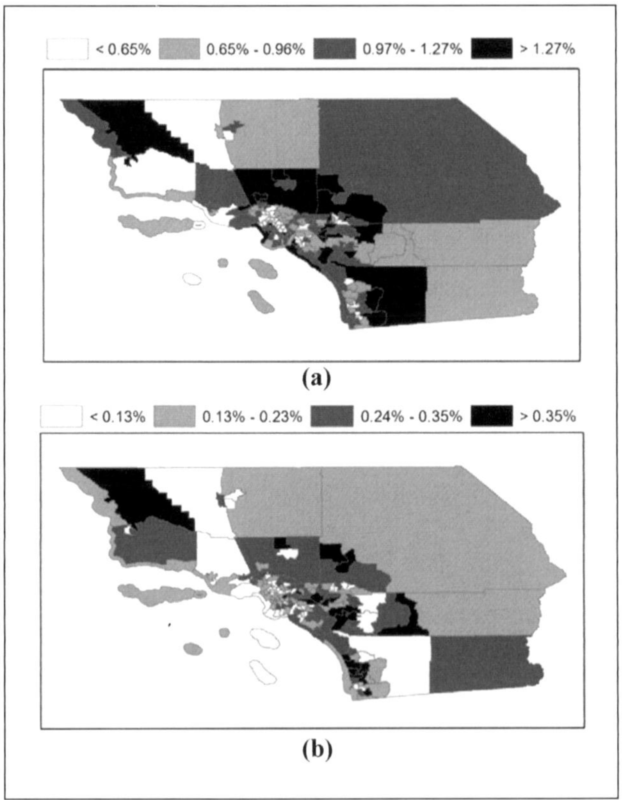

Figure 3. Concentration of resident (*a*) ESF#1 and (*b*) ESF#2 workers (in % of the total working population).

highest mean PGAs occur along the fault, through Imperial, Riverside, San Bernardino, Los Angeles, and Kern counties. Several PUMAs in northeast Los Angeles County are also subject to higher mean PGAs.

The maximum mean PGA in a PUMA is 0.70*g*, in southwest San Bernardino County (PUMA ID = 0607108, San Bernardino [West]). San Bernardino County claims seven of the 10 PUMAs with the top 10 highest mean PGAs in the Ardent Sentry scenario. Los Angeles (LA) County claims the other three of the top 10. The two highest mean PGAs in LA County are 0.62*g* and 0.48*g*, in north central LA County (PUMA IDs = 0603704 and 0603703, Palmdale and Lancaster, respectively). Meanwhile, the 10th highest mean PGA (at 0.46*g*) is in central LA County (PUMA ID = 0603738, El Monte & South El Monte).

Result 2: Proposed SOC & NAICS Codes for ESF#1 & ESF#2, and Comparison with the General Working Population

Table 1 shows the SOC and NAICS codes which the authors propose as corresponding to those workers and industry sectors (both private and public) who would perform the Transportation (ESF#1) and Communications (ESF#2) emergency support functions.

Italic entries indicate the six-digit specialized industry NAICS codes which are not directly available as filtering values in the 2011–2015 ACS PUMS. Thus, the corresponding two-digit sector or three-digit subsector codes are marked with one asterisk "*". For example, while the authors deem the Highway, Street, and Bridge Construction NAICS (237310) to be within the scope of ESF#1, the ACS PUMS merely indicates if a worker is in the Construction sector (NAICS = 23). Thus, Table 1 includes "23XXXX Construction*" among the NAICS codes for ESF#1.

Using these sector and subsector codes widens the scope of industries included in an ESF. When combined with SOCs in which many workers would be employed by both public and private entities, this widening of industry scope consequently includes many private-sector workers which the authors deem to be beyond the scope of the ESF of interest. One example SOC is Civil Engineers (17-2051).

Thus, Table 1 marks certain SOCs with "(p)" to indicate SOC codes for which only public-sector workers are considered for selection. In the 2011–2015 ACS PUMS public-sector workers are those for whom the Class of Worker variable COW is 3, 4, or 5 – local, state, or federal government employee, respectively. This measure introduces the assumption that public-sector workers in these particular SOC codes are more likely to be critical to an emergency support function than private-sector workers. While imperfect, the authors present it as one possible solution to this issue.

Figures 3 and 4, and Table 2 compare ESF#1 and ESF#2 workers against the general working population and against each other.

Figure 3 depicts the concentration of resident ESF#1 and ESF#2 workers in the study area, in terms of percentage of the total working population residing in the PUMA. While the grayscale color ramps for both are identical in the figure, the legend indicates that overall concentrations of ESF#2 workers are lower than those of ESF#1 workers.

The PUMA with the highest concentration of resident ESF#1 workers is in southeast Orange County (PUMA ID = 0605915, Rancho Santa Margarita (East) & Ladera Ranch). Among the top 10 PUMAs with the highest ESF#1 worker concentrations, Orange County claims two PUMAs, Los Angeles County claims three, San Bernardino County claims two, and San Diego, Ventura, and Riverside Counties claim one each. These concentrations range from 1.5% to 2%.

The PUMA with the highest concentration of resident ESF#2 workers is in central San Diego County (PUMA

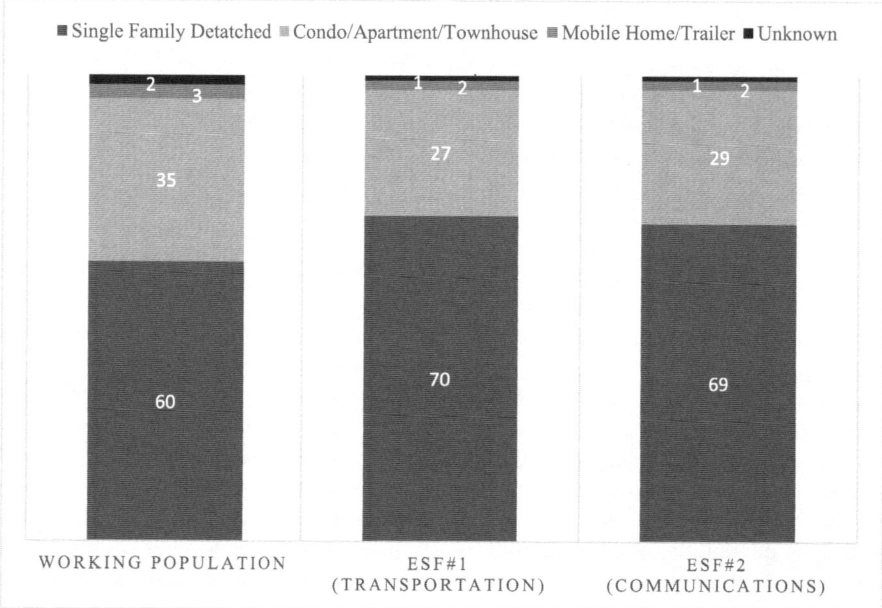

Figure 4. Distribution of housing unit building types for the total working population (left), ESF#1 workers (center), and ESF#2 workers (right) in Southern California, 2011–2015.

ID = 0607308, San Diego (Northeast/Rancho Bernardo) & Poway). San Diego County claims three of the top 10 the PUMAs in ESF#2 worker concentration, followed by San Bernardino and Riverside Counties (two each), and Orange, San Luis Obispo, and Los Angeles counties (one each). These concentrations range from 0.4% to nearly 0.6%.

Lancaster (PUMA ID = 0603703) and Victorville & Adelanto (PUMA ID = 0607102) are two PUMAs that are in the top 10 for both ESF#1 and ESF#2 worker concentrations. The former (Lancaster) also ranks among the top 10 PUMAs with the highest mean PGA.

Figure 4 shows the distribution of housing unit building types for the total working population, ESF#1 workers, and ESF#2 workers in Southern California. For all three building types, the percentage of workers with unknown housing unit building types is less than 3%. Furthermore, for each of the three building type categories, the percentage of workers who live in mobile homes is less than 2%. Mobile homes as a class of building are particularly vulnerable to earthquakes (*25*, *26*).

At a glance, ESF#1 and ESF#2 workers are more similar to each other than to the general working population in their distribution between Single Family Detached and Condo, Apartment, and Townhouse (multifamily) housing. For example, whereas 70% of ESF#1 and 69% of ESF#2 workers live in Single Family Detached homes, only 60% of the total working population live in such homes.

Table 2 presents the household income distribution of the general working population, ESF#1 workers, and ESF#2 workers at the 10th, 25th, 50th (median), 75th, and 90th percentiles. The authors chose household income as opposed to the individual worker's income since the household income better reflects the resources that would be available to these workers in the event of an earthquake or fire. ESF#1 and ESF#2 household incomes are higher than those of the general working population from the 10th to the 75th percentiles – including the median household income. However, the 90th percentile ESF#1 worker's household income is less than either the 90th percentile ESF#2 worker or the 90th percentile household income in the general working population. This paper does not attempt to determine if these differences are statistically significant.

Result 3: Means and Standard Deviations of PGA Exposure, and t-Statistics for Hypothesis Tests

The top three rows of Table 3 summarize the means and standard deviations of PGA exposure for ESF#1 and non-ESF#1 workers, and for ESF#2 and non-ESF#2 workers. The bottom two rows present the resulting *t*-statistic values to test the two null hypotheses $H_0^{ESF\#1}$ and $H_0^{ESF\#2}$. For both *t*-statistics, the value is large enough

Table 3. Mean and Standard Deviation Exposure to PGA for ESF#1 and Non-ESF#1 Workers, and for ESF#2 and Non-ESF#2 Workers

	Worker Category			
	ESF#1	Non-ESF#1	ESF#2	Non-ESF#2
Total count *(n)*	102,991	10,695,776	22,360	10,776,407
Mean PGA exposure experienced in $g(\bar{x})$	0.193*g*	0.190*g*	0.188*g*	0.190*g*
Standard deviation of PGA exposure in g(s)	0.141*g*	0.136*g*	0.133*g*	0.137*g*
Hypothesis test	$H_0^{ESF\#1}: \mu_{ESF\#1} = \mu_{non-ESF\#1}$		$H_0^{ESF\#2}: \mu_{ESF\#2} = \mu_{non-ESF\#2}$	
t-statistic	+ 5.67**		−2.18*	

*$p < .05$. **$p < .01$. ***$p < .001$.

to reject the null hypothesis, although to differing levels of confidence. Namely:

1. ESF#1 workers are exposed to *higher* mean PGAs than non-ESF#1 workers ($p < .01$)
2. ESF#2 workers are exposed to *lower* mean PGAs than non-ESF#2 workers ($p < .05$)

ESF#1 workers are more exposed to PGAs in their PUMAs of residence than ESF#2 workers. In an earthquake like the 2015 Ardent Sentry scenario, ESF#1 workers may be more affected by damage to their homes than ESF#2 workers are. They could thus face more difficulty in returning to work and restoring transportation infrastructure.

This discrepancy may be due to the higher concentration of ESF#1 workers in southwest San Bernardino and northern Los Angeles counties. The dark PUMAs of Figure 3*a* roughly correspond with the PUMAs of higher mean PGA in Figure 2. Of the PUMAs with the top 30 highest mean PGAs, 27 are within San Bernardino and LA Counties. Meanwhile, among the PUMAs with the top 30 highest concentrations of ESF#1 workers, 14 are in those two counties. In contrast, only nine San Bernardino and LA county PUMAs are among the PUMAs with the highest ESF#2 worker concentrations.

Conclusion

This study aims to make two contributions. First, it proposes using SOC and NAICS codes to identify those worker and industry categories that would most likely participate in restoring damaged transportation and communications infrastructure. This study lists those SOC and NAICS codes which the authors believe most closely correspond to the Transportation (ESF#1) and Communications (ESF#2) emergency support functions of the *National Response Framework* (*3*).

Second, this study proposes a method to determine whether any category of workers is exposed to higher PGAs in their PUMAs of residence. This method determines whether the mean exposure to PGA of a worker category (e.g., ESF#1 or ESF#2) is equal to or different from the mean PGA exposure of the rest of the working population. For the 2015 Ardent Sentry scenario, ESF#1 workers are found to be exposed to higher PGAs than non-ESF#1 workers, while ESF#2 workers are exposed to lower PGAs than non-ESF#2 workers. For both cases, the differences in PGA exposure are statistically significant.

For practitioners planning for disasters, awareness of worker categories that are disproportionately affected could facilitate pre-event response and recovery planning efforts. For example, public agencies can mitigate these potential worker shortfalls with investments in mutual aid agreements and worker preparedness. Mutual aid agreements prior to a disaster allow an affected jurisdiction to request resources (e.g., equipment, worker teams) from unaffected neighboring jurisdictions. Worker preparedness measures would depend upon jurisdiction.

These results are subject to limitations. First, PUMAs are geographically large, with some PUMAs in this study larger than 1000 square miles. Since PGA can vary considerably even within a PUMA, these large sizes can distort the calculation for mean PGA in the PUMA. However, PUMAs were the finest geographic resolution for which SOC and NAICS data could be obtained for all worker categories and still achieve a sufficient sample size.

Second, while a San Andreas earthquake like the Ardent Sentry scenario is often used for planning purposes, large earthquakes could occur on other Southern California faults (e.g., the Northridge blind thrust fault). Ultimately, 2015 Ardent Sentry is only one scenario earthquake in the USGS catalog (*20*). Including scenario earthquakes of other magnitudes, locations, and return periods can result in a more complete seismic hazard profile of mean PGA exposure for ESF#1 and ESF#2 workers.

Lastly, the NAICS codes available within the 2011–2015 ACS PUMS are only a subset of the full 2012 NAICS code structure from the Bureau of Labor

Statistics. Consequently, in this study, several six-digit specialized industry NAICS codes are not directly available in the PUMS data. Out of necessity, this study uses the next-lowest-level NAICS code available (e.g., four-digit industry group or three-digit subsector code) and restricts the Class of Worker variable to public employees only.

There is potential for applying this methodology elsewhere. For example, since transportation has interdependencies with energy infrastructure, one can extend this methodology to ESF#12 (Energy) workers. This methodology could be extended to all 14 ESFs, or similarly, even to the six recovery support functions of the *National Disaster Recovery Framework* and to the Department of Homeland Security's sixteen CISs. In these cases, the scope of the study would extend beyond transportation and cover other areas of emergency management.

Appendix

Frequency Table Calculations

Equations 1 and 2 obtain the mean and standard deviation via a frequency table (*27*):

$$\bar{x} = \frac{\sum_{i=1}^{k} f_i x_i}{(n-1)} \tag{1}$$

$$s = \sqrt{\sum_{i=1}^{k} \frac{f_i (x_i - \bar{x})^2}{(n-1)}} \tag{2}$$

where

i = index for PUMA identifier

k = total number of PUMAs

n = total number of workers residing in study area

There are several worker categories of interest possible, e.g., ESF#1, ESF#2, non-ESF#1, or non-ESF#2 workers.

x_i = mean PGA (in fraction of g) of PUMA i, calculated in Step 1

f_i = worker count in PUMA i for the worker categories of interest

$\bar{x}$ = mean PGA (% of g) to which workers are exposed in this category

s = standard deviation of the PGA (% of g) exposure of the workers in this category

Table 4 depicts an example frequency table to calculate the mean and standard deviation of ESF#1 workers' exposure to PGAs across nine PUMAs in the study area. Estimates of the mean and standard deviation of PGA exposure for ESF#1, ESF#2, non-ESF#1, or non-ESF#2 workers across all 156 PUMAs of the study area are presented in Table 3. To perform these calculations, the authors use SAS version 9.4.

t-Test Hypothesis Testing

This study uses the t-test formula in Equation 3:

$$t = \frac{(\bar{x}_1 - \bar{x}_2) - (\mu_1 - \mu_2)}{\sqrt{\frac{s_1^2}{n_1} + \frac{s_2^2}{n_2}}} = \frac{(\bar{x}_1 - \bar{x}_2)}{\sqrt{\frac{s_1^2}{n_1} + \frac{s_2^2}{n_2}}} \tag{3}$$

where

$\bar{x}_1$ = estimated mean PGA exposure of either ESF#1 or ESF#2 workers, calculated in Step 3

$\bar{x}_2$ = estimated mean PGA exposure of either non-ESF#1 or non-ESF#2 workers, calculated in Step 3

μ_1 = true mean PGA exposure of either ESF#1 or ESF#2 workers

μ_2 = true mean PGA exposure of either non-ESF#1 or non-ESF#2 workers

n_1 = number of ESF#1 or ESF#2 workers in the estimate

n_2 = number of non-ESF#1 or non-ESF#2 workers in the estimate, respectively

Note that if testing the null hypothesis $\mu_1 = \mu_2$, then the term $\mu_1 - \mu_2 = 0$. Since the counts of ESF#1 and

Table 4. Sample Frequency Table to Calculate the Mean and Standard Deviation of ESF#1 Workers' PGA Exposure

PUMA ID	(x_i) Mean PGA (g)	(f_i) Count of ESF#1 workers residing in PUMA	$f_i x_i$	$f_i(x_i - \mu)^2$
0602500	0.07	729	49.71	13.65
0602901	0.08	461	35.67	7.51
0602902	0.12	901	108.32	6.48
0602903	0.11	642	70.92	5.74
0602904	0.14	242	34.94	0.89
0602905	0.09	603	52.26	8.45
0603701	0.32	1131	356.40	13.71
0603702	0.21	1493	313.25	0.03
0603703	0.48	913	437.25	68.49
			$\bar{x}$ = 0.21g	s = 0.13g

ESF#2 are sufficiently large ($n > 100$), the following t-statistic critical values (two-tailed) would permit the rejection of these null hypotheses: $t > \pm 1.96$ for a 95% confidence level ($p < .05$), and $t > \pm 2.58$ for 99% confidence ($p < .01$). Table 3 summarizes the outcome of these two hypothesis tests.

Author Contributions

The authors confirm contribution to the paper as follows: study conception and design: Pierre Auza, R Jayakrishnan, Yuko Nakanishi; data collection: Diana Lavery; analysis and interpretation of results: Pierre Auza, Diana Lavery; draft manuscript preparation: Pierre Auza. All authors reviewed the results and approved the final version of the manuscript.

References

1. *Standing Committees by Mode and Topic*. Transportation Research Board. Washington, D.C. http://www.trb.org/AboutTRB/StandingCommitteesMT.aspx. Accessed July 15, 2017.
2. *Transportation Systems Sector-Specific Plan*. U.S. Department of Homeland Security and U.S. Department of Transportation, Washington, D.C., 2015. https://www.dhs.gov/publication/nipp-ssp-transportation-systems-2015.
3. *National Response Framework*. 3rd ed. Federal Emergency Management Agency, Washington, D.C., June 2016. https://www.fema.gov/media-library/assets/documents/117791.
4. *Emergency Support Function #1 – Transportation*. Federal Emergency Management Agency, June 2016. https://www.fema.gov/media-library-data/1470148635327-75b99900ae83949a9c5577c1dc99ccdd/ESF_1_Transportation_20160705_508.pdf.
5. Wallischeck, E. *GPS Dependencies in Transportation: An Inventory of Global Positioning System Dependencies in the Transportation Sector, Best Practices for Improved Robustness of GPS Devices, and Potential Alternative Solutions for Positioning, Navigation and Timing*. Publication DOT-VNTSC-NOAA-16-01. Volpe National Transportation Systems Center and National Oceanic and Atmospheric Administration, 2016. http://ntl.bts.gov/lib/60000/60400/60433/DOT_VNTSC_NOAA_16_01.pdf.
6. *Recommended Practice: Securing Control and Communications Systems in Transit Environments*. APTA-SS-CCS-RP-001-10. American Public Transportation Association, Washington, D.C. http://www.apta.com/resources/standards/Documents/APTA-SS-CCS-RP-001-10.pdf.
7. Countermeasures Assessment and Security Experts LLC and Western Management and Consulting LLC. *NCHRP Report 221/TCRP Report 67: Protection of Transportation Infrastructure from Cyber Attacks: A Primer*. Transportation Research Board of the National Academies, Washington, D.C., 2015.
8. *Federal Automated Vehicles Policy*. National Highway Traffic Safety Administration, Washington, D.C., September 2016. https://www.transportation.gov/AV.
9. *Market Report: State ITS Trends*. The Intelligent Transportation Society of America (ITS America), Washington, D.C., December 2013. http://itsamerica.org/state-dot-its-trends/. Accessed July 15, 2017.
10. *2013 Deployment Tracking Survey Results*. The Intelligent Transportation Society of America (ITS America), Washington, D.C., December 2013. http://www.itsdeployment.its.dot.gov/Default.aspx. Accessed July 15, 2017.
11. *Emergency Support Function #2 – Communications*. Federal Emergency Management Agency, Washington, D.C., June 2016. https://www.fema.gov/media-library-data/1473679033823-d7c256b645e9a67cbf09d3c08217962f/ESF_2_Communications_FINAL.pdf.
12. CoreLogic. *California Earthquake Risk Report*. CoreLogic, Irvine, CA, 2016. http://www.corelogic.com/about-us/researchtrends/california-earthquake-risk-report.aspx#.WNGvePkrJEY.
13. Jones, L. M., R. Bernknopf, D. Cox, J. Goltz, K. Hudnut, D. Mileti, S. Perry, D. Ponti, K. Porter, M. Reichle, H. Seligson, K. Shoaf, J. Treiman, and A. Wein. *The ShakeOut Scenario*. U.S. Geological Survey Open-File Report 2008-1150 and California Geological Survey Preliminary Report 25. 2008. https://pubs.usgs.gov/of/2008/1150/.
14. Southworth, F., J. Hayes, S. McLeod, and A. Strauss-Wieder. *NCFRP Report 30: Making US Ports Resilient as Part of Extended Intermodal Supply Chains*. Transportation Research Board of the National Academies, Washington, D.C., 2014.
15. Lupa, M. R, R. W. Eash, P. Mescher, J. T. von Brown, and C. A. Clymer. Rail Freight Commodity Model for Iowa: Methodology & First Steps. Presented at 94th Annual Meeting of the Transportation Research Board, Washington, D.C., 2015.
16. Oliveira-Neto, F. M., S. M. Chin, H. L. Hwang, and B. Peterson. *Methodology for Estimating Ton-Miles of Goods Movements for U.S. Freight Multimodal Network System*. Oak Ridge National Laboratory, Oak Ridge, TN, 2013, p. 15
17. Oliveira-Neto, F. M., S. M. Chin, and H. L. Hwang. Aggregate Freight Generation Modeling: Assessing Temporal Effect of Economic Activity on Freight Volumes with Two-Period Cross-Sectional Data. *Transportation Research Record: Journal of the Transportation Research Board*, 2012. 2285: 145–154.
18. Cook, K. Ports, Impact Assessments, and Emergency Management. Presented at 96th Annual Meeting of the Transportation Research Board, Washington, D.C., 2017.
19. Sheehan, T. St. Louis County, Minnesota, June 2012. Floods: Critical Connections of Low-Volume Roads. Presented at 94th Annual Meeting of the Transportation Research Board, Washington, D.C., 2015.
20. *SCLEGACY (Scenario Catalog): Southern California Legacy Catalog*. United States Geological Survey (USGS). https://earthquake.usgs.gov/scenarios/catalog/sclegacy/. Accessed July 15, 2017.
21. Hagihara, K. U.S. 3rd Fleet Participates in Exercise Ardent Sentry 2015. *Navy News Service*, May 5, 2015.

http://www.navy.mil/submit/display.asp?story_id = 87132. Accessed July 15, 2017.
22. *M 7.8 Scenario Earthquake - Ardent Sentry 2015 Scenario.* United States Geological Survey (USGS). https://earthquake.usgs.gov/scenarios/eventpage/sclegacyardentsentry2015_se.
23. *M 7.8 Scenario Earthquake – Shakeout2 Full Scenario.* United States Geological Survey (USGS). https://earthquake.usgs.gov/scenarios/eventpage/sclegacyshakeout2full_se.
24. *2011–2015 ACS PUMS Data Dictionary.* U.S. Census Bureau, January 19, 2017. https://www2.census.gov/programs-surveys/acs/tech_docs/pums/data_dict/PUMS_Data_Dictionary_2011-2015.pdf.
25. Wein, A., L. Johnson, and R. Bernknopf. Recovering from the ShakeOut Earthquake. *Earthquake Spectra*, Vol. 27, No. 2, May 2011, pp. 521–538.
26. Maison, B., and K. Cobeen. Analytical Study of Mobile Home Response to the 2014 South Napa Earthquake. *Earthquake Spectra*, Vol. 32, No. 1, 2016, pp. 1–22.
27. *Mean & Standard Deviation Calculator for Frequency Table & Grouped Data.* KNOWPAPA.COM. http://knowpapa.com/sd-freq/. Accessed July 15, 2017.

The Standing Committee on the Logistics of Disaster Response and Business Continuity (ABR20) peer-reviewed this paper (18-04708).

Article

TRR
JOURNAL OF THE TRANSPORTATION RESEARCH BOARD

Transportation Research Record
2018, Vol. 2672(1) 125–137

DOI: 10.1177/0361198118773896
journals.sagepub.com/home/trr
SAGE

Crisis Communication Patterns in Social Media during Hurricane Sandy

Arif Mohaimin Sadri[1], Samiul Hasan[2], Satish V. Ukkusuri[3], and Manuel Cebrian[4]

Abstract

Hurricane Sandy was one of the deadliest and costliest of hurricanes of the past few decades. Many states experienced significant power outage; however, many people used social media to communicate while having limited or no access to traditional information sources. Using machine learning techniques, this study explored the evolution of various communication patterns and determined user concerns that emerged over the course of Hurricane Sandy. The original data included ~52M tweets coming from ~13M users between October 14, 2012 and November 12, 2012. A topic model was run on ~763K tweets from the top 4,029 most frequent users who tweeted about Sandy at least 100 times. Some 250 well-defined communication patterns based on perplexity were identified. Conversations of the most frequent and relevant users indicate the evolution of numerous storm-phase (warning, response, and recovery) specific topics. People were also concerned about storm location and time, media coverage, and activities of political leaders and celebrities. Also presented is each relevant keyword that contributed to one particular pattern of user concerns. Such keywords would be particularly meaningful in targeted information-spreading and effective crisis communication in similar major disasters. Each of these words can also be helpful for efficient hash-tagging to reach the target audience as needed via social media. The pattern recognition approach of this study can be used in identifying real-time user needs in future crises.

In 2012, coastal residents of New York and New Jersey experienced a massive storm surge produced by late-season Hurricane Sandy that caused disastrous affects in the mid-Atlantic and northeastern United States across the Atlantic basin (*1*). Some examples include: $50Bn in property damage, 72 fatalities, at least 147 direct deaths, destruction of 570,000 buildings, cancellation of 20,000 flights, 8.6M power outages in 17 states, 230,000 cars destroyed as a result of flooding, and thousands of people displaced from their homes (*2–5*). Whereas many previous studies primarily explored evacuation logistics (*6–11*), the focus has now shifted to mitigation of and resilience to natural hazards (*12*). Disaster resilience with a view to limiting such adverse impacts is now a national imperative at all levels, and more efficient policies are required by the government to allow people to be less dependent on federal resources (*13*).

As such, vulnerable communities need to respond to any form of disaster with sufficient preparation (*14–17*), and effective crisis communication is one major aspect of this. This includes systematic planning, collection, organization, and circulation of relevant awareness information to the target audience, reaching out to every individual in a community (*18–20*). During an emergency, people may obtain weather information from traditional media such as radio or television, and social media such as Facebook, Twitter, or the internet. Social media platforms, uniquely different from traditional ones, can help disseminate targeted information during a crisis. However, it is critical to efficiently harness the large-scale and rich information from these communication sources (*21*). Many studies have acknowledged the value of social media data in an emergency response (*22–27*), community interactions (*28, 29*), crisis informatics (*30–38*), and predicting real-world actions/events (*39, 40*).

During Hurricane Sandy, social media also played an important role in crisis communication. New York and New Jersey residents used smart-phones to access social media, as they had limited access to traditional

[1]Moss School of Construction, Infrastructure and Sustainability, Florida International University, Miami, FL
[2]Department of Civil, Environmental, and Construction Engineering, University of Central Florida, Orlando, FL
[3]Lyles School of Civil Engineering, Purdue University, West Lafayette, IN
[4]Media Lab, Massachusetts Institute of Technology, Cambridge, MA

Corresponding Author:
Address correspondence to Arif Mohaimin Sadri: sadri.buet@gmail.com

information sources (*41*). Social media communications via social media continued during and after the storm in areas without power (*42*). Studies have identified authorities, news media, and peers as emergency warning sources (*43*), and psychological and social factors are very important in translating hazard warning information into a collective decision (*6*, *44–47*). For example, local authorities, peers, local and national media, and the internet significantly influence evacuation decision-making (*45*). Moreover, individuals were more likely to evacuate if they relied on social media for weather-related information during Sandy (*7*).

Twitter users can post a brief message (maximum 140 characters) and follow other users in a network setting. Twitter thus exhibits the characteristics both of a social network and an informational network (*48*). Whereas the social network aspects of Twitter provide access to geographically and personally relevant information, the information network can help to control global information contagion (*40*). These specific features make Twitter particularly useful for effective information dissemination during crises, as evidenced in many empirical studies: service characteristics (*24*, *33*), retweeting activity (*49*, *50*), situational awareness (*51*, *52*), online communication of emergency responders (*53*, *54*), text classification and event detection (*30*, *31*, *35*, *55*, *56*), devising sensor techniques for early awareness (*57*), quantifying human mobility (*58*, *59*), classifying urban activity and lifestyle patterns (*60–63*), and disaster relief efforts (*64*).

This paper uses machine learning techniques to analyze the evolution of communication patterns, and determines various user concerns that emerged over the course of Hurricane Sandy. The original data included ~52M tweets coming from ~13M users between October 14, 2012 and November 12, 2012. The topic model is run on ~763K tweets from the top 4,029 most frequent users who tweeted about Sandy at least 100 times. Some 250 well-defined communication patterns were identified based on perplexity. Conversations of the most frequent and relevant users indicate the evolution of numerous storm-phase (warning, response, and recovery) specific topics. People were also concerned about storm location and time, media coverage, and activities of political leaders and celebrities. Also presented is each relevant keyword that contributed to one particular pattern of user concern.

Data Description and Analyses

In this study, raw data (~52M tweets, ~13M users, Oct 14–Nov 12, 2012) obtained from Twitter were analyzed. The detailed steps involved in data collection can be found in Kryvasheyeu et al. (*57*). The data include a text database with user and text identifiers, texts, and some additional useful information. The network database includes the relationship graphs of active users, that is, the list of followees for each user (*38*). These were reconstructed using Twitter API. Only a minor fraction of the texts (~1.35%) are geo-tagged by Twitter. For relevance, user activity was assessed on the basis of number of tweets (~11.83M) in the data that included the word 'sandy', and that co-appeared with other words, after filtering out ~46.45M tweets that were in English; that is, non-English tweets were removed.

First, the texts were analyzed that evolved in the tweets of the top 4,029 most frequent users who tweeted at least 100 times about Sandy (AF $\geq$ 100), producing ~763K tweets and ~95M words. After removing punctuation, English stop-words, repeated words such as "http" and "@" from these tweets, it was found that that the 50 most frequent words contributed ~30% of all the words in the texts. The highly frequent words are shown as a word cloud in Figure 1, where the frequencies of these words can be seen. The words "sandy" and "hurricane" are the top two most frequent words. The word "storm" is also at the top of the list. Interestingly, the word "storm" was more popular on October 23, 2012, shortly after the tropical depression was formed (Table 1). Later on, "hurricane" and "sandy" became more frequent, close to the landfall on October 29. The word "superstorm" was also popular. The appearance of "sandy" before the tropical depression was formed (October 22) suggests those tweets having "sandy" out of context. The labeling of a tropical storm should be given importance, as it is strongly connected to how information disseminates later. Any given word can be a part of a specific topic evolving over time. Also, each word can be a part of multiple topics. The text database needs to be carefully analyzed to infer aggregate and user-specific topic patterns.

Also analyzed were the texts that evolved in the tweets of the top 157,622 most frequent users who tweeted at least ten times about Sandy (AF $\geq$ 10), producing ~3.9M tweets. After removing the punctuation, English stop-words, repeated words such as "http" and "@" from these tweets, it was found that these 100 most frequent words (Figure 1) contribute ~37.5% of all the words in the texts. Any given word can be a part of a specific topic which evolved over time. Also, each word can be a part of multiple topics. The text database needs to be carefully analyzed to infer aggregate and user-specific topic patterns. The highly frequent words are shown as a heatmap over time in Figures 2 and 3, where the frequencies of these words can be seen. The words "sandy" and "hurricane" are the top two most frequent words. The word "storm" is also at the top of the list. Interestingly, the word "storm" was more popular on October 23, 2012 shortly after the tropical depression was formed. Later

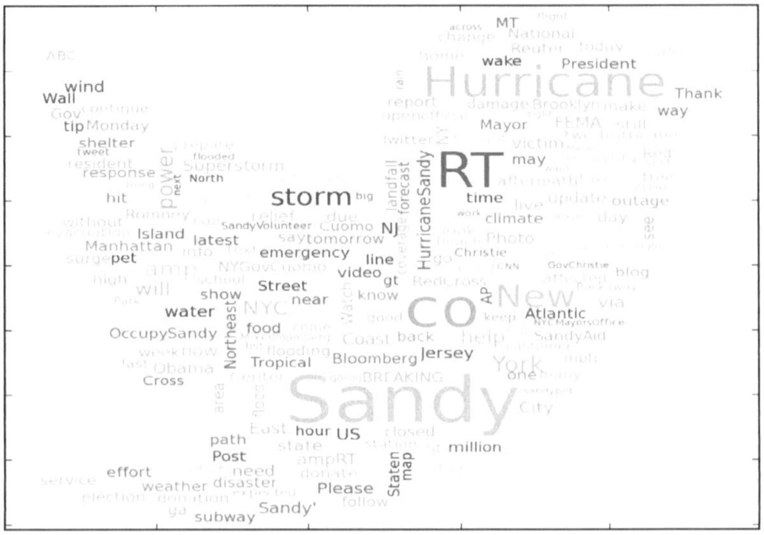

Figure 1. Patterns obtained from the text analysis (AF $\geq$ 100; 4,029 nodes/users and 763,000 tweets).

Table 1. Communication Patterns: Warning and Recovery Phase

Topic	No. of topics	Pattern #	Relevant keywords
Warning phase			
Storm prediction	13	W1–13	model, track, show, predict, impact, data, confirm, intense, image, satellite, latest, radar, nasa, loop, powerful, view, cuba, jamaica, hurricane, might, become, frankenstorm, increase, storm, surge, lifethreatening, flood, graphics, make, landfall, expect, along, threaten, become, merge, tropical, strength, downgrade, form, cyclone, gain, weaken, regain, region, hit, path, across, map, crisis, gouge, noaa, see, direct
Storm watch	6	W14–19	warn, storm, watch, effect, issue, tropical, huge, hit, eye, keep, unclear, video, space, show, nasa, station, timelapse, national, hurricane, center, forecast, nhc, advisory, cone
Preparedness	4	W34–37	tip, disaster, children, care, preparedness, parent, kid, prepare, residents, urge, plan, officials, get, ready, try, serious, make, sure, check, neighbor, elderly
User concern	8	W38–45	affect, thoughts, prayers, everyone, pray, god, love, hope, feel, worry, call, shit, take, seriously, aim, place, care, action, warn, stay, safe, keep, please, inside, tune, inform, impact, wake, destructive, know, need, want, anyone, let, call, someone, put, risk, really, info, visit, check, control, rumor, share
Weather condition	9	W20–28	wind, rain, high, heavy, strong, snow, mph, move, category, sustain, mb, pressure, water, level, tide, feet, rise, surge, record, dangerous, outside, inside, scary, gust, far, miles, force, storm, center, field, wide, air, drop, eye, low, central, inch, report, blizzard, part
Previous hurricane	2	W29–30	review, katrina, unprepared, usgov, irene, tragedy, vulnerability
Storm cause	3	W31–33	blame, gay, marriage, preacher, global, warming, climate, change, cause
Recovery phase			
Disaster relief	16	RC1–16	concert, victims, raise, relief, benefit, sandyaid, occupysandy, sandyvolunteer, sandyrelief, sandyhelp, mutualaid, donate, redcross, help, efforts, fund, support, million, show, need, church, volunteer, help, pls, collect, please, recover, rt, spread, word, families, charity, ask, offer, free, aid, provide, displace, want, supply, join, dayofgiving, fundraise, disaster, federal, fema, blood, drive, fdr, food, meals, emergency, serve, response
Recovery	10	RC17–26	reopen, jfk, airport, help, survivors, victims, impact, cash, recovery, efforts, begin, fuel, frustration, please, need, help, family, claim, pay, insurance, assistance, fema, tax, health, workers, mental, cut, learn, lessons, get, back, work, normal, finally, replacement, recover, add, businesses
Damage/Aftermath	7	RC27–33	damage, cost, billion, estimate, economic, losses, aftermath, wake, wrath, fury, devastation, devastate, fee, waive, bank, ever, history, record, largest, worst, biggest, sales, auto, industry, affect, buy, please, help, areas, concern, postsandy, expose

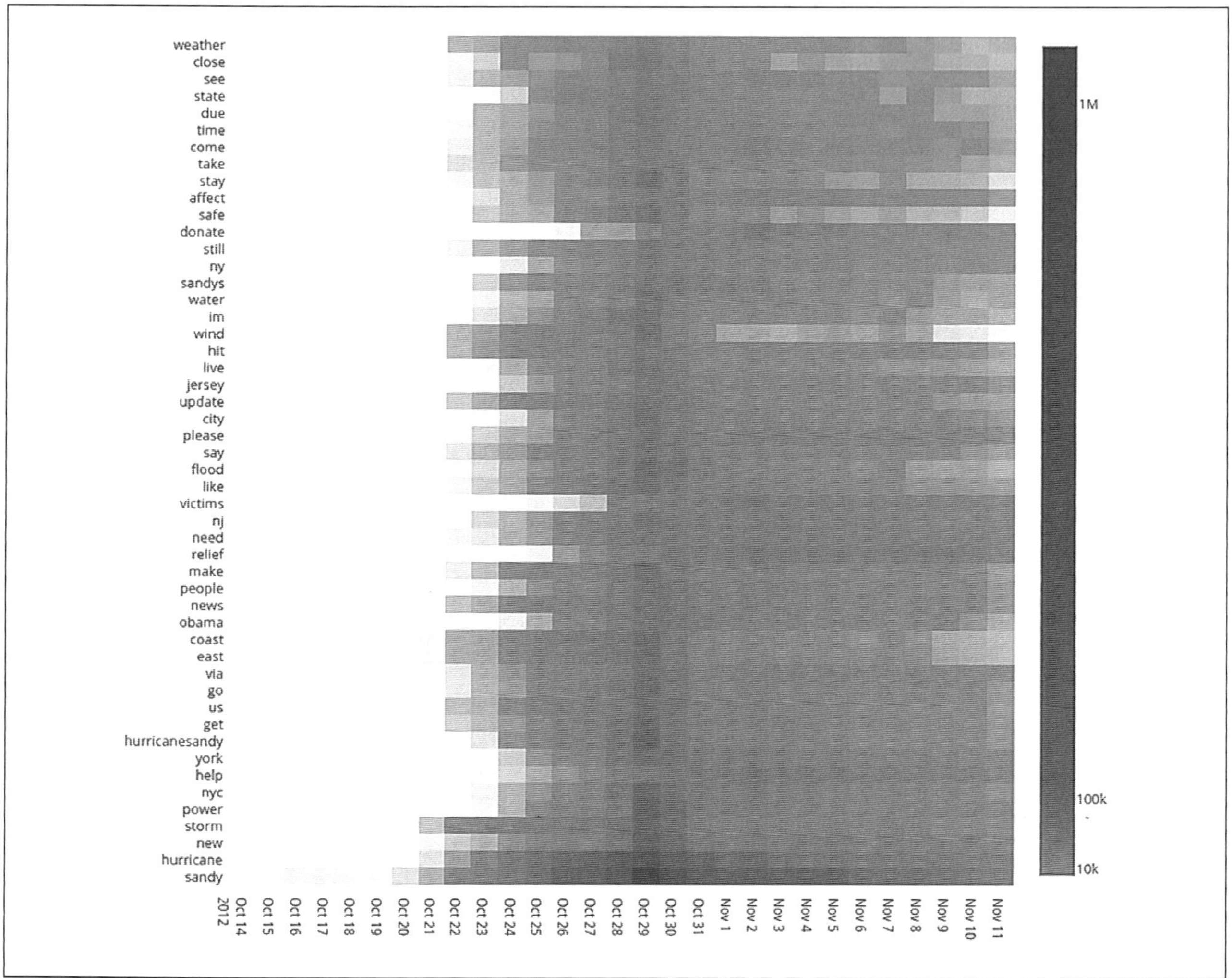

Figure 2. Word appearance over time (word frequency rank: 1–50).

on, "hurricane" and "sandy" became more frequent, close to the landfall on October 29. The word "superstorm" was also popular. The appearance of "sandy" before the tropical depression was formed (October 22) suggests those tweets having "sandy" out of context. The labeling of a tropical storm should be given importance, as it is strongly connected to how information disseminates later. The other words in the list can be broadly classified into several categories. Words such as "east," "coast," "stay," "safe," "update," "watch," "path," "prepare," and "emergency" appeared more frequently before the storm. Some words, for example, "power," "water," "food," "gas," and "home" were more prominent close to or shortly after the landfall. Considerably after the landfall, topics including "help," "relief," "victims," "donate," "affect," "please," "fema," "without," and "shelter" appeared to emerge. Words such as "weather," "state," "time," "still," "sandys," "hit," "ny," "nj," and "york" were fairly uniformly distributed both before and after the landfall.

Model Description and Formulation

The crisis communication patterns and their evolution in these networks were based on topic models that are well established in the machine learning literature. A topic model is a generative model that follows a probabilistic process for generating documents based on a set of straightforward probabilistic sampling rules. This process explains how words in a document can be generated based on some latent variables, that is, topics that evolve in a document. The overall procedure includes the following steps (*65*): (i) select a distribution over topics to make a new document, (ii) for each word in the document, select a topic randomly and then a word from that topic following that distribution. The process can also be reversed and the set of topics that generate the collection of documents can be obtained. The model fitness of such generative models should find the best set of latent variables (i.e., topics of the documents) based on which the observed data can be reasonably explained (i.e., words in the

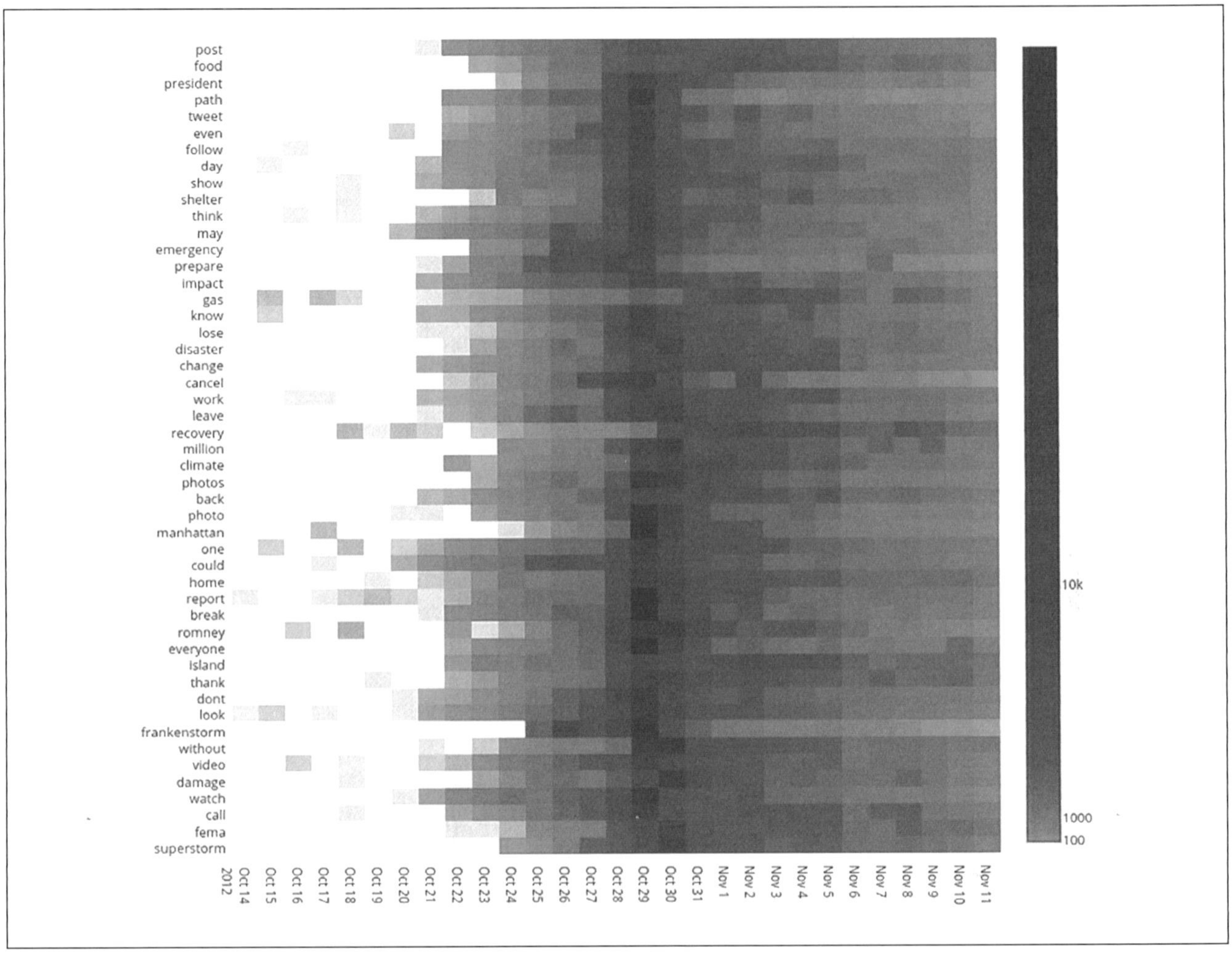

Figure 3. Word appearance over time (word frequency rank: 51–100).

documents). The generative process, as discussed above, does not assume any specific ordering of the words ("bag-of-words" assumption in natural language processing) in a document (65), and the frequency of word appearance in a document is the only information that is relevant. The ordering of the words can be useful at times; however, this is not captured by topic models. A topic model can also be applied to other types of discrete data.

Formally speaking, the problem of identifying various social interaction topics is to determine K latent patterns through ϕ_k for $k \in \{1, 2, \ldots K\}$ and each topic ϕ_k is a distribution of different words. A word is defined as the basic unit of data to be selected from a set of possible words of size W, a user tweets about N topics, and the user can contribute to the collection of M tweets. The generative process is summarized below (see Figure 4*a*):

1. For each topic, $k \in \{1, 2, \ldots K\}$, a distribution over words is selected

$$\phi^{(k)} \sim \text{Dirichlet}(\beta)$$

2. For each tweet, $m \in \{1, 2, \ldots M\}$,
   (a) A distribution over topics is selected
   $$\theta^{(m)} \sim \text{Dirichlet}(\alpha)$$
   (b) For each word i in tweet m
      i. Select a topic $z_i \sim \text{Multinomial}\left(\theta^{(m)}\right)$; $z_i \in \{1, 2, \ldots K\}$
      ii. From topic z_i, a word is selected

$$w_i \sim \text{Multinomial}\left(\phi^{(z_i)}\right);\ w_i \in \{1, 2, \ldots W\}$$

Now, given M tweets, K topics over W unique words, the main objectives of the inference of topic pattern classifications are:

i. Find the probability of a word w given each topic k, $P(w|z = k) = \phi_k^w$ where $P(w|z = k)$ is

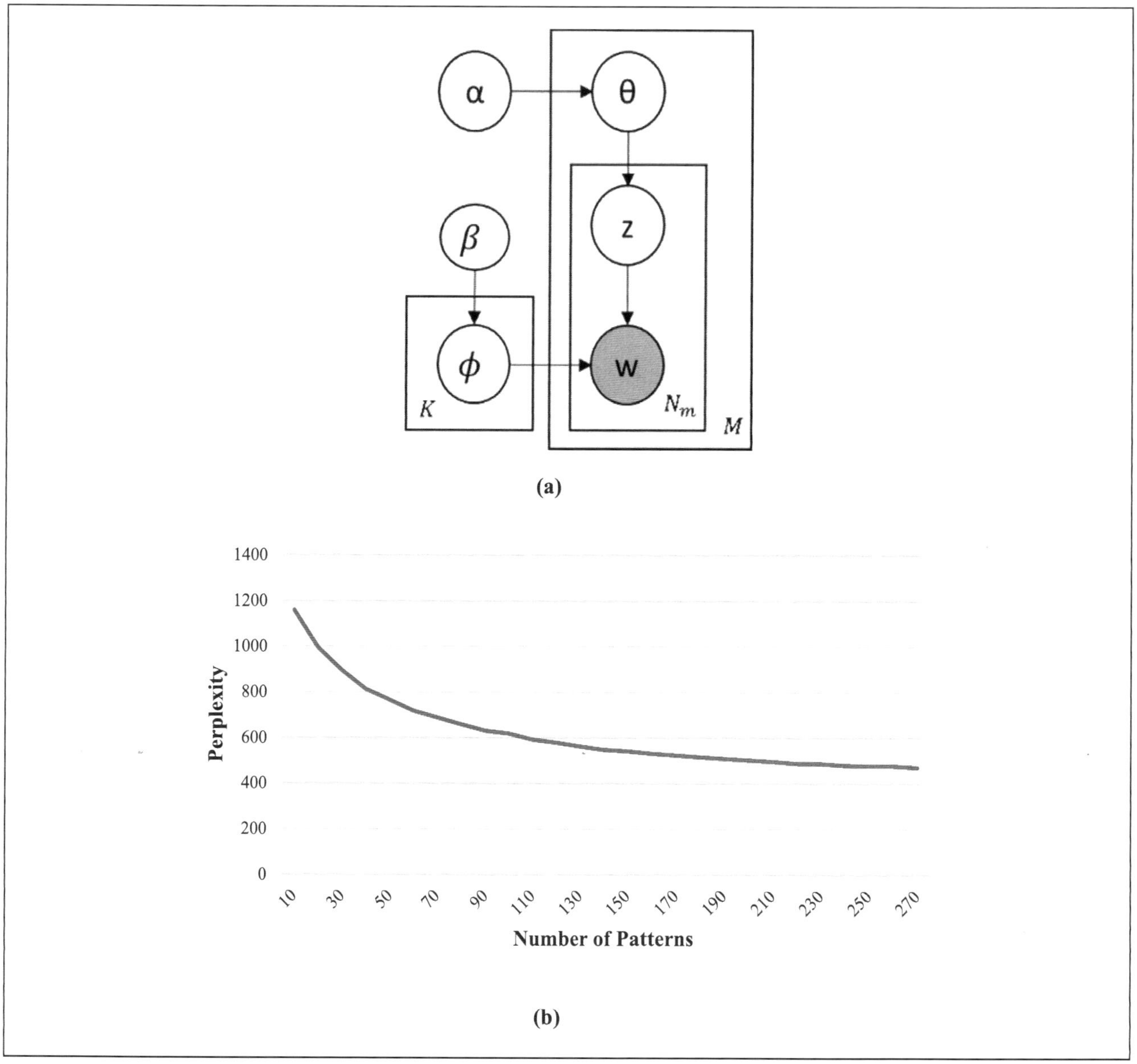

Figure 4. (*a*) Probabilistic generative process of topic models and (*b*) perplexity analysis.

represented with K multinomial distributions ϕ over words of size W

ii. Find the probability of a topic k for a word in tweet m, $P(z = k|m) = \theta_m^k$ where $P(z|m)$ is represented with M multinomial distributions θ over K topics.

The above model views the topic pattern as a probability distribution over words and tweeting activities as a mixture of these patterns. From K topics, the probability of **i**th word in a given tweet m is:

$$P(w_i|m) = \sum_{j=1}^{K} P(w_i|z_i = j)P(z_i = j|m) \qquad (1)$$

Here, z_i is the latent variable referring to the pattern from which the **i**th word is drawn, $P(w_i|z_i = j)$ indicates the probability of word w_i under the **j**th pattern and $P(z_i = j|m)$ is the probability of selecting a word from pattern **j** in the tweet m. Intuitively, $P(w|z)$ determines the importance of a word in forming a pattern and $P(z|m)$ determines the prevalence of the pattern in different tweets. The complete model of pattern generation follows:

$$w_i|z_i, \phi^{(z_i)} \sim \text{Multinomial}\left(\phi^{(z_i)}\right)$$

$$\phi \sim \text{Dirichlet}(\beta)$$

$$z_i|\theta^{(m)} \sim \text{Multinomial}\left(\theta^{(m)}\right)$$

$$\theta \sim \text{Dirichlet}(\alpha)$$

Here, α and β are hyper-parameters for the prior distributions of θ and ϕ, respectively. Dirichlet prior distributions are assumed, which are conjugate to the multinomial distributions.

The joint distribution of words and patterns $P(\boldsymbol{w}, z)$ written as:

$$P(\boldsymbol{w}, z) = P(\boldsymbol{w}|z)P(z) \tag{2}$$

The first term can be written as (*66*):

$$P(\boldsymbol{w}|z) = \left(\frac{\Gamma(W\beta)}{\Gamma(\beta)^W}\right)^K \prod\nolimits_{k=1}^{K} \frac{\prod_w \Gamma(n_k^w + \beta)}{\Gamma\left(n_k^{(\cdot)} + W\beta\right)} \tag{3}$$

Here n_k^w is the number of times word w is assigned to pattern k and $n_k^{(\cdot)} = \sum_{w-1}^{W} n_k^w$.

The second term can be written as:

$$P(z) = \left(\frac{\Gamma(K\alpha)}{\Gamma(\alpha)^K}\right)^M \prod\nolimits_{m=1}^{M} \frac{\prod_k \Gamma(n_m^k + \alpha)}{\Gamma\left(n_m^{(\cdot)} + K\alpha\right)} \tag{4}$$

Here n_m^k is the number of times a word from tweet $m$is assigned to pattern k and $n_m^{(\cdot)} = \sum_{k-1}^{K} n_m^k$. A pattern can be assigned to a word using the following conditional distribution (*66*):

$$P(z_i = k|z_{-i}, \boldsymbol{w}) \propto \frac{n_{-\mathrm{i,k}}^{w_i} + \beta}{n_{-\mathrm{i,k}}^{(\cdot)} + W\beta} \frac{n_{-\mathrm{i,m_i}}^{k} + \alpha}{n_{-\mathrm{i,m_i}}^{(\cdot)} + K\alpha} \tag{5}$$

Here, n_{-i} is the count excluding the current pattern assignment of z_i. The first ratio expresses the probability of word w_i in pattern k and the second ratio expresses the probability of pattern k in the tweet m.

Parameter Estimation and Model Selection

There are various approximation techniques for estimating the parameters of this model (*66*, *67*). Here the Gibbs sampling approach proposed by Griffiths and Steyvers was used; the algorithm can be found in detail in Griffiths and Steyvers (*66*). Only a brief description of the approach is provided here. To estimate the model parameters, a Markov chain Monte Carlo (MCMC) procedure is used. In MCMC, samples are taken from a Markov chain constructed to converge to a target distribution. In the model developed here, each state of the chain is the assignment of a pattern to a word and the transition from one state to another follows a specific rule based on Gibbs sampling approach (*68*). In this procedure, the next state is reached by sampling the variables from a conditional distribution which specifies the distribution of the variables conditioned on the current assignment of all other variables and the observations. The parameters of Latent Dirichlet Allocation (LDA) representing the hidden patterns, can be computed as:

$$\hat{\phi}_k^w = \frac{n_k^w + \beta}{n_k^{(\cdot)} + W\beta};\ \hat{\theta}_m^k = \frac{n_m^k + \alpha}{n_m^{(\cdot)} + K\alpha} \tag{6}$$

The model selection was based on *Perplexity*—a metric to measure the predictive capacity of the model to infer the unseen data in each run. In machine learning, *Perplexity* is a commonly used metric to report the performance of a probabilistic model that refers to the average likelihood of obtaining a test data set given a set of model parameters. *Perplexity* can be defined as the exponential of the negative of average predictive likelihood of test data given a model (*66*). In this study, the algorithm was run for different number of topic patterns (K) and perplexity in each run was computed. Next, the optimal number of patterns was selected based on perplexity values. For a given set of words $\{\boldsymbol{w_m}\}$ and $m \in D^{test}$ given a model $\mathcal{M}$, *Perplexity* of a test data set can be defined as:

$$Perplexity = \exp\left[-\frac{\sum_{m=1}^{M} \log p(\boldsymbol{w_m}|\mathcal{M})}{\sum N_m}\right] \tag{7}$$

where N_m is the number of words in each topic m and $p(\boldsymbol{w_m}|\mathcal{M})$ can be derived from Eq. (1). Figure 4*b* indicates that there exist 250 well-defined patterns in the data as there is no significant change in perplexity after that.

Crisis Communication Patterns

Patterns of Warning Phase

Tables 1–3 present the details on each type of the communication pattern along with the associated keywords. The top ten words contributing to each pattern are tabulated in the Appendix (Tables 1–8) in Sadri, in the decreasing order of their occurrence probability from left to right (*69*). The topic model was applied to ~763K tweets including ~95M words from the 4,029 users who were active at least 100 times (AF $\geq$ 100), demonstrating their high degree of appearance and relevance to Sandy. Most discussions during the warning phase (before October 29, 2012), listed in Table 1, are related to storm prediction (W1–13), storm watch (W14–19), and storm preparedness (W34–37). Storm prediction topics consist of words such as NASA, NOAA, satellite, predict, path, model, track, form, regain, weaken, and other similar words. Conversations on storm watch are expressed through keywords including warn, watch, issue, advisory, and forecast. People were also

Table 2. Communication Patterns: Response Phase

Topic	No. of topics	Pattern #	Relevant keywords
Response phase			
Gas/fuel	3	RP1–3	fuel, spill, diesel, oil, gasoline, near, shut, gas, line, station, ration, price, shortage, wait
Food/water	2	RP4–5	food, water, need, supply, blanket, batteries, distribute, clothe, ice, drink, pump, bottle, boil, safe, sewage
Power outage	11	RP6–16	power, without, million, customers, still, remain, restore, utility, phone, charge, cell, internet, mobile, battery, explosion, dark, safety, generator, caution, blackout, tip, cold, house, electricity, outage, without
Transportation	9	RP17–25	flight, travel, tour, airlines, subway, transit, mta, commuter, bike, walk, bus, service, train, cancel, suspend, system, line, resume, suspend, traffic light, break, stop, wait, port, metro, authority, bridge, tunnel, open, close, station, boat, empty, flood
Local officials	3	RP26–28	officials, governor, cuomo, christie, obama
Evacuation	1	RP29	evacuation, zone, order, mandatory
Infrastructure	3	RP30–32	statue, liberty, nuclear, power, plant, world, trade, center
Death tolls	5	RP33–37	dead, death, rise, toll, destruction, homeless, kill, people, many, lot, bahamas, haiti, caribbean, fear, video
Hospital	1	RP38	hospital, evacuate, medical, patient, nurse, fail, baby
Rescue	2	RP39–40	police, people, nypd, rescue, fire, trap, loot, report, ship, miss, sink
Fire	1	RP41	home, fire, destroy, burn
Flood	2	RP42–43	flood, street, swim, severe, underwater
Trees/Debris	2	RP44–45	cleanup, clean, clear, debris, house, tree, fall, home, block, line, wire
Pets/Animals	2	RP46–47	pet, shelter, rescue, animal, find, help, dog, cat
Crime	1	RP48	worry, crime, strand, unplug
Stock market	1	RP49	stock, market, trade, close, exchange, open, close
FEMA	3	RP50–52	fema, response, respond, call, 911, register, insurance, damage, shelter, zip, code
First responder	6	RP53–58	national, guard, navy, aid, response, assist, soldier, army, support, carrier, american, red, cross, donation, million, relief, first, responder, team, supply, military, truck, community, aid, group
Event cancellation	4	RP59–62	cancel, due, event, class, marathon, runner, suspend, office, postpone, date
Work/School closure	2	RP63–64	school, close, reopen, remain, due, office

concerned about storm preparedness and words such as "get," "ready," "try," "prepare," "parent," "children," "care," and "neighbor" contributed to such topics. Several topics are specific to user concern (W38–45), in which people expressed their thoughts and prayers for the people who stayed along the path of the hurricane. Some topics were related to updates on weather condition, incorporating "wind," "rain," "snow," "water," "heavy," "strong," and several weather measurement units. People also expressed their concern about previous hurricanes such as Katrina and Irene as Sandy was approaching. The causality of a major hurricane like Sandy was also a major topic of concern (W31–33). Some claimed global warming and climate change to be responsible, some supposed gay marriage to be a source of storm formation.

Patterns of Response Phase

Turning to topics that evolved during the response phase (during or immediately after the landfall on October 29, 2012), it was found that people were mainly concerned about basic needs such as gas/fuel (RP1–3), food/water (RP4–5), and the adverse impacts caused by significant power outage in different states (RP6–16). Several discussions were on transportation-related topics (RP17–25) as expressed by "flight," "travel," "airlines," "subway," "bus," "train," "metro," "station," "bridge," "tunnel," "mta," "cancel," "service," "suspend," "resume," "line," and other similar words (Table 2). People also talked about how local officials such as state governors responded (RP26–28) to the situation, and whether or not an evacuation order is issued (RP 29). Major infrastructure such as the Statue of Liberty, World Trade Center, and Nuclear Power Plant were also part of many interactions (RP30–32). The severity of death tolls was highlighted in several topics (RP33–37). Hospitals undertook precarious evacuations (RP38) and several rescue efforts were made by the police department (RP39–40). Concerns about the impact of fire (RP41), flood (RP42-43), clearing debris (RP44–45), and pets/animals (RP46–47) also emerged. People also talked about different ongoing crimes (RP48) and the status of the stock market (RP49). First responders such as FEMA (RP50–52), National Guard, Navy, American Red Cross, and Army also appeared in several topics (RP53–58). Some people discussed the cancellation of

Table 3. Communication Patterns: Location, Time, News & Media, Politics, Leaders & Celebrities, and Others

Topic	No. of topics	Pattern #	Relevant keywords
Location			
East Coast	3	L1–3	eastern, east, coast, us
NY/NJ	4	L4–7	ny, nj, new, york, city, jersey, state
NY	3	L8–10	new, york, yorkers, newyork, nyc, nysandy
NJ	2	L11–12	nj, new, jersey, sandynj, njsandy
Other states	9	L13–21	li, long, island, longisland, staten, statenisland, borough, rhode, si, ct, connecticut, ctsandy, va, virginia, vasandy, nc, north, carolina, delaware, philadelphia, philly, sandycenpa, md, maryland, mdsandy
River	1	L22	river, hudson, bank
Street address	1	L23	st, ave, cars, water, flood, avenue, street
Politics, Leaders & Celebrities			
President Obama	3	PC1–3	obama, gop, govt, america, president, barack, statement, speak, brief, deliver
Others	4	PC4–7	romney, mitt, bill, clinton, government, capital, mayor, endorse
US Election	5	PC8–12	presidential, election, us, race, politics, gop, washington, vote, poll, voters, early, election2012
Celebrities	4	PC13–16	karl, rove, lindsay, lohan, dina, lohan, jim, cantore
News & Media			
News media	16	NM1–16	cnn, los, angeles, times, sky, news, nyt, business, abc, daily, mail, cbs, usa, today, box, break, post, blog, report, washington, huffington, yahoo, cover, buzzfeed, bbc, wsj, wall, street, journal, weather, channel, national, coverage, forbes, reuters
Social media	9	NM17–25	social, media, internet, share, friends, photo, video, app, download, mobile, tweet, send, message, text, receive, facebook, fb, page, status, topic, top, twitter, follow, info, information, picture, image, pics, instagram
Broadcasting	5	NM26–30	live, update, coverage, watch, brief, press, conference, hold, stream, tv, online, broadcast, youtube, chat, report, call, ask, question, answer, reporters
Time			
Time	6	T1–6	oct, 29, 30, 2012, pm, tonight, afternoon, night, week, year, last, next, hours, days, minutes, past, next
Others			
Others	33	O1–33	hurricane, sandy, dont, think, cant, u, cant, look, one, superstorm, say, good, ppl, like, get, give, hear, us, go, yet

major events (RP59–62) and closure of offices and schools (RP63–64).

Recovery Phase and Phase-Independent Patterns

The recovery phase of the storm primarily included discussions about numerous disaster relief efforts such as Occupy Sandy and popular hashtags like "sandyaid," "sandyrelief," "sandyhelp," and "sandyvolunteer," among others (RC1–16). Some discussions were specific to the recovery process itself when people talked about survivors and victims of the storm and lessons learned from Hurricane Sandy (RC17–26). Economic losses, devastating damage, and other aftermaths caused by Sandy (RC17–26) were also discussed in this particular phase of the storm (Table 1). Various topics were phase independent and specific to location, time, media coverage, political leaders, and celebrities, as listed in Table 3. When the location category is considered, it was found that many patterns represent user concern about East Coast, New York, and New Jersey (L1–12). These were the areas that experienced the major impact of Sandy. People also communicated about other states/areas slightly affected or close to the path of Sandy. These included Long Island, Staten Island, Rhode Island, Connecticut, Virginia, North Carolina, Delaware, Philadelphia, and Maryland (L13–21). Other specific locations included the Hudson River (L22) and any given street address that was flooded (L23). Time-related topics were also prominent (T1–6) that included day and year of Sandy's landfall (October 29, 2012) in addition to other specific time units.

Media played a salient role during Hurricane Sandy. Numerous patterns were part of the media coverage, both traditional news media (NM1–16) and easily accessible social media (NM17–25). Traditional media sources included CNN, *LA Times*, Sky News, *NYT*, ABC News, *Daily Mail*, CBS, *USA Today*, *Washington Post*, *Huffington Post*, BBC, *Forbes*, Reuters, and *Wall Street Journal*. On the other hand, Facebook, Twitter, and

Instagram were among the popular social media topics as people talked about several social media activities such as "tweet," "status," "page," "share," "message," "follow," and so on. Some topics represented solely different broadcasting steps such as live updates, live coverage, press briefings, and others (NM26–30). Political leaders such as President Obama (PC1–3), Mitt Romney, and Bill Clinton (PC4–7) appeared in several discussion topics. Some interactions were on the 2012 US Election and its relevance to Hurricane Sandy (PC8–12). Celebrity activities during that period were also prominent on several occasions. For example, Karl Rove, Jim Cantore, Lindsay Lohan, and Dina Lohan appeared in a few conversations (PC13–16). Some other topics were uncategorized (O1–33).

Discussion

Conversations of most frequent and relevant users ($\sim$763K tweets from top 4,029 highly active users; AF $\geq$ 100) indicate evolution of numerous topics at different phases of the storm such as warning, response, and recovery. People were also concerned about phase-independent topics specific to location, time, media coverage, political leaders, and celebrity activities. The top 50 most frequent words contribute up to $\sim$30% of all words in those tweets; however, each word can be part of any specific topic or multiple topics. Warning phase topics were related to storm prediction, storm watch and advisory, storm preparedness, user concerns about the storm, previous hurricane experience, and causes responsible for storm formation such as climate change and global warming. Communications during the response phase may include basic needs of people such as food, water, gas, power, and so on. People also express concern on major infrastructure and different transportation facilities such as subway, metro, train, bus, airlines, and others. Many interactions include how local officials responded to the storm threat and ordered mandatory evacuations. Topics included first responders such as FEMA, hospitals, pets/animals, fire, flooding, event cancellation, rescue operations, work or school closure. People primarily discussed several disaster relief and fundraising efforts in the recovery phase. In addition, possible aftermath and lessons learned from such a major disaster also emerged during this phase. While some communication patterns are location (affected states/areas) and time specific, some were specific to media coverage, including both traditional news media and social media. Political leaders, the 2012 US election, and celebrity activities were among other topics of discussion.

The pattern recognition approach of this study can be used in identifying real-time user needs in future crises. For example, if users express their concern on social media about locating nearby shelters, this will emerge if the model is applied. As such, the analyses of the text using the proposed modeling framework would enable emergency response officials to know the evolution of concerns/sentiment over time, thereby allowing agencies to prioritize response and enabling better resource allocation. As part of the emerging patterns through texts, such predictive models allow users to identify shelter locations, route congestion, and other real-time information about the impacts of the disaster, thus facilitating better decision-making. This is evident in some of the Sandy patterns predicted by the proposed model that appeared in the response phase (RP46–47 and RP50–52 in Table 2).

Introductory hashtags are common these days in social media. The keywords that constitute a particular pattern of user concern, as presented in this study, would help emergency managers to use such keywords as hashtags at different phases (warning, response, and recovery) of major hurricanes and help vulnerable population receive more crisis information via social media. Future studies should capture the complex contagion of Hurricane Sandy in more detail; for example, how user concerns spread through the network of Twitter users (followee–follower versus direct user mentions) and the rate of information propagation, content relevance, and risk profiling. Subgraphs of geo-tagged users ($\sim$2% in the raw data) may help to understand the mobility pattern and spatial correlation with their network activity.

Acknowledgments

The authors are grateful to National Science Foundation for the grant CMMI-1131503 and CMMI- 1520338 to support the research presented in this paper. The second author is supported by the U.S. Department of Transportation University Transportation Centers Program under the project "Disaster Analytics: Disaster Preparedness and Management through Online Social Media," for which the author is grateful.

Author Contributions

All authors contributed to the study conception and design, data collection, analysis and interpretation of results, and draft manuscript preparation. All authors reviewed the results and approved the final version of the manuscript.

References

1. Blake, E. S., T. B. Kimberlain, R. J. Berg, J. Cangialosi, and J. L. Beven, II. Tropical Cyclone Report: Hurricane Sandy. *National Hurricane Center*, Vol. 12, 2013, pp. 1–10.
2. Abramson, D. M., and I. Redlener. Hurricane Sandy: Lessons Learned, Again. *Disaster Medicine and Public Health Preparedness*, Vol. 6, No. 4, 2012, pp. 328–329.
3. Halverson, J. B., and T. Rabenhorst. Hurricane Sandy: The Science and Impacts of a Superstorm. *Weatherwise*, Vol. 66, No. 2, 2013, pp. 14–23.

4. Meyer, R. J., Baker, J., Broad, K., Czajkowski, J., & Orlove, B. The Dynamics of Hurricane Risk Perception: Real-Time Evidence from the 2012 Atlantic Hurricane Season. *Bulletin of the American Meteorological Society*, Vol. 95, No. 9, 2014, pp. 1389–1404.
5. Gladwin, H., B. Morrow, and J. Lazo. *Communication and Understanding of Hurricane Sandy Storm Surge Forecast and Warning Information*. NSF CMMI 1322088. Florida International University Natural Hazards Workshop, Florida International University, Miami, FL, 2013, http://www.academia.edu/4316097/Communication_and_Understanding_of_Hurricane_Sandy_Storm_Surge_Forecast_and_Warning_Information.
6. Sadri, A. M., S. V. Ukkusuri, and H. Gladwin. Modeling Joint Evacuation Decisions in Social Networks: The Case of Hurricane Sandy. *Journal of Choice Modelling*, Vol. 25, 2017, pp. 50–60.
7. Sadri, A. M., S. V. Ukkusuri, and H. Gladwin. The Role of Social Networks and Information Sources on Hurricane Evacuation Decision Making. *Natural Hazards Review*, Vol. 18, No. 3, 2017, p. 04017005.
8. Sadri, A. M., S. V. Ukkusuri, and P. Murray-Tuite. A Random Parameter Ordered Probit Model to Understand the Mobilization Time during Hurricane Evacuation. *Transportation Research Part C: Emerging Technologies*, Vol. 32, 2013, pp. 21–30.
9. Sadri, A. M., S. V. Ukkusuri, P. Murray-Tuite, and H. Gladwin. How to Evacuate: Model for Understanding the Routing Strategies during Hurricane Evacuation. *Journal of Transportation Engineering*, Vol. 140, No. 1, 2013, pp. 61–69.
10. Sadri, A. M., S. V. Ukkusuri, P. Murray-Tuite, and H. Gladwin. Analysis of Hurricane Evacuee Mode Choice Behavior. *Transportation Research Part C: Emerging Technologies*, Vol. 48, 2014, pp. 37–46.
11. Sadri, A. M., S. V. Ukkusuri, P. Murray-Tuite, and H. Gladwin. Hurricane Evacuation Route Choice of Major Bridges in Miami Beach, Florida. *Transportation Research Record: Journal of the Transportation Research Board*, 2015. 2532: 164–173.
12. Sadri, A. M., S. V. Ukkusuri, S. Lee, R. Clawson, D. Aldrich, M. S. Nelson, J. Seipel, and D. Kelly. The Role of Social Capital, Personal Networks, and Emergency Responders in Post-Disaster Recovery and Resilience: A Study of Rural Communities in Indiana. *Natural Hazards: Journal of the International Society for the Prevention and Mitigation of Natural Hazards*, Vol. 90, No. 3, 2017, pp. 1377–1406.
13. Cutter, S. L., J. A. Ahearn, B. Amadei, P. Crawford, E. A. Eide, G. E. Galloway, M. F. Goodchild, H. C. Kunreuther, M. Li-Vollmer, and M. Schoch-Spana. Disaster Resilience: A National Imperative. *Environment: Science and Policy for Sustainable Development*, Vol. 55, No. 2, 2013, pp. 25–29.
14. Diffenbaugh, N. S., and C. B. Field. Changes in Ecologically Critical Terrestrial Climate Conditions. *Science*, Vol. 341, No. 6145, 2013, pp. 486–492.
15. Guikema, S. D. Infrastructure Design Issues in Disaster-Prone Regions. *Science*, Vol. 323, No. 5919, 2009, pp. 1302–1303.
16. Kennedy, D. Science, Terrorism, and Natural Disasters. *Science*, Vol. 295, No. 5554, 2002, pp. 405–405.
17. Press, F., and R. M. Hamilton. Mitigating Natural Disasters. *Science*, Vol. 284, No. 5422, 1999, p. 1927.
18. Cutter, S. L., and C. Finch. Temporal and Spatial Changes in Social Vulnerability to Natural Hazards. *Proceedings of the National Academy of Sciences*, Vol. 105, No. 7, 2008, pp. 2301–2306.
19. Helbing, D. Globally Networked Risks and How to Respond. *Nature*, Vol. 497, No. 7447, 2013, pp. 51–59.
20. Vespignani, A. Predicting the Behavior of Techno-Social Systems. *Science*, Vol. 325, No. 5939, 2009, pp. 425–428.
21. Lazer, D., A. S. Pentland, L. Adamic, S. Aral, A. L. Barabasi, D. Brewer, N. Christakis, N. Contractor, J. Fowler, and M. Gutmann. Life in the Network: The Coming Age of Computational Social Science. *Science*, Vol. 323, No. 5915, 2009, p. 721.
22. Bagrow, J. P., D. Wang, and A.-L. Barabasi. Collective Response of Human Populations to Large-Scale Emergencies. *PLoS One*, Vol. 6, No. 3, 2011, p. e17680.
23. Hughes, A. L., and L. Palen. Twitter Adoption and Use in Mass Convergence and Emergency Events. *International Journal of Emergency Management*, Vol. 6, No. 3–4, 2009, pp. 248–260.
24. Li, J., and H. R. Rao. Twitter as a Rapid Response News Service: An Exploration in the Context of the 2008 China Earthquake. *The Electronic Journal of Information Systems in Developing Countries*, Vol. 42, No. 1, 2010, pp. 1–22.
25. Van Hentenryck, P. Computational Disaster Management. *Proc., 23rd International Joint Conference on Artificial Intelligence (IJCAI 2013)*, Beijing, China, AAAI Press/International Joint Conference on Artificial Intelligence, Menlo Park, Calif., 2013, pp. 12–19.
26. Wang, D., Y.-R. Lin, and J. P. Bagrow. Social Networks inEmergency Response. In *Encyclopedia of Social Network Analysis and Mining* (A. Reda and J. Rokne eds.), Springer, 2014. pp. 1904–1914.
27. Watts, D., M. Cebrian, and M. Elliot. Dynamics of Social Media. In *Public Response to Alerts and Warnings Using Social Media: Report of a Workshop on Current Knowledge and Research Gaps* (N. R. Council, ed.), The National Academies Press, Washington, D.C., 2013.
28. Sadri, A. M., S. Hasan, and S. V. Ukkusuri. Joint Inference of User Community and Interest Patterns in Social Interaction Networks. *arXiv preprint arXiv:1704.01706*, 2017.
29. Sadri, A. M., S. Hasan, S. V. Ukkusuri, and J. E. S. Lopez. Analyzing Social Interaction Networks from Twitter for Planned Special Events. *arXiv preprint arXiv:1704.02489*, 2017.
30. Caragea, C., N. McNeese, A. Jaiswal, G. Traylor, H.-W. Kim, P. Mitra, D. Wu, A. H. Tapia, L. Giles, and B. J. Jansen. Classifying Text Messages for the Haiti Earthquake. *Proc., 8th International Conference on Information Systems for Crisis Response and Management (ISCRAM2011)*, Lisbon, Portugal, Citeseer, Vol. 1, 2011.
31. Earle, P. S., D. C. Bowden, and M. Guy. Twitter Earthquake Detection: Earthquake Monitoring in a Social World. *Annals of Geophysics*, Vol. 54, No. 6, 2012, 708–715.

32. Freeman, M. Fire, Wind and Water: Social Networks in Natural Disasters. *Journal of Cases on Information Technology*, Vol. 13, No. 2, 2011, pp. 69–79.
33. Guy, M., P. Earle, C. Ostrum, K. Gruchalla, and S. Horvath. Integration and Dissemination of Citizen Reported and Seismically Derived Earthquake Information via Social Network Technologies. In *International Symposium on Intelligent Data Analysis*, Springer, Berlin, Heidelberg, 2010. pp. 42–53.
34. Pickard, G., W. Pan, I. Rahwan, M. Cebrian, R. Crane, A. Madan, and A. Pentland. Time-Critical Social Mobilization. *Science*, Vol. 334, No. 6055, 2011, pp. 509–512.
35. Sakaki, T., M. Okazaki, and Y. Matsuo. Earthquake Shakes Twitter Users: Real-Time Event Detection by Social Sensors. *Proc., 19th International Conference on World Wide Web*, Raleigh, N.C., ACM, New York, 2010, pp. 851–860.
36. Skinner, J. Natural Disasters and Twitter: Thinking from Both Sides of the Tweet. *First Monday*, Vol. 18, No. 9, 2013. http://dx.doi.org/10.5210/fm.v18i9.4650.
37. Ukkusuri, S., X. Zhan, A. Sadri, and Q. Ye. Use of Social Media Data to Explore Crisis Informatics: Study of 2013 Oklahoma Tornado. *Transportation Research Record: Journal of the Transportation Research Board*, 2014. 2459: 110–118.
38. Sadri, A. M., S. Hasan, S. V. Ukkusuri, and M. Cebrian. Understanding Information Spreading in Social Media during Hurricane Sandy: User Activity and Network Properties. *arXiv preprint arXiv:1706.03019*, 2017.
39. Korolov, R., J. Peabody, A. Lavoie, S. Das, M. Magdon-Ismail, and W. Wallace. Actions are louder than Words in Social Media. *Proc., 2015 IEEE/ACM International Conference on Advances in Social Networks Analysis and Mining 2015*, Paris, France, ACM, New York, 2015, pp. 292–297.
40. Kryvasheyeu, Y., H. Chen, N. Obradovich, E. Moro, P. Van Hentenryck, J. Fowler, and M. Cebrian. Rapid Assessment of Disaster Damage Using Social Media Activity. *Science Advances*, Vol. 2, No. 3, 2016, p. e1500779.
41. Kaufman, S., C. Qing, N. Levenson, and M. Hanson. *Transportation During and After Hurricane Sandy*. Rudin Center for Transportation Policy & Management, New York, 2012, pp. 1–34.
42. Floating Sheep. *Floating Sheep*. http://www.floatingsheep.org/2012/10/the-urban-geographies-of-hurricane.html. Accessed July 29, 2016.
43. Drabek, T. E. Social Processes in Disaster: Family Evacuation. *Social Problems*, Vol. 16, No. 3, 1969, pp. 336–349.
44. Gladwin, H., J. K. Lazo, B. H. Morrow, W. G. Peacock, and H. E. Willoughby. Social Science Research Needs for the Hurricane Forecast and Warning System. *Natural Hazards Review*, Vol. 8, No. 3, 2007, pp. 87–95.
45. Lindell, M. K., J.-C. Lu, and C. S. Prater. Household Decision Making and Evacuation in Response to Hurricane Lili. *Natural Hazards Review*, Vol. 6, No. 4, 2005, pp. 171–179.
46. Lindell, M. K., and C. S. Prater. A Hurricane Evacuation Management Decision Support System (EMDSS). *Natural Hazards*, Vol. 40, No. 3, 2007, pp. 627–634.
47. Lindell, M. K., and C. S. Prater. Critical Behavioral Assumptions in Evacuation Time Estimate Analysis for Private Vehicles: Examples from Hurricane Research and Planning. *Journal of Urban Planning and Development*, Vol. 133, No. 1, 2007, pp. 18–29.
48. Myers, S. A., A. Sharma, P. Gupta, and J. Lin. Information Network or Social Network?: The Structure of the Twitter Follow Graph. *Proc., 23rd International Conference on World Wide Web*, Seoul, Korea ACM, New York, 2014. pp. 493–498.
49. Kogan, M., L. Palen, and K. M. Anderson. Think Local, Retweet Global: Retweeting by the Geographically-Vulnerable during Hurricane Sandy. *Proc., 18th ACM Conference on Computer Supported Cooperative Work & Social Computing*, Vancouver, B.C., ACM, New York, 2015. pp. 981–993.
50. Starbird, K., and L. Palen. Pass It On?: Retweeting in Mass Emergency. *Proc., 7th International Community on Information Systems for Crisis Response and Management*, Seattle, Washington, May 2–5, 2010.
51. Power, R., B. Robinson, J. Colton, and M. Cameron. Emergency Situation Awareness: Twitter Case Studies. In *International Conference on Information Systems for Crisis Response and Management in Mediterranean Countries*, Springer, Dordrecht, 2014. pp. 218–231.
52. Vieweg, S., A. L. Hughes, K. Starbird, and L. Palen. Microblogging during Two Natural Hazards Events: What Twitter May Contribute to Situational Awareness. *Proc., SIGCHI Conference on Human Factors in Computing Systems*, Atlanta, Ga., ACM, New York, 2010, pp. 1079–1088.
53. Hughes, A. L., L. A. St Denis, L. Palen, and K. M. Anderson. Online Public Communications by Police & Fire Services during the 2012 Hurricane Sandy. *Proc., 32nd Annual ACM Conference on Human Factors in Computing Systems*, Toronto, Ontario, ACM, New York, 2014, pp. 1505–1514.
54. St Denis, L. A., L. Palen, and K. M. Anderson. Mastering Social Media: An Analysis of Jefferson County's Communications during the 2013 Colorado Floods. *Proc., 11th International ISCRAM Conference*, ISCRAM, University Park, Pa., 2014. pp. 737–746.
55. Imran, M., S. M. Elbassuoni, C. Castillo, F. Diaz, and P. Meier. Extracting Information Nuggets from Disaster-Related Messages in Social Media. *Proc., ISCRAM*, Baden-Baden, Germany, 2013.
56. Kumar, S., X. Hu, and H. Liu. A Behavior Analytics Approach to Identifying Tweets from Crisis Regions. *Proc., 25th ACM Conference on Hypertext and Social Media*, Santiago, Chile, ACM, New York, 2014, pp. 255–260.
57. Kryvasheyeu, Y., H. Chen, E. Moro, P. Van Hentenryck, and M. Cebrian. Performance of Social Network Sensors During Hurricane Sandy. *PLoS One*, Vol. 10, No. 2, 2015, p. e0117288.
58. Wang, Q., and J. E. Taylor. Quantifying Human Mobility Perturbation and Resilience in Hurricane Sandy. *PLoS One*, Vol. 9, No. 11, 2014, p. e112608.
59. Wang, Q., and J. E. Taylor. Resilience of Human Mobility Under the Influence of Typhoons. *Procedia Engineering*, Vol. 118, 2015, pp. 942–949.
60. Hasan, S., and S. V. Ukkusuri. Urban Activity Pattern Classification Using Topic Models from Online Geo-

Location Data. *Transportation Research Part C: Emerging Technologies*, Vol. 44, 2014, pp. 363–381.
61. Hasan, S., and S. V. Ukkusuri. Location Contexts of User Check-Ins to Model Urban Geo Life-Style Patterns. *PLoS One*, Vol. 10, No. 5, 2015, p. e0124819.
62. Hasan, S., S. V. Ukkusuri, and X. Zhan. Understanding Social Influence in Activity-Location Choice and Life-Style Patterns Using Geo-Location Data from Social Media. *Frontiers in ICT*, Vol. 3, 2016, p. 10.
63. Hasan, S., X. Zhan, and S. V. Ukkusuri. Understanding Urban Human Activity and Mobility Patterns Using Large-Scale Location-Based Data from Online Social Media. *Proc., 2nd ACM SIGKDD International Workshop on Urban Computing*, Chicago, Ill., ACM, New York, 2013. p. 6.
64. Gao, H., G. Barbier, R. Goolsby, and D. Zeng. *Harnessing the Crowdsourcing Power of Social Media for Disaster Relief. IEEE Intelligent Systems*, Vol. 26, No. 3, pp.10–14.
65. Steyvers, M., and T. Griffiths. Probabilistic Topic Models. *Handbook of Latent Semantic Analysis*, Vol. 427, No. 7, 2007, pp. 424–440.
66. Griffiths, T. L., and M. Steyvers. Finding Scientific Topics. *Proceedings of the National Academy of Sciences*, Vol. 101, No. Suppl. 1, 2004, pp. 5228–5235.
67. Blei, D. M., A. Y. Ng, and M. I. Jordan. Latent Dirichlet Allocation. *Journal of Machine Learning Research*, Vol. 3, 2003, pp. 993–1022.
68. Robert, C. *The Bayesian Choice: From Decision-Theoretic Foundations to Computational Implementation*. Springer Science & Business Media, Berlin, Germany, 2007.
69. Sadri, A. M. *Social Network Influence on Ridesharing, Disaster Communication and Community Interaction*. PhD dissertation. Purdue University, West Lafayette, Indiana, USA, 2016.

The Standing Committee on Logistics of Disaster Response and Business Continuity (ABR20) peer-reviewed this paper (18-04173).

The authors are solely responsible for the facts and accuracy of the information presented in the paper.

Article

TRR
JOURNAL OF THE TRANSPORTATION RESEARCH BOARD

Transportation Research Record
2018, Vol. 2672(1) 138–147

DOI: 10.1177/0361198118797510
journals.sagepub.com/home/trr
SAGE

Application of Novel Recovery Techniques to Enhance the Resilience of Transportation Networks

Ning Zhang[1], Alice Alipour[1], and Laura Coronel[1]

Abstract

Resilience is an important characteristic of the transportation system. It reflects the network's ability to mitigate shocks, provide alternatives, and rapidly recover to a target performance level. Earthquakes can cause the transportation network to experience severe disruptions that significantly reduce network resilience. To prevent long-term closures after earthquakes, the development of innovative approaches for their rapid restoration is necessary. This paper uses the recent developments in accelerated bridge construction (ABC) techniques as a means to enhance the rapid recovery of the system. ABC techniques often come with increased initial construction costs. In most cases, the additional cost is offset by the improvement of the network performance. To examine the efficiency of ABC techniques on the rapid recovery and on the network performance after earthquakes, the direct and indirect costs during the entire recovery period were analyzed and the relationship between network recovery time and network performance was estimated under different construction techniques. Additionally, the effect of using the incentive method to reduce the repair time was studied. The results show that the use of ABC techniques and the incentive method have great potential to minimize the transportation network's indirect losses, improve network performance, increase the network's resilience by decreasing recovery time, and justify the additional initial costs associated with these techniques.

For society to function properly, it is essential to maintain the transportation infrastructure at high performance levels. Major disruption to the functionality of the network can lead to detrimental effects to the surrounding community. Seismic events can severely damage different components of the network, result in closures, and hamper access for emergency vehicles and the distribution of goods, and affect everyday community activities. For these reasons, it is paramount for transportation networks to remain functional or rapidly return to acceptable levels of functionality after a seismic event.

To date, there is neither a common agreement within the research community about the definition of resilience nor an agreed-upon measurement system for resilience (*1–4*). Brunaeu et al. defined resilience as the reduction in (1) probability of failure, (2) consequences from failures, and (3) recovery time (*5*). Based on this, four dimensions of resilience were defined: robustness, redundancy, resourcefulness, and rapidity (R4). Robustness is the ability of the structure or network to endure stresses and ensure minimal loss; redundancy is the ability to provide alternative routes to compensate for the decrease in a network's capacity; resourcefulness is the ability prioritize restoration actions and to identify an action depending on the resources available; rapidity is the speed at which recovery can reach a target value.

Although resilience is a multi-dimensional property, most of the studies to date have focused primarily on robustness. This includes multitudes of approaches to bolster bridges against earthquakes. However, it is impractical to strengthen every part of a network due to the wide array of possible events (both intensity and frequency) and limited budgetary resources (*6*, *7*). Couple that with the sheer size of the transportation network and it becomes evident that transportation agencies must tolerate some failure in the system and plan for faster recovery to achieve enhanced resiliency. With continued recognition of the value of recovery, methods are required that allow for the rapid recovery of the transportation system. Considering that bridges are thought

[1]Department of Civil, Construction and Environmental Engineering, Iowa State University, Ames, IA

Corresponding Author:
Address correspondence to Alice Alipour: alipour@iastate.edu

to be the most vulnerable components of the transportation system in seismic events, one aspect of recovery could be focused on the development of techniques that would result in the faster restoration of damaged bridges.

Accelerated bridge construction (ABC) is a construction technique that reduces onsite construction time when building new or replacing old bridges (*8*). ABC techniques can be categorized as foundation and wall elements, rapid embankment constructions, prefabricated bridge element and systems (PBES), self-propelled modular transporters, slide-in constructions, longitudinal launching, horizontal sliding or skidding, and fast-tracked contracting. ABC techniques combined with the "A" + "B" (bid items + critical duration) bidding method (*9*) may contribute to shorter duration of restoration activities. "A" means to minimize bridge repair costs, and "B" aims to reduce and control bridge closure time. Though effective, ABC techniques and the incentive method are costly and ABC techniques cannot universally be applied to every bridge type because of the technical limitations on prefabrication and onsite construction. Therefore, a holistic network-level analysis is required to implement ABC techniques properly throughout the network to decrease the recovery time. To measure the efficiency of the aforementioned accelerated recovery techniques, the time to reach a target threshold was defined as a measure of resilience. At the same time, a series of techniques (i.e., ABC, Incentives, ABC + Incentives) that could potentially speed up the restoration processes was simulated, and performance measures such as travel delay and opportunity loss were used to evaluate the effectiveness of the proposed techniques in enhancing resilience of the system.

Focusing on earthquakes as the extreme events to be considered, there are different aspects that need to be contemplated for the risk and consequence analysis: earthquakes vary spatially in intensity and frequency, the effects they have on different infrastructure components are different, and there are dependencies between the loads that adjacent components experience. For large infrastructure systems that are regionally distributed, the loading intensity needs to be characterized at a great number of locations. Researchers have used seismic risk analysis to describe the degree of damage that could occur after an earthquake (*10–13*). For example, Werner et al. developed REDARS software for seismic risk analysis (*14*). In addition to seismic risk analysis, several research efforts have focused on designing efficient retrofit strategies before disasters to enhance network resilience through enhancing robustness (*15–17*). A few studies have focused on measuring network resilience using topology-based approaches and assessing post-event losses during the recovery time (*18–22*).

In each of the listed research areas, appropriate performance measures must be employed to evaluate the post-event functionality of the network and identify the required recovery strategies. Three commonly used measures are flow capacity, travel time, and network connectivity. Flow capacity is independent of origin–destination (O-D) data or the behavior of drivers, and can be used to directly quantify the extent of damage to the network after an earthquake. This measure represents the largest possible flow between the origin and destination nodes without exceeding the capacity of corresponding links. Travel time is another functionality measure that has been widely employed to estimate the performance level of a damaged network. The calculation of this measure requires O-D data, which can be accessed from surveys or mathematical models. The third functionality measure is network connectivity, which can be defined as the ability to get from one node to another. The connectivity measure is mostly suitable for immediate post-earthquake emergency response and disregards network capacity, travel time, and trip length.

Any disruption in the functionality of the network may hinder economic activities in the region and result in socio-economic losses. Indirect losses, such as those related to excessive traffic delay and opportunity loss (*19*), are more difficult to estimate compared with direct losses due to bridge damage. Furthermore, the state of the system after an earthquake changes over time as the restoration strategies are implemented. Although losses caused by delays are relatively simple to calculate based on the value of time assigned to each of the constituents, losses due to the constraints that affect traffic flow are more complicated to quantify. For instance, loss of accessibility could create problems for employees and may result in loss of productivity and increased overheads. Such impacts are more noticeable when an earthquake occurs in industrial or urban areas.

In the current study, the bridge/road network of the San Francisco Bay Area was developed: the bridges were assigned to different categories, and their vulnerability was measured through fragility functions. A specific scenario earthquake event was considered, and the distribution of seismic intensity was measured at the bridge locations. Damage to the bridges was translated into loss of functionality of the links. A traffic analysis model was established to quantify the direct and indirect costs including drivers' delay and opportunity loss. Considering the direct impact of the closure durations on the indirect costs of the project, application of accelerated construction techniques such as ABC and innovative contracting methods (such as incentives) were used to enhance the speed of recovery. Network performance and costs with and without ABC techniques or the innovative contracting method were compared.

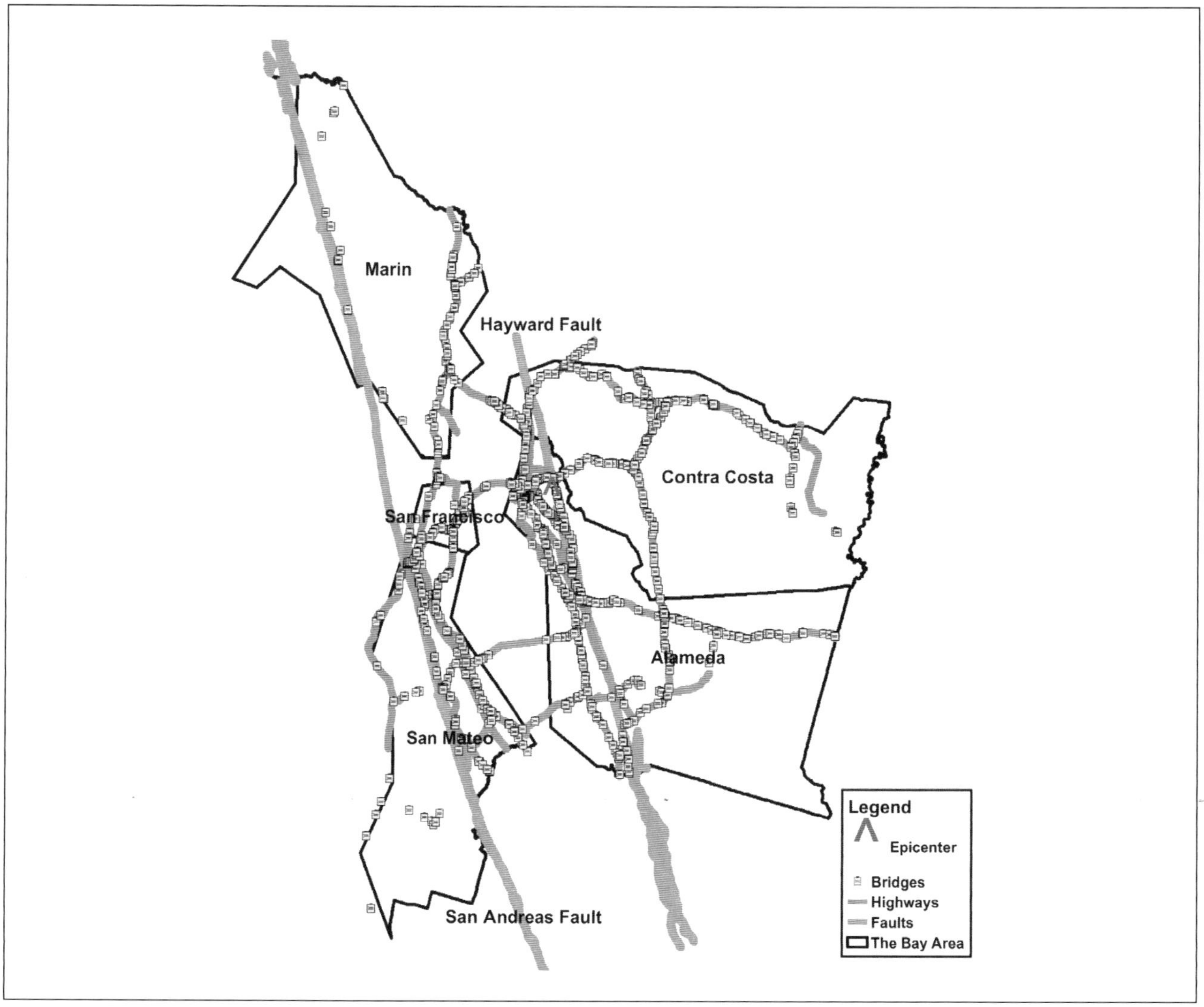

Figure 1. Map of San Francisco Bay Area with the two major faults and location of bridges and highways.

Fragility Analysis of the Components of the Test Bed Network

To demonstrate the capabilities of the developed approach, the San Francisco Bay Area network was tested. This area has high seismicity, a large population, and high dependency on the transportation network. The San Andreas Fault and the Hayward Fault are located in the region and have been the source of historical earthquakes that have resulted in damage to the transportation network. In the current study, an earthquake on Hayward Fault was simulated. Figure 1 shows the locations of the faults with respect to the highways and bridges in the test bed. This earthquake scenario was chosen to present the methodology framework. For a satisfactory decision-making process, a suite of earthquakes in the target region should be considered. The earthquake in the current case spread through the Hayward Fault and resulted in different levels of seismic intensities at the bridges' locations. The peak ground acceleration (PGA) is the intensity measure at the location of the bridges (Equation 1):

$$ln(PGA_H) = -3.512 + 0.904M - 1.328 ln\sqrt{{d_{\text{sitesource}}}^2 + (0.149e^{0.647M})^2} + c[1.125 - 0.112 ln d_{\text{sitesource}} - 0.0957M] \quad (1)$$

where PGA_H is the peak ground acceleration on horizontal direction; the $d_{\text{sitesource}}$ is the shortest distance between the epicenter and the location of each bridge (Equation

Table 1. Definition of Bridge and Link Damage States

Bridge damage BD_i	Definition	Post-event utility of bridge	Repairing requirement	Expected closure time	Critical relationship	Link damage LD_i	Link residual capacity
0	None	Normal	None	0	$SD_i > F_1$	0	100%
1	Minor	Slight damage	Inspect, adjust, patch	<3 days	$F_2 < SD_i \leq F_1$	1	100%
2	Moderate	Repairable damage	Repair components	<3 weeks	$F_3 < SD_i \leq F_2$	2	25%
3	Extensive	Irreparable damage	Rebuild components	<3 months	$F_4 < SD_i \leq F_3$	3	0
4	Complete	Irreparable damage	Rebuild structure	≥3 months	$SD_i \leq F_4$	4	0

2); ε is a random error term related to the $ln(PGA_H)$; and $c = 1$, if the fault type is strike slip, $c = 0$ otherwise:

$$d_{\text{sitesource(i)}} = \sum_{n=0}^{N} \frac{1}{2}\left[d_{\text{sitestart(i)}} - (n-1)d_{\text{startend(i)}} \cos\alpha\right] \quad (2)$$

where $d_{\text{sitestart(i)}}$ is the distance between bridge i and the fault rupture's start point; $d_{\text{startend(i)}}$ is the fault's rupture length; N is the number of rupture pieces; and α is the angle between bridge i to the rupture's start point and the rupture's start point to its end.

Using Equation 1, the seismic intensity measures were estimated at each bridge's location. The fragility estimates were then used to evaluate the extent of the damage to the bridges. Fragility curves present the cumulative probabilities of failures as a function of the ground motion intensity (PGA) (Equation 3) (*23–25*):

$$F(a_{\text{ik}}) = \Phi\left[\frac{ln\left(\frac{a_{\text{ik}}}{c_{\text{k}}}\right)}{\xi_k}\right] \quad (3)$$

where $F(a_{\text{ik}})$ is the cumulative probability of bridge i under damage state k; Φ is the standardized log-normal cumulative distribution function; a_{ik} is the value of PGA_H of bridge i under damage state k; and c_k, ζ_k are the median and log-standard deviations respectively, maximizing the likelihood function of damage state k.

The definition of the damage states was extracted from the HAZUS-MH (2012) (*26*) and is categorized as no damage, minor, moderate, extensive, and complete damage. When a bridge can potentially be damaged (SD_i) with a specific PGA_H value, the bridge damage state BD_i can be predicted based on the relationships in Table 1. Here F_1, F_2, F_3, and F_4 are values from the fragility curve at the PGA_H value point, representing the lower and upper bounds of the variable damage states. Bridges are sensitive to ground motion and are usually the most seriously disrupted elements of the transportation networks. The link damage state, LD_i, is defined as being equal to the damage state of the worst damaged bridge on the link.

Network-Level Assessment

The network analysis method used in the current study is based on the four step model (*27*). It provides an algorithm to determine the equilibrium of traffic flows in the network, and consists of four steps: trip generation, trip distribution, mode choice, and trip assignment. To calculate the trip rate for a specific trip purpose, the gravity model and the reduction model were considered (*27, 28*). The developed model in this paper is dynamic as it considers the travel fluctuations in the post-event travel demand. This is because after a major earthquake some travelers forgo their trips, due to a multitude of reasons such as extensive damage to the transportation network and buildings that house different activities. To make the travel demands more representative after an earthquake, Shinozuka et al. generated a relationship between the intensity of the seismic event and a reduction rate that could be applied on trip distribution to estimate the O-D data after earthquakes (*28*). Equations 4 and 5 are introduced to provide for reduced traffic demands:

$$Att_{\text{i,post}} = Att_{\text{i,pre}}(1 - \xi_{\text{i}}) \quad (4)$$

$$Pro_{\text{i,post}} = Pro_{\text{i,pre}}(1 - \xi'_{\text{i}}) \quad (5)$$

where $Att_{\text{i,post}}$, and $Pro_{\text{i,post}}$ are the post-event reduced trip attraction and production from node i; $Att_{\text{i,pre}}$ and $Pro_{\text{i,pre}}$ are the original trip attraction and production from node i; and ξ_i and ξ_i' are the reduction rates based on the intensity of the earthquake experienced at each node.

Figure 2 presents the procedure used for the network analysis.

It is assumed that network repairs start right after the occurrence of earthquakes. As the bridges are restored, the link capacities and traffic demands increase. Because of these dynamic traffic characteristics, the transportation analysis model needs to be run repeatedly in set time-frames until the whole network is restored to a target functionality level. The target functionality in this study is considered as 100% restoration of service in the network.

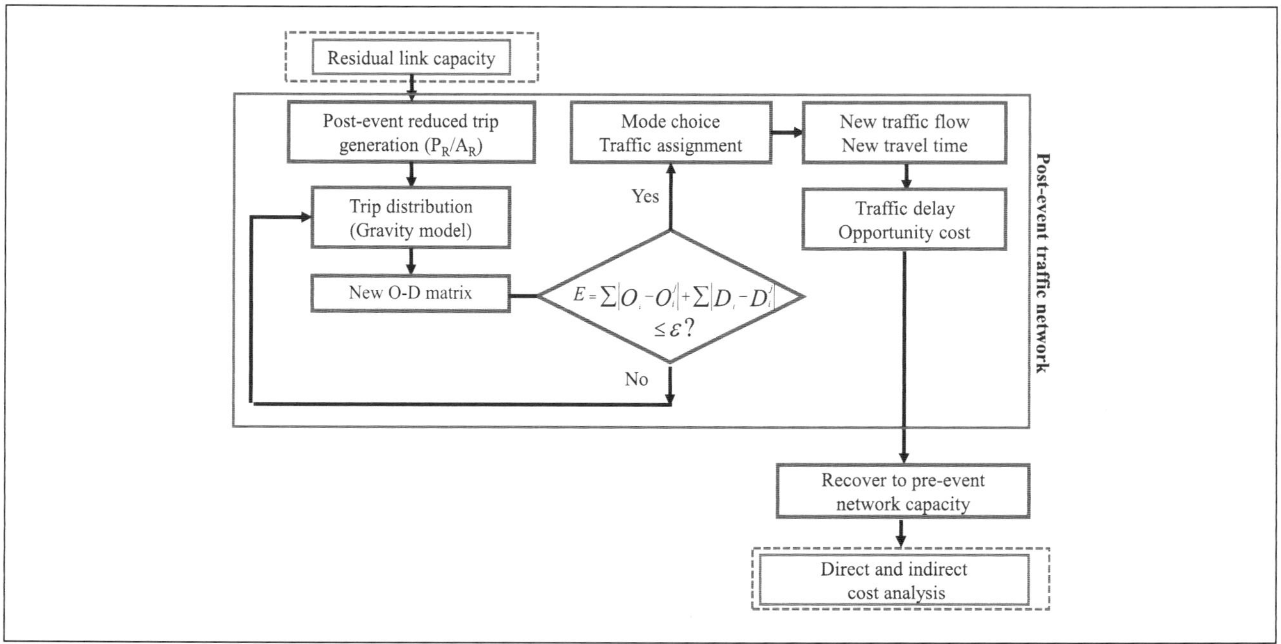

Figure 2. Post-event traffic network model.

Network Performance Measures

The consequences of network disruption can be (a) the direct costs associated with the replacement of an asset (assuming that the downtime is negligible), (b) the indirect costs to the users of the affected asset (such as increased travel time or longer detours in medium-range downtime), or even (c) the economic costs to society given a longer duration of disruption in a transportation system serving a community (for cases in which downtime is extended). The direct costs consist of conventional construction costs, the additional cost of ABC techniques, and the cost of the incentive method (Equation 6). The indirect cost consists of the traffic delay and the opportunity cost (equations 7 to 9):

$$C_{\text{direct}} = C_{\text{con}} + Z_1 A_{\text{ABC}} + Z_2 A_{\text{inc}} \tag{6}$$

where C_{direct} is the total direct cost of the whole recovery period; C_{con} is the conventional construction cost; A_{ABC} is the additional cost of using ABC techniques; and A_{inc} is the cost of using incentive methods.

If any ABC technique is adopted, $Z_1 = 1$, otherwise $Z_1 = 0$. If the incentive method is applied, $Z_2 = 1$, otherwise $Z_2 = 0$:

$$C_{\text{indirect}} = C_{\text{del}} + C_{\text{opp}} \tag{7}$$

$$C_{\text{del}} = \left(\sum_i f'_i t'_i - \sum_i f_i t_i\right) \tag{8}$$

$$C_{\text{opp}} = \left\{\sum_i \sum_j \left[\frac{(T^{\text{p}}_{\text{ij}} - T'^{\text{p}}_{\text{ij}})\left(t'_{\text{ij}} - t_{\text{ij}}\right)}{2}\right]\right\} V_{\text{time}} \tag{9}$$

where C_{indirect} is the total indirect cost; f_i and f'_i are the pre- and post-event traffic flows on link i; t_i and t'_i are the pre- and post-event travel times on link i; T^{p}_{ij} and $T'^{\text{p}}_{\text{ij}}$ are the pre- and post-event trips on type p from node i to node j; t_{ij} and t'_{ij} are the pre- and post-event travel times from node i to node j; and V_{time} is a monetary value that the unit travel time values. This is assumed to be \$12/hour for the state of California.

The use of ABC techniques for recovery is limited to the critical bridges that experience high levels of damage and cause the worst traffic cost after earthquakes. There are many network characteristics in graph theory such as node degree, average shortest path, and clustering coefficient that could be used to prioritize nodes or subgraphs. In this study betweenness centrality was used to rank the most important elements in a network system (*29*):

$$BC_l = \sum_{i \neq j \neq l} \frac{SP^{(l)}_{\text{ij}}}{SP_{\text{ij}}} \tag{10}$$

where BC_l is the betweenness centrality of link l; $SP^{(l)}_{\text{ij}}$ is the number of shortest paths between node i and j through link l; and SP_{ij} is the total number of shortest paths between node i and j.

Network performance during the recovery time is evaluated by the residual percentage of the network performance. Loss is defined as the ratio of the total traffic cost to the initial total vehicle hours traveled (VHT) of the network system (Equation 11):

$$P_{\mathrm{it}} = 1 - \frac{Tcost_{\mathrm{it}}}{VHT_{\mathrm{Initial}}} \qquad (11)$$

where P_{it} is the residual network performance with measure i at time t during recovery time; $Tcost_{\mathrm{it}}$ is the total traffic cost with measure i at time t during its recovery time; and VHT_{Initial} is the VHT before earthquakes.

To present the post-event network performance and the recovery condition through using ABC techniques and the incentive method, the current paper simulated four bridge repair strategies: conventional construction (Conventional), using ABC techniques for the top 30% of the most critical bridges (0.3ABC), using incentives on the top 30% of the most critical bridges (0.3Incentive) and using both ABC techniques and the incentive method for the top 30% of the most critical bridges (0.3ABC + 0.3Incentive). Additionally, the conventional scenario means the network is repaired using conventional construction techniques. Incentivization means the contractor is rewarded if projects are finished ahead of schedule and penalized if any delays take place.

Results and Discussion

Under an earthquake with a moment magnitude of 7.5, the level of intensity measures were estimated at each bridge's location. By comparing these values with the bridges' fragility curves, the bridge and link damage states were calculated. Figure 3 shows the state of damage in the bridges and links under one of the simulations. After the earthquake and before the initiation of the restoration processes, the average number of bridges in damage state 4 (DS4) and 3 (DS3) was 575 and 89, respectively. As the restoration of bridges takes place, the number of damaged bridges decreases until the network is fully recovered. Figure 4 shows the recovery process for the bridges in different damage states and for the bridges in DS4 under different scenarios of recovery. With the implementation of ABC techniques and the incentive method, the number of damaged bridges in DS4 decreases at a faster rate.

According to the California Department of Transportation (*30*), the additional construction cost of ABC techniques is approximately 30% of the initial construction cost. The incentive cost varies depending on the bridge types and the construction company. Therefore, the total direct cost of each scenario should be the sum of conventional construction costs, the additional cost of implementing ABC techniques, and the cost of incentives. The total direct cost for conventional construction and the scenario using ABC techniques and incentives on the top 30% of bridges ranges from $1,363.7 million to $1,614.5 million. Figure 5 shows the state of damage in bridges and links 200 days and 400 days after start of the recovery process.

Figure 6 shows the continuous network performance improvement during the recovery time. Immediately after the earthquake, the network is heavily damaged and loses nearly 75% of its performance. Using the ABC techniques and the incentive method, the network

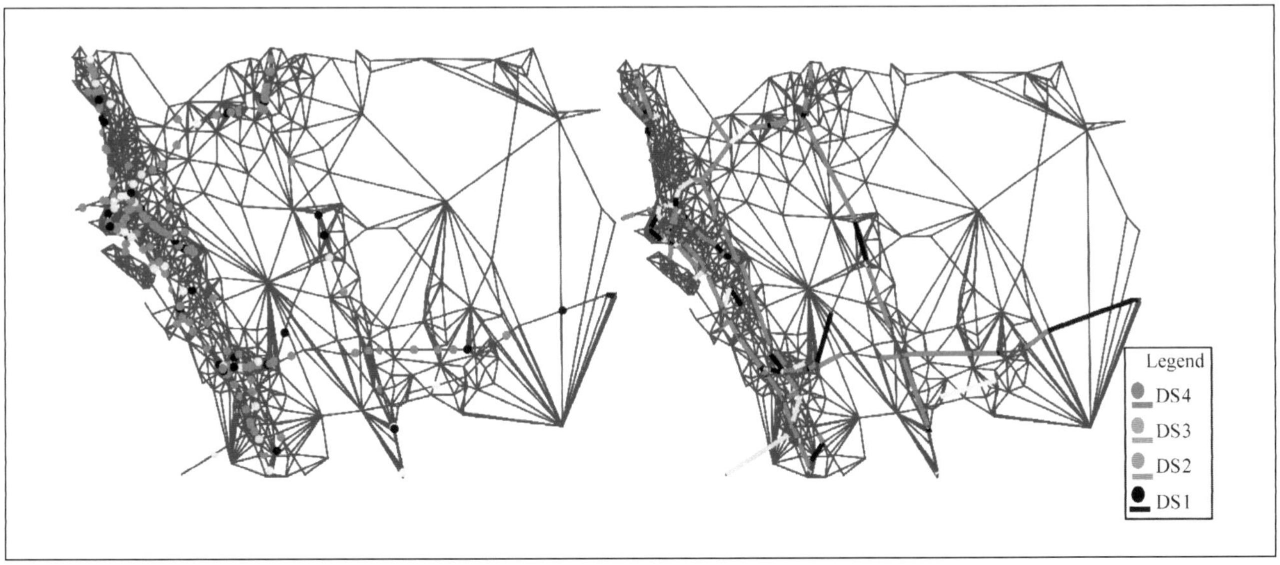

Figure 3. Bridge (left) and link damage (right) states after earthquake.

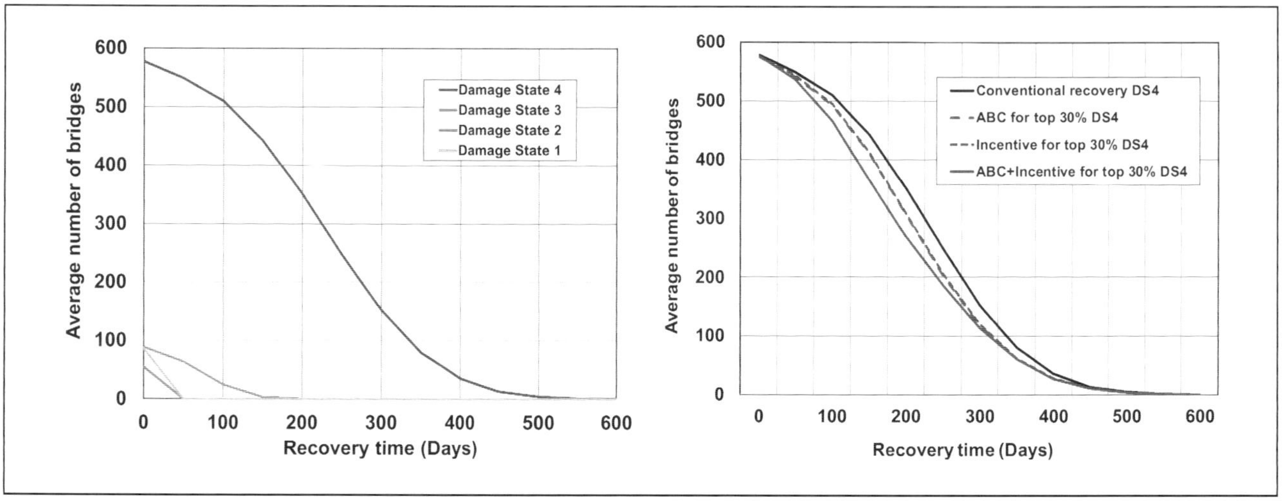

Figure 4. Number of bridges in each damage state (left) and under different recovery scenarios (right) as restoration takes place.

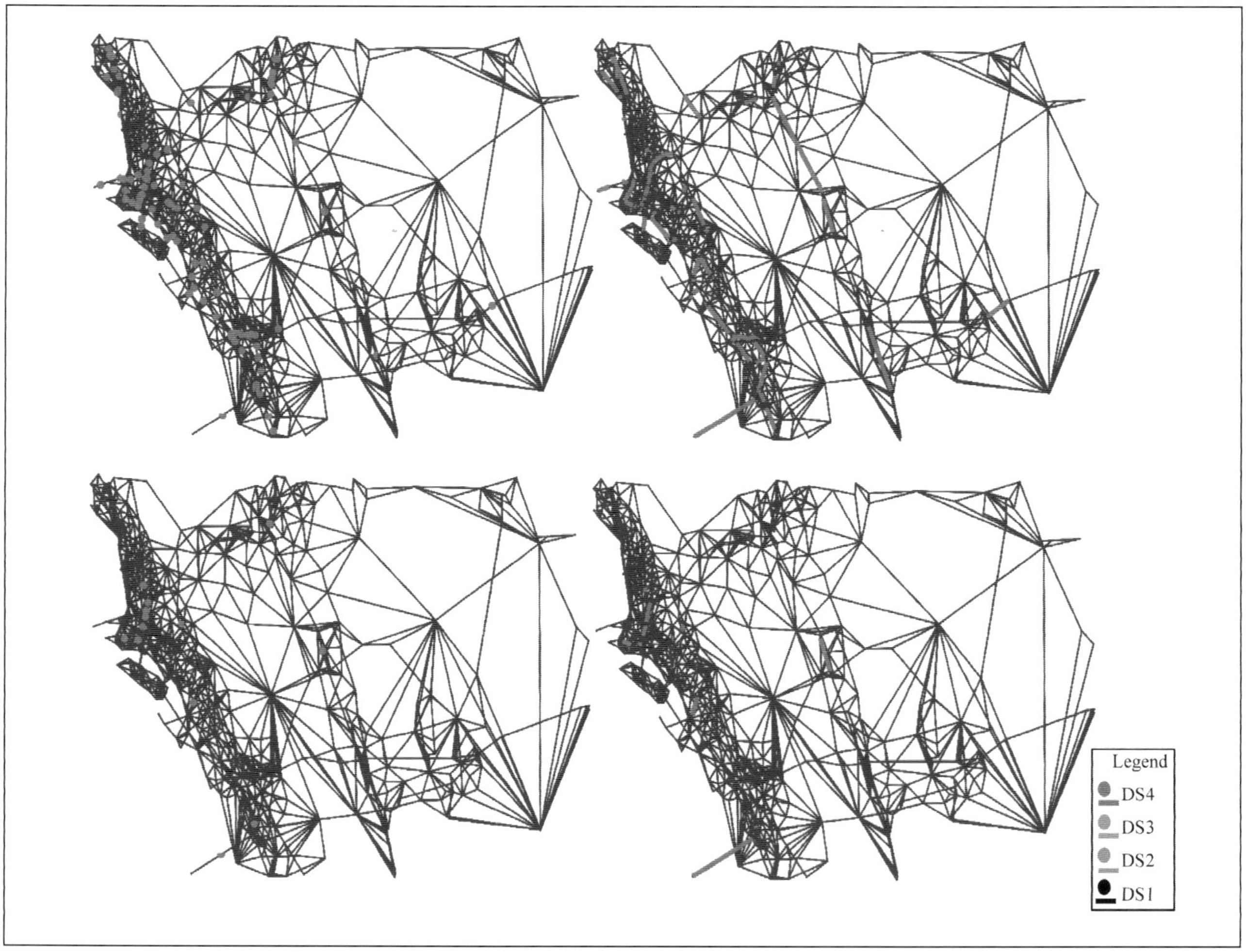

Figure 5. Bridge and link damage states at 200 days (top) and 400 days (bottom) after the start of recovery.

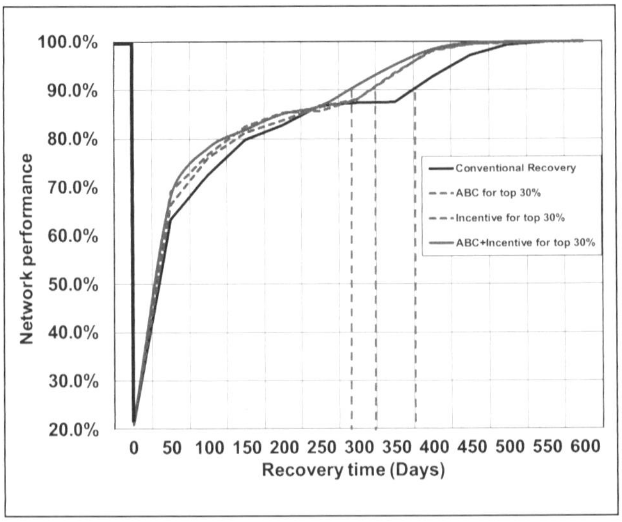

Figure 6. Network performance during recovery process with different repair strategies.

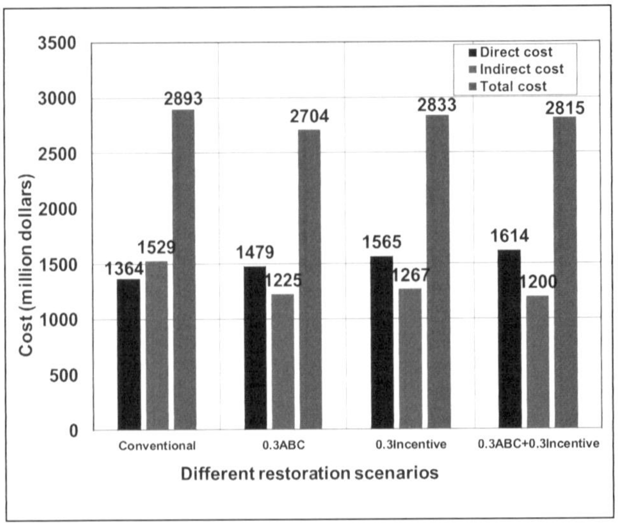

Figure 7. Direct, indirect, and total costs associated with different restoration scenarios.

performance recovers faster than the one with conventional techniques. Additionally, Figure 6 shows that conventional construction techniques require 375 days to achieve a 90% network performance. Though using both ABC and incentives for the top 30% of bridges takes less than 280 days. The time that is saved through the recovery process helps decrease the indirect costs associated with the network closures. Figure 7 presents the data for direct, indirect, and total costs for the different scenarios.

Applicability of the Method to Other Regions

Application of the developed methodology could be extended to other regions exposed to dangers other than seismic hazards. The procedure would require identification of the hazard, the effect it has on the components of the system (e.g., bridges or roadways), and the intensity and frequency of the hazard. Then the selected assets should undergo a vulnerability analysis process. Similar to the procedure provided in this paper, vulnerability could be measured in terms of the fragility of the component as it is exposed to different hazard intensity levels. A similar set of information would be required to develop the transportation network model of the study area and conduct the traffic analysis pre- and post-event (*7*). For analysis of direct and indirect costs, the values specific to the study region need to be adopted. The ABC application should follow the standard practices relevant to the study region. An example extension of topography-based analysis (compared with flow-based analysis in this paper) was established by the second author for the New York City region under the effects of a hurricane-induced storm surge (*18*).

Conclusion

Resilience represents the capacity of the system to absorb shocks (robustness), provide alternatives (redundancy), and recover rapidly (resourcefulness and rapidity). However, the majority of efforts to bolster the resilience of transportation systems to extreme events have set off on a path that primarily emphasizes robustness. This paper aimed to approach the problem of enhancing the resilience of the transportation network from a different perspective, that is, to accept that not all vulnerabilities can be mitigated and that a true resilient system should be able to recover rapidly despite possible failures. To achieve this goal the use of ABC techniques as a means to speed up recovery processes was proposed in this paper. It has been shown that ABC techniques combined with incentivizing processes can decrease the total repair time by up to 60%, which greatly improves rapid network recovery.

Acknowledgments

The research study results reported in this paper were partially sponsored by the Accelerated Bridge Construction University Transportation Center at Florida International University with Iowa State University and University of Nevada-Reno as partner universities. The authors acknowledge and thank the sponsors for their support.

Author Contributions

The authors confirm contribution to the paper as follows: study conception and design: AA; data collection: NZ, AA; analysis and interpretation of results: NZ, AA; draft manuscript preparation: NZ; parts of literature review: LC. All authors reviewed the results and approved the final version of the manuscript.

References

1. Holling, C. S. Resilience and Stability of Ecological Systems. *Annual Review of Ecology and Systematics*, Vol. 4, 1973, pp. 1–23.
2. Godschalk, D. R., A. Rose, E. Mittler, K. Porter, and C. T. West. Estimating the Value of Foresight: Aggregate Analysis of Natural Hazard Mitigation Benefits and Costs. *Journal of Environmental Planning and Management*, Vol. 52, No. 6, 2009, pp. 739–756. https://doi.org/10.1080/09640560903083715.
3. Folke, C. Resilience: The Emergence of a Perspective for Social–Ecological Systems Analyses. *Global Environmental Change*, Vol. 16, No. 3, 2006, pp. 253–267.
4. McCarthy, I. P., M. Collard, and M. Johnson. Adaptive Organizational Resilience: An Evolutionary Perspective. *Current Opinion in Environmental Sustainability*, Vol. 28, Suppl. C, 2017, pp. 33–40.
5. Bruneau, M., S. E. Chang, R. T. Eguchi, G. C. Lee, T. D. O'Rourke, A. M. Reinhorn, M. Shinozuka, K. Tierney, W. A. Wallace, and D. von Winterfeldt. A Framework to Quantitatively Assess and Enhance the Seismic Resilience of Communities. *Earthquake Spectra*, Vol. 19, No. 4, 2003, pp. 733–752.
6. Douglas, E., J. Jacobs, K. Hayhoe, L. Silka, J. Daniel, M. Collins, A. Alipour, B. Anderson, C. Hebson, E. Mecray, and R. Mallick. Progress and Challenges in Incorporating Climate Change Information into Transportation Research and Design. *Journal of Infrastructure Systems*, Vol. 23, No. 4, 2017, p. 4017018.
7. Alipour, A. Enhancing Resilience of Bridges to Extreme Events by Rapid Damage Assessment and Response Strategies. *Transportation Research Record: Journal of the Transportation Research Board*, 2017. 2604: 54–62.
8. Culmo, M. P. *Accelerated Bridge Construction - Experience in Design, Fabrication and Erection of Prefabricated Bridge Elements and Systems*. U.S. Department of Transportation, Federal Highway Administration, McLean, VA, 2011.
9. Prefabricated Bridge Elements and Systems. *PBES Cost Study: Accelerated Bridge Construction Success Stories*. Federal Highway Administration, Washington, D.C., 2006.
10. Adachi, T., and B. Ellingwood Impact of Infrastructure Interdependency and Spatial Correlation of Seismic Intensities on Performance Assessment of a Water Distribution System. Presented at 10th International Conference on Applications of Statistics and Probability in Civil Engineering, Tokyo, Japan, 2007.
11. Jayaram, N., and J. Baker. Efficient Sampling and Data Reduction Techniques for Probabilistic Seismic Lifeline Risk Assessment. *Earthquake Engineering & Structural Dynamics*, Vol. 39, No. 10, 2010, pp. 1109–1131.
12. Park, J., P. Bazzurro, and J. W. Baker. Modeling Spatial Correlation of Ground Motion Intensity Measures for Regional Seismic Hazard and Portfolio Loss Estimation. In J. Kanda, T. Takada, and H. Furuta (eds.), *Proc. 10th International Applications of Statistics and Probability in Civil Engineering Conference*, Tokyo, Japan, CRC Press, Boca Raton, FL, 2007. http://citeseerx.ist.psu.edu/viewdoc/download?doi = 10.1.1.452.9728&rep = rep1&type = pdf.
13. Shinozuka, M., Y. Murachi, X. Dong, Y. Zhou, and M. J. Orlikowski. Effect of Seismic Retrofit of Bridges on Transportation Networks. *Earthquake Engineering and Engineering Vibration*, Vol. 2, No. 2, 2003, pp. 169–179.
14. Werner, S. D., C. E. Taylor, S. Cho, J. Lavoie, C. Huyck, C. Eitzel, H. Chung, and R. T. Eguchi. *Redars 2 Methodology and Software for Seismic Risk Analysis of Highway Systems*. MCEER, Buffalo, NY, 2006.
15. Liu, C., Y. Fan, and F. Ordóñez. A Two-Stage Stochastic Programming Model for Transportation Network Protection. *Computers & Operations Research*, Vol. 36, No. 5, 2009, pp. 1582–1590.
16. Fan, Y., and C. Liu. Solving Stochastic Transportation Network Protection Problems using the Progressive Hedging-Based Method. *Networks and Spatial Economics*, Vol. 10, No. 2, 2010, pp. 193–208.
17. Chen, L., and E. Miller-Hooks. Resilience: An Indicator of Recovery Capability in Intermodal Freight Transport. *Transportation Science*, Vol. 46, No. 1, 2011, pp. 109–123.
18. Testa, A. C., M. N. Furtado, and A. Alipour. Resilience of Coastal Transportation Networks Faced with Extreme Climatic Events. *Transportation Research Record: Journal of the Transportation Research Board*, 2015. 2532: 29–36.
19. Furtado, M., and A. Alipour. Cost Assessment of Highway Bridge Network Subjected to Extreme Seismic Events. *Transportation Research Record: Journal of the Transportation Research Board*, 2014. 2459: 29–36.
20. Alipour, A., and B. Shafei. Assessment of Postearthquake Losses in a Network of Aging Bridges. *Journal of Infrastructure Systems*, Vol. 22, No. 2, 2016, p. 4015023.
21. Alipour, A., and B. Shafei. Seismic Resilience of Transportation Networks with Deteriorating Components. *Journal of Structural Engineering*, Vol. 142, No. 8, 2016, p. C4015015.
22. Testa, A., M. Furtado, and A. Alipour. Seismic Resilience Assessment of Highway Networks: A Topology-Based Approach. *Proc. of the 2nd European Conference on Earthquake Engineering and Seismology, The Eurpoean Association for Earthquake Engineering*, Istanbul, Turkey, 2014.
23. Alipour, A. Life-Cycle Performance Assessment of Highway Bridges under Multi-Hazard Conditions and Environmental Stressors. PhD Dissertation, University of California, Irvine, CA, 2010.
24. Alipour, A., B. Shafei, and M. S. Shinozuka. Capacity Loss Evaluation of Reinforced Concrete Bridges Located in Extreme Chloride-Laden Environments. *Structure and Infrastructure Engineering*, Vol. 9, No. 1, 2013, pp. 8–27.
25. Shinozuka, M., M. Q. Feng, H. Kim, T. Uzawa, and T. Ueda. *Statistical Analysis of Fragility Curves*. Technical Report MCEER-03-0002. State University of New York, Buffalo, NY, 2003, p. 192.
26. Federal Emergency Management Agency [FEMA] Hazus-MH–MH 2.0. Earthquake Model Technical Manual. FEMA, Mitigation Division, Washington, D.C., 2012.

27. Ortuzar, J., D. de, and L. G. Willumsen. *Modelling Transport*. John Wiley & Sons Ltd, Chichester, UK, 1994.
28. Shinozuka, M., Y. Zhou, S. Kim, Y. Murachi, S. Banerjee, S. Cho, and H. Chung. *Socio-Economic Effect of Seismic Retrofit Implemented on Bridges in the Los Angeles Highway Network*. Report No. CA F/CA/SD-2005/03. California Department of Transportation, Sacramento, CA, 2005.
29. Rodrigue, J. P., C. Claude, and S. Brian. *The Geography of Transport Systems*, 4th ed. Routledge, New York, NY, 2013.
30. Furtado, M. N., and A. Alipour. Enhancement of Post-Earthquake Resilience of Bridge Networks with Prioritization and Accelerated Repair Strategies. In *Structures Congress 2014*, Boston, MA, pp. 2014–2024. https://doi.org/10.1061/9780784413357.177.

The Transportation Systems Resilience Section (ABR00) peer-reviewed this paper (18-06760).

Article

TRR
JOURNAL OF THE TRANSPORTATION RESEARCH BOARD

Transportation Research Record
2018, Vol. 2672(1) 148–160

DOI: 10.1177/0361198118799169
journals.sagepub.com/home/trr
SAGE

Evaluation of the Traffic Impacts of Mass Evacuation of Halifax: Flood Risk and Dynamic Traffic Microsimulation Modeling

MD Jahedul Alam[1], Muhammad Ahsanul Habib[2], Kevin Quigley[3], and Tim L. Webster[4]

Abstract

This paper presents a comprehensive evaluation of traffic impacts of a mass evacuation of the Halifax Peninsula under several flooding scenarios. Flood extent and associated damages to the transport network are identified through digital elevation modeling that intersects with the Halifax stream and transport networks. The resulting flood scenarios inform a traffic microsimulation model that uses a dynamic traffic assignment-based microsimulation approach and simulates the evacuation of 34,808 evacuees estimated from the Halifax Regional Transport Network Model. The simulation results suggest that flooding of the links by 7.9 m flood reduces alternative evacuation routes by 31.2%. It takes 15 hours to evacuate 83% of evacuees while the remaining 17% are not accommodated in the network due to reduced network capacity. The number of vehicles in the network has peaked at 13,000 in this flooding scenario. An evaluation of network performance reveals a sustained congestion prevailing from 4th to 7th hour of the evacuation. The novelty of this study is that it develops a comprehensive tool of flood risk and dynamic traffic microsimulation modeling to offer an in-depth evaluation of potential impacts during evacuation. The results will help emergency professionals in evacuation planning and making emergency decisions.

Extreme weather events are becoming more frequent and flooding is the most common type of natural disasters among them in Canadian cities. Examples include the Toronto flood of 2013, the Halifax winter storm of 2015, and Hurricane Juan in 2003. Regardless of the type of disaster, evacuation is considered an effective way to save lives when the affected zone is not safe for habitation. Safely evacuating citizens to secure destinations is critical during any natural or human-made disaster. City evacuation is complex and a highly uncertain undertaking. Particularly, it is hard to know how traffic would flow through the disrupted transport network, and how severe the degree of resulting traffic congestion would be during an evacuation. Traffic congestion is an important determinant to predict the time to evacuate people during an emergency. Moreover, depending on the type and severity of the disasters, not all network links may be recommended for use during an evacuation. For example, during flooding, network links are likely subjected to complete or partial inundation which decreases the options for evacuation gateways as well as the network capacity. Interruptions to vehicle paths lead traffic to bottlenecks and degrades the overall network system efficiency.

In this context, coastal cities such as Halifax, the capital of Nova Scotia, face considerable challenges. The Halifax Peninsula, a small land area of 3.3 × 7.5 km^2 that serves as the city's central business district, has only 5 exit and entry points. As downtown Halifax is the financial hub of the province of Nova Scotia, its population nearly doubles during work hours. Halifax is also on

[1]Department of Civil and Resource Engineering, Dalhousie University, Halifax, NS, Canada
[2]School of Planning, and Department of Civil and Resource Engineering, Dalhousie University, Halifax, NS, Canada
[3]School of Public Administration, Dalhousie University, Halifax, NS, Canada
[4]Applied Geomatics Research Group, Nova Scotia Community College, Lawrencetown, NS, Canada

Corresponding Author:
Address correspondence to Muhammad Ahsanul Habib: ahsan.habib@dal.ca

a hurricane path, and extreme events such as hurricanes are common threats to the Atlantic coast of Canada. The region has experienced natural disasters in the past that have affected its critical infrastructure and transport network. In Halifax, most commuters use two bridges connecting twin cities Halifax and Dartmouth which are vulnerable to extreme weather events. Moreover, the region has low-lying transportation infrastructure such as the Bedford Highway and the Armdale Rotary. These major exit points for the Peninsula would be likely to be affected during floods. Thereby, Halifax provides an interesting test bed for developing a comprehensive framework to assess flood risk-related network disruptions and associated challenges in a mandatory mass evacuation.

Therefore, the main objectives of this study are (i) to develop a flood risk model to identify the potential network links that are prone to inundation during floods of different extremes in the Halifax Peninsula, and (ii) to develop a dynamic traffic microsimulation model to predict the traffic flow through uninterrupted origin–destination vehicle paths during an evacuation. The flood model informs the traffic microsimulation model by conveying flood-related damages to the existing Halifax transport network. The dynamic traffic simulation modeling utilizes a dynamic traffic assignment (DTA) module to address the routing policies, queues and spillback in the evacuation traffic stream. The traffic assignment step uses the business as usual (BAU) traffic demand and evacuation demand obtained from the Halifax Regional Transport Network Model (*1*).

Literature Review

Research on mass evacuation has been gaining attention in recent years to reduce loss of lives and properties during an emergency condition. Additionally, many recent disaster events have emphasized the importance of efficient and effective evacuation planning. The tsunami that resulted from the Tohoku earthquake in 2011 hit the coast in only 15 minutes (*2*). In total 386,739 people were displaced due to the Fukushima Daiichi nuclear disaster caused by the same tsunami (*3*). Natural disasters are inherently rare events; they occur with potentially catastrophic consequences (e.g., Hurricane Katrina, Hurricane Irma, Indian Ocean Tsunami). Thus, it is hard in practice to gain knowledge about the extent of the disaster-related damages to the transport network and the subsequent impacts on the overall evacuation process in disaster response. Nevertheless, a long continuous rising trend in sea level warrants awareness of the extent and timing of coastal floods and associated damages. In this regard, high-resolution data resources of surface elevation are required for the analysis of the flood extent and associated damages and/or for other large-scale modeling applications. With advancement in technology, the Light Detection and Ranging (LiDAR) is becoming a well-recognized data source for developing accurate surface models and provides a potential platform to develop a digital elevation model (DEM) (*4*). With the aid of a DEM, flooding limits can be mapped and analyzed in high resolution. Forbes et al. (*5*) developed future extreme water level scenarios for the Halifax Regional Municipality (HRM) during floods by considering the rising mean sea level, land subsidence, storm surge, wave set up and run up, and harbour seiche. However, a hydraulically connected flood risk model is of paramount importance to predict network interruptions due to flooding at different extremes. This model then informs the microsimulation model by conveying the flood-related damages to the transport network.

Simulation-based traffic network models are very effective for formulating, implementing, evaluating and optimizing critical transport operation in a stochastic transport network (*6*). These models can represent the actual traffic environment over a large network. They demonstrate an efficient traffic flow forecasting capability in critical traffic conditions, including evacuating people to safe destinations. The available traffic simulation models use either static or DTA procedures to load traffic into the network. The benefit of using DTA-based traffic simulation models is that this type of model is efficient to estimate time-dependent differential link flow in the network and to generate time-dependent network performances in fine-grained detail (*7*). The evacuation process is complex and traffic flow is critical during the evacuation. Therefore, advanced models including DTA-based microsimulation model are required to capture the complex interactions among vehicles, and to execute real-time driver behavior considering all spatial attributes (e.g., paths, lane changing) and temporal aspects (e.g., time-dependent congestion) under extensively congested traffic conditions in the network.

Much effort has been invested in developing traffic simulation models to predict evacuation time and traffic congestion spillback during an evacuation in the network. Sheffie et al. (*8*) was the first study to develop the Network Emergency Evacuation (NETVAC) to estimate the evacuation time in the event of an emergency at several nuclear power plant stations in the U.S.A. Their model determined evacuees' route choices based on the network familiarity and myopic view which can be captured by the speed on the outbound links. Dynamic Network Evacuation (DYNEV) was developed by KLD Associates (*9*) to estimate the network capacity and evacuation demand for nuclear power plant areas. Lieberman and Xin (*10*) developed a macroscopic traffic simulation model that includes a kinematic flow model, a lane assignment model, and can regulate inflow into the links. Their model was used in DYNEV-II for large-scale evacuations, such as the evacuation of 300,000 evacuees utilizing an available highway network of 3,000 links, 850 origins and 75 destinations. Though the network anticipated congestion initially, after 5 hours from the start of the evacuation the congestion disappeared from the emergency planning zone (EPZ). Shao et al. (*11*) developed an evacuation model within the VISSIM platform to test an evacuation process for vehicles in the Beijing National Stadium's parking lots, and to estimate the evacuation time for all vehicles. Several evacuation studies (*12*, *13*) have evaluated strategies involving limited access to some facilities and roads to improve both total evacuation time and the needed time to only evacuate the population within the most dangerous areas. However, most studies are experimental in nature or deal with limited network simulation. Existing studies (*14–16*) mainly focus on the aftereffects of natural disasters such as hurricane, flood etc. But there is a clear gap in prediction modeling that identifies traffic impacts at city scale. In addition, calibration and validation of the simulation model is also challenging due to lack of sufficient field traffic data (*17*). This study proposes a process to calibrate and validate driving behavior and route choice parameters utilizing a Latin Hypercube Sampling (LHS) technique.

This study fills the gap in literature by developing a GIS-based hydraulically connected flood risk model that generates fine-scale network disruptions and informs the microsimulation modeling about network damages during the displacement of citizens to safe destinations. Moreover, this study examines how the degree of network damage is related to the flooding scenarios of different water levels and how this affects network connectivity.

Methodology

This study develops a comprehensive framework to test different evacuation scenarios in the Halifax Peninsula transport network. Figure 1 shows a sequential modeling framework where the dynamic traffic microsimulation model recognizes the interrupted network links informed by the flood risk model and utilizes BAU traffic demand and evacuation demand obtained from the Halifax Regional Transport Network Model.

Spatial Flood Risk Model

Data Used. The flood risk analysis requires the development of inventories of major Halifax transport network links (Figure 2) impacted under various flooding scenarios. Data used to develop the flood risk model include stream (water) network provided by the NS Topographic Database 2012 and LiDAR data obtained from the HRM. The LiDAR data is used to develop the DEM for the region. The road network data for Halifax is extracted from the National Road Network (NRN) Database 2014. All spatial data are projected in North American Datum (NAD) of 1983, Universal Transverse Mercator (UTM) zone 20 North. The vertical datum is referenced to the Canadian Geodetic Vertical Datum of 1928 (CGVD28).

Flood Layer Generation. Flood layers are generated using an Applied Geomatics Research Group (AGRG) proprietary tool that raises flood water on a flat plane (known as "still-water") to inundate low areas. As the tool requires an elevation model with correct connectivity, all culverts or bridges within the study area need to be represented as low areas in the DEM. First, the GIS-based tool is used to generate an initial culvert database. The culverts are then used to notch the DEM to allow a path for the water to move, resulting in a hydraulically connected DEM. The flood levels (*5*) for five scenarios selected for analysis are listed below:

- Scenario 1 (2.9 m): Water level 2.9 m relative to CGVD28, if Hurricane Juan occurred today on a higher high water at large tide (HHWLT)
- Scenario 2 (3.9 m): Water level 3.9 m relative to CGVD28, if Hurricane Juan occurred today at HHWLT + conservative 100-year sea level rise of 1 m
- Scenario 3 (7.9 m): Water level 7.9 m relative to CGVD28, if Hurricane Juan occurred today at HHWLT + extreme 100-year sea level rise of 5 m
- Scenario 4 (15 m): Water level 15.0 m relative to CGVD28, if moderate Tsunami occurs
- Scenario 5 (30 m): Water level 30.0 m relative to CGVD28, if large Tsunami occurs

Three of the scenarios are related to the water level seen during Hurricane Juan, which made landfall in the HRM as a Category 2 hurricane on September 29, 2003 and produced a maximum recorded water level of 2.1 m CGVD28 (*4*). Had Hurricane Juan occurred on HHWLT, the water level would have been 2.9 m CGVD28. If Hurricane Juan were to happen on HHWLT after a very conservative sea level rise of 1 m in 100 years, the water level would be 3.9 m CGVD28, and so on. The first three scenarios are selected for microsimulation modeling, as the peninsula becomes almost an island under the last two scenarios which might require an evacuation through airlift.

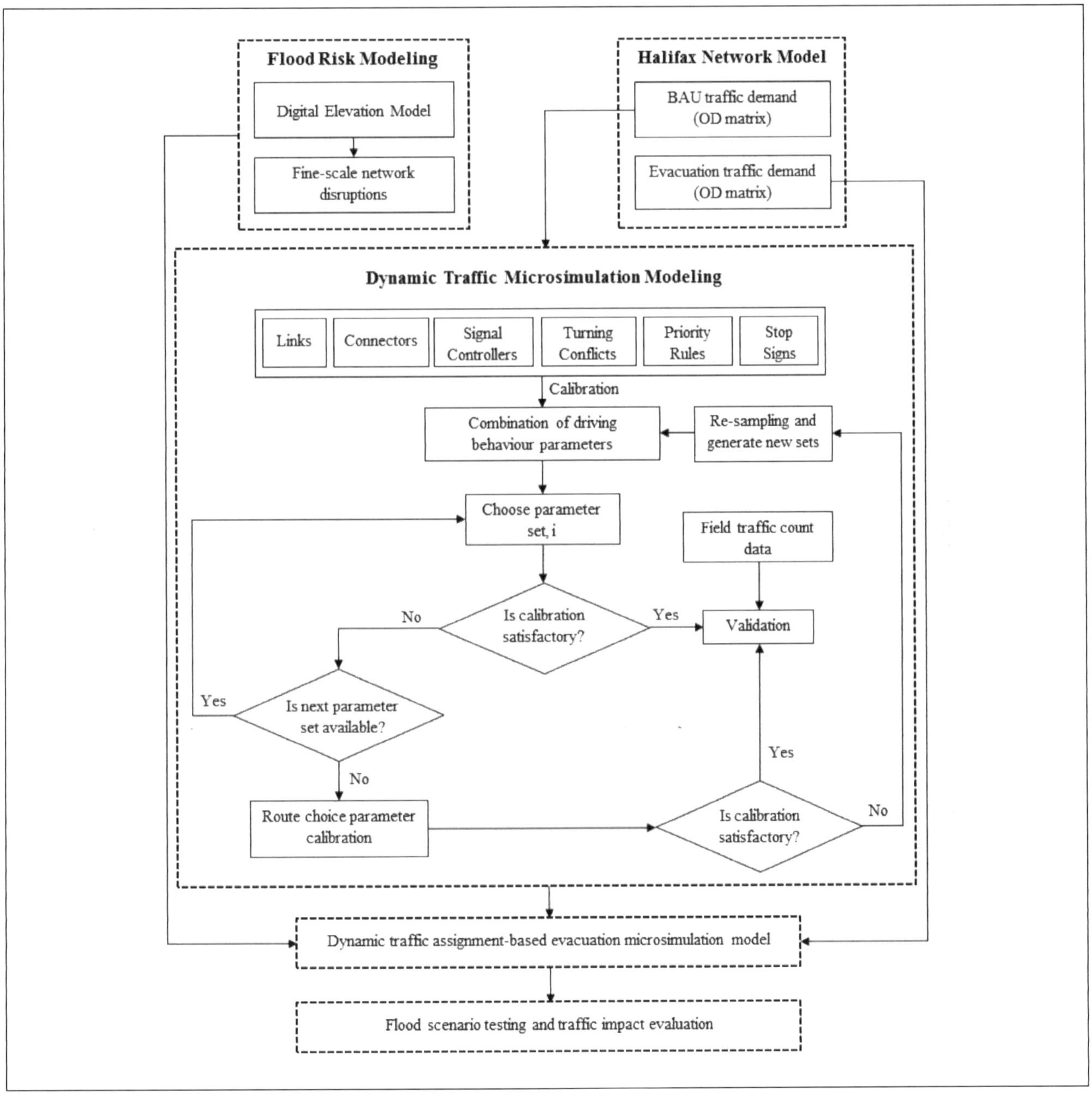

Figure 1. A sequential modeling framework for evacuation of the Halifax peninsula.

Flood Depth Analysis. The depth of water over the land for each of the flooding scenarios is determined by subtracting the flood water level surface from the DEM. This analysis results in depth rasters that provide a depth value for each 2 m pixel. Some areas contained extensive flooding over roads (>2 km inundated). Thus, reporting the exact depth of water over each meter of road would produce extensive results that would require further summarization and analysis. Therefore, it was determined that water depth would be divided into 6 categories as below:

- Depth class 1: water depth $\leq$ 0.5 m
- Depth class 2: water depth > 0.5 m–1.0 m
- Depth class 3: water depth > 1.0 m–2.0 m
- Depth class 4: water depth > 2.0 m–5.0 m
- Depth class 5: water depth > 5.0 m–10.0 m
- Depth class 6: water depth > 10.0 m

Flood Risk Model Results. The simulation of flood water level within the developed GIS-based flood risk model for each flood extent is shown in Figure 2. The Halifax peninsula has five exits as shown in Figure 2: exit 1,

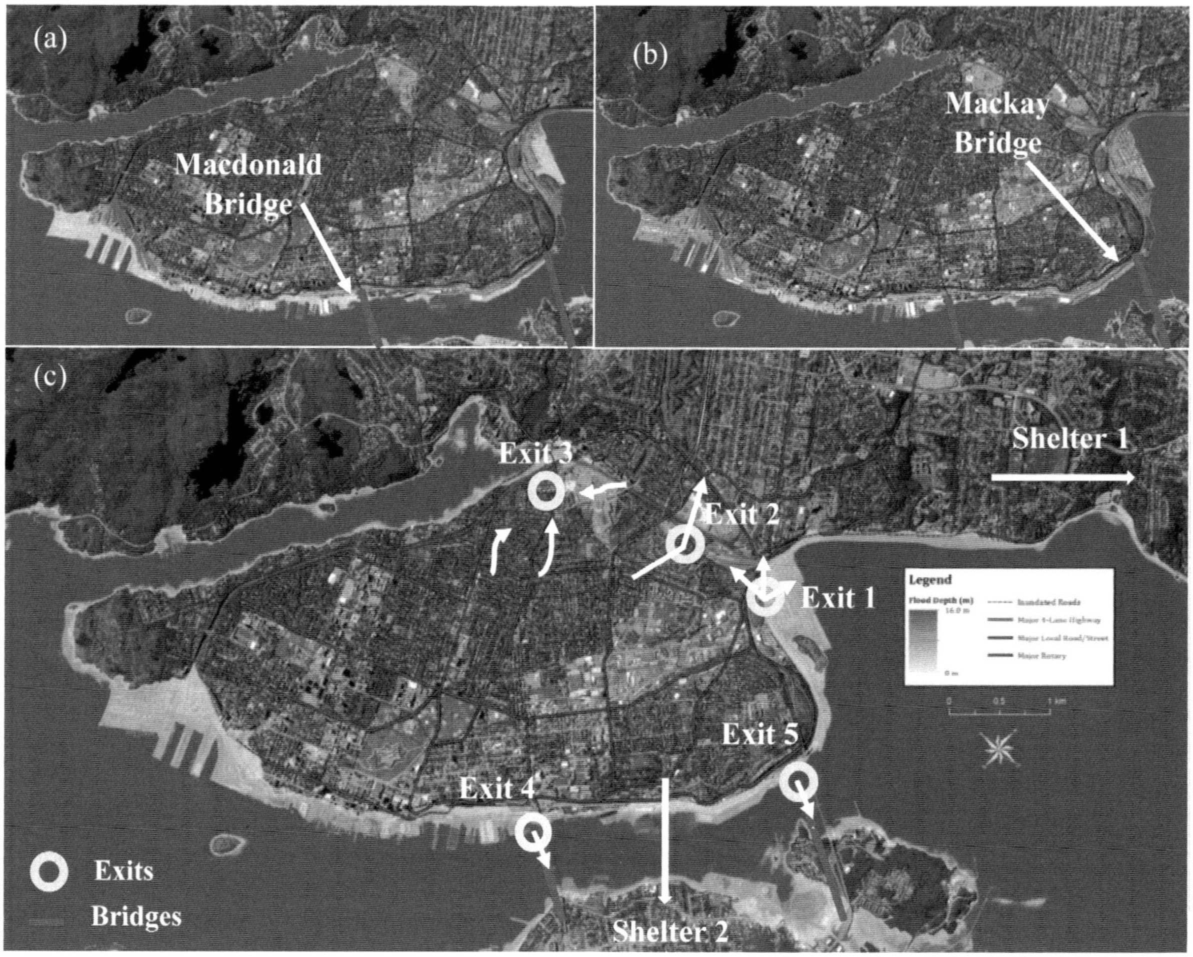

Figure 2. Flooding extent and depth of flooding in the Halifax peninsula under (*a*) 2.9 m, (*b*) 3.9 m, and (*c*) 7.9 m flooding scenarios.

Bedford Highway; exit 2, Highway 102, exit 3, Armdale Rotary; exit 4, Macdonald Bridge; and exit 5, Mackay Bridge.

The major street flooding occurs near exit 3 on Quinpool Road. The rotary (exit 3) is the most susceptible piece of road on the peninsula, with over 43% of the rotary impacted under 3.9 m flood scenario and over 88% impacted under 7.9 m flood scenario.

In addition, Bedford Highway (exit 1) is also inundated by 47.8% of its total length in the 7.9 m flood scenario. The length of each road within the study area impacted by floods is shown in Table 1.

Dynamic Traffic Microsimulation Modeling

Network Model and Data Used. This study has developed a microscopic traffic simulation model for Halifax within the VISSIM platform. The study area (Figure 2) includes two bridges, the Macdonald Bridge and the Mackay Bridge, that connect Halifax and Dartmouth across the Harbor. In total, the coded network has 1,310 links and connectors, including, Highway 102, Highway 118, the Armdale Rotary, and all arterial roads within the study area. Halifax Geodatabase 2012, Google Earth Image, and Google Street View are utilized to properly set up all the network elements in their true geographic locations. Major signalized intersections are coded with actual phase and split time through building the signal controller in the simulator. Signal time data is obtained from the Public Work Traffic Study of HRM, 2014. In total, 41 major- and 12 stop-sign-controlled intersections with 1,930 resolved turning conflicts are included in the network. The study area consists of 19 zones within the peninsula which are considered for evacuation.

The BAU traffic flow from 0600 to 1000 is simulated for calibration and validation purposes where the simulation hour between 0600–0700 is used as warming period. The model is calibrated and validated for the hours

Table 1. Identified Flooded Links Based on the Flood Risk Model Results

Name of roads	Flooding scenario water level (m CGVD28)	Length of roads flooded (m)	Original length of roads (m)	% of roads impacted by flooding
Armdale Rotary	2.9	234.6	1324.3	17.7
	3.9	576.1		43.5
	7.9	1175.1		88.7
Barrington Street	2.9	0.0	7782.2	0.0
	3.9	162.4		2.1
	7.9	660.5		8.5
Bedford Highway	2.9	0.0	3974.6	0.0
	3.9	0.0		0.0
	7.9	1898.9		47.8
Chebucto Road	2.9	0.0	2201.2	0.0
	3.9	0.0		0.0
	7.9	79.0		3.6
Joseph Howe Drive	2.9	0.0	2873.3	0.0
	3.9	0.0		0.0
	7.9	149.2		5.2
Kearney Lake Road	2.9	0.0	269.4	0.0
	3.9	0.0		0.0
	7.9	110.2		40.9
Quinpool Road	2.9	245.4	2420.7	10.1
	3.9	398.4		16.5
	7.9	464.3		19.2

between 0700–0900. Traffic volume data obtained from the Public Work Traffic Study of HRM, 2014 is used for the validation of the microsimulation model.

Calibration of Driving Behavior Parameters. This study calibrates three car-following parameters of the Wiedemann 74 car-following model. The parameters include (i) parameter 1-average standstill distance (ax), (ii) parameter 2-additive part of safety distance (bx_add), and (iii) parameter 3-multiplicative part of safety distance (bx_mult) (*18–20*). These three parameters together give a car-following distance according to the following equations (*19*):

$$d = ax + bx \tag{1}$$

where

d is safety distance;

ax is average standstill distance;

bx adjusts time requirement values which can be written as:

$$bx = (bx\_add + bx\_mult^{*}z)^{*}\sqrt{v} \tag{2}$$

where

z is a value of range [0, 1], which is normally distributed around 0.5 with a standard deviation of 0.15;

v is vehicle speed. and

Literature suggests that the values of the parameters range between 1–3 m for ax (*17*), and 0–3 for bx_add and bx_mult used in Cobb Parkway model calibration (*21*).

The combination of the values of the parameters plays an important role to replicate actual traffic flow in the simulator. Very often, the number of combinations of the parameters is so large that it is difficult to evaluate possible scenarios within the feasible time limit. If there are n variables, each having z number of subdivisions within its range, the total number of possible combinations is z^n. Supposing $n = 4$ and $z = 10$, this gives a very large combination number ($10^4 = 10000$) to evaluate. Therefore, this study uses a Latin hypercube sampling (LHS) technique to control the number of parameter combinations and to reduce the simulation time and cost. This is a statistical method in which the ranges of the values of n variables are subdivided into equally probable intervals, z. Next, sample points are selected from the interval to fill a Latin square grid, which contains pair variables along two axes and only one element in each row and each column. Thereby, the advantage of using this method is that it does not need more samples for more variables and the full range of the distribution can be sampled consistently. The LHS process in this study has identified 13 parameter combinations.

Each combination is considered for simulation. The goodness-of-fit of the model is measured in terms of R^2 value which measures the closeness between the simulated and observed traffic count. The combination, which gives a value 1 for ax, 0.6 for bx_add, and 0.7 for bx_mult exhibits better network performance compared to other combinations in terms of R^2 value. However, this study also performs route choice parameter

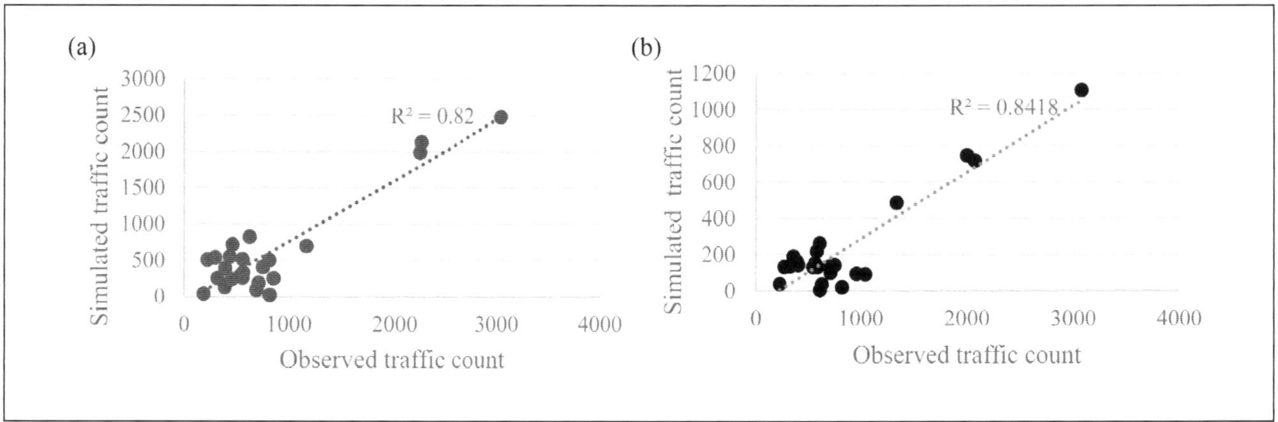

Figure 3. Validation of microsimulation model for (*a*) 0700–0800 and (*b*) 0800–0900.

calibration for further improvement of the traffic flow in the simulator.

Route Choice Parameter Calibration. For further improvement of the traffic flow at a few local network links, route choices are calibrated in the simulator. Route choice calibration in VISSIM can only be done by adding link surcharges (i.e., cost components) to modify assignment results. Links that attract more traffic volume than observed can be assigned a positive surcharge, thus reducing their desirability. In the absence of well-defined guidelines on the relationship between the surcharge value and traffic divergence, a surcharge value of 30 is considered as the starting value based on experience. Through several iterations, surcharge value 50 for 6 links, 100 for 7 links, 150 for 2 links, 200 for 6 links, and 500 for 3 links have been imposed. Improvement is identified by comparing the simulated traffic flow with the observed traffic count at these links.

Validation of Microsimulation Model. This study has conducted a traffic volume-based validation using traffic count data obtained from the multiple data sources as mentioned earlier. The exit points and several nearby intermediate intersections totaling 21 locations are validated in terms of minimum deviation between the observed and simulated traffic counts. The goodness-of-fit of the model is evaluated by determining R^2 for the hours of 0700–0800 and 0800–0900. The validation results in Figure 3*a* and 3*b* suggest R^2 values of 82% and 84% respectively for these two hours.

DTA-Based Evacuation Microsimulation Model and Evacuation Cases

The Halifax Regional Transport Network Model yields 34,808 trips completed in the Halifax Peninsula between 0600–1000. The time 0900–1000 is considered as the ending hour of the morning peak period and the head count is the maximum after this four-hour peak period. This study selects two shelters due to their sufficient accommodation capacity and safe distance from the Halifax Peninsula. The shelters include (i) Shelter 1- Charles P. Allen High School, and (ii) Shelter 2- Nova Scotia Community College (NSCC) Akerley campus (Figure 2). It is assumed that 20% of the total evacuees will take other shelters, for instance, the homes of relatives and friends. The rest can be accomodated within the selected shelters. In total, four evacuation cases are considered within the simulation model:

- Case 1 (No links flooded): No links are affected by any flood scenario
- Case 2 (2.9 m): 0.11% links of the modeled network are affected
- Case 3 (3.9 m): 0.27% links of the modeled network are affected
- Case 4 (7.9 m): 1.1% links of the modeled network are affected

Results and Discussion

Overall Evacuation Results

The simulation results in Table 2 regarding case 4 suggest that the available vehicle paths between the zones and the shelters are disrupted by 31.2% with respect to case 1. The more intense the flood-related network damages, the fewer alternative routes exist for evacuees. The results suggest that the capacity of the Halifax Peninsula transport network is limited to accommodate evacuation demand in the network if flood water level is 7.9 m (case 4) and above unless countermeasures are applied. Only 83% of evacuees can be admitted into the network and evacuated within 15 hours from the beginning of the

Table 2. Overall Evacuation Results for the Halifax Peninsula Transport Network

Scenarios	Case 1: No links flooded	Case 2: 2.9 m	Case 3: 3.9 m	Case 4: 7.9 m
Loss of connectivity, %	0.0	23.3	23.7	31.2
Total departure, %	100	100	100	83
Completed evacuation, %	99	99	98	83
Evacuation time, hr	11	13	15	15
Increment in evacuation time w.r.t case 1, hr	–	2	4	4

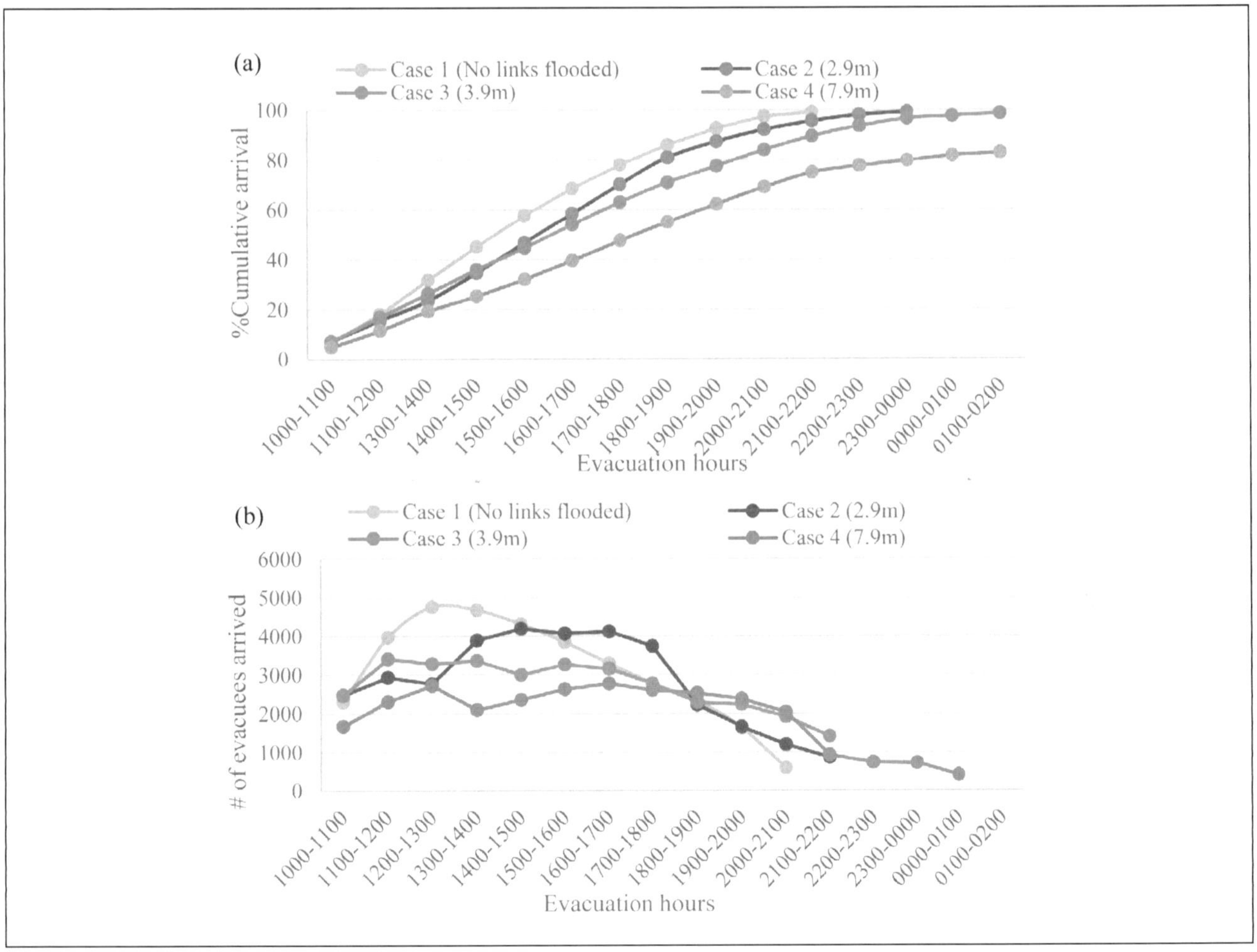

Figure 4. (*a*) Cumulative percent arrival; (*b*) hourly arrival rate of evacuees at shelters during the evacuation of the Halifax peninsula under different flood scenarios.

evacuation under this flooding condition. Traffic congestion in the network peaks when 17% of evacuees are not yet in the network in this case. A prioritizing and staged evacuation strategy could help to evacuate all evacuees under this circumstance. In the case of flood scenarios that demonstrate lower water level rise (2.9 m and 3.9 m), almost all the evacuees can be evacuated within an evacuation time equal to or less than 15 hours.

Arrival of Evacuees at Shelters

Figure 4 presents (*a*) cumulative percent arrival of evacuees, and (*b*) the hourly arrival rate of evacuees at shelters for all the cases considered in this study. The results in Figure 4*a* suggest that to complete a 50% evacuation it takes 5.5 hours to 7.5 hours with the increase in water level from 2.9 m in case 2 to 7.9 m in case 4. In the case of a 7.9 m flood scenario, the evacuees are distributed over

Figure 5. (*a*) Traffic congestion in the network during evacuating the Halifax Peninsula; (*b*) long traffic queue across the Macdonald Bridge which is exit 4; (*c*) traffic flow on major arterial roads approaching exit 2 causes severe traffic congestion.

15 hours as the network has limited discharge capacity from the beginning of the evacuation due to flooding of the key links in the network. It can be concluded that the arrival rate of evacuees at shelters is more uniform when network has less route options. In this case, congestion arises from the beginning of the evacuation in the network. Moreover, it is evident that overall evacuation rate decreases with the increase in flood water level as network interruptions increase. In case 2, parts of Armdale Rotary (exit 3) on the west of the Halifax Peninsula (Figure 2) are affected. This interrupts traffic operations of several routes from the Peninsula to shelter 1. The damage spreads to the east in case 3. During the flood of 7.9 m water level (case 4), Bedford Highway (exit 1) located on the west of the peninsula is also added in the list of flooded links. This scenario presents an immensely disrupted transport network for an evacuation. Figure 5 demonstrates the traffic flows and congestion spillback during the evacuation under a 7.9 m flood scenario. Exits 1 and 3 are completely flooded in this scenario.

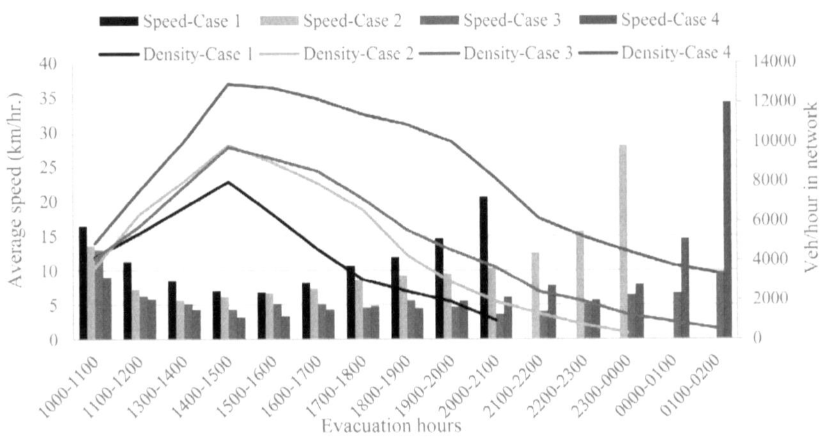

Figure 6. Vehicle density and average speed profile during evacuation of the Halifax Peninsula under different flood scenarios.

Network Performance

Vehicle Density and Average Speed in the Network. This study also examined the average speed and the number of vehicles present in the network for each evacuation case. Flood-related damages to the network links cause an increased number of vehicles to stay in the network. Figure 6 shows that the number of vehicles in the network remains high over a longer duration in case 4 since there are fewer alternative evacuation paths. However, the congestion level in terms of the number of vehicles in the network in different evacuation cases depends on the arrival rate of evacuees to their destinations. For example, the number of vehicles in the network varies in the range of 8,000–13,000 from case 1 to case 4. As indicated earlier, case 4 has the lowest arrival rate; the number of vehicles in the network has peaked at 13,000 in this case. These vehicles create congested traffic conditions and thereby reduce the average speed in the network.

Hourly Network Performance Evaluation. Table 3 presents an in-depth evaluation of the network performance during the evacuation of the Halifax Peninsula. The traffic analysis results are tabulated for each of the four evacuation cases at one-hour interval within the respective total evacuation times. Delays per vehicle increase with the increase in flood water level from case 2 to case 4 as does total delay time. Initially, in each case, delay increases rapidly and reaches a maximum within 5–7 hours from the beginning of the evacuation. This result indicates that vehicles accumulate and create congested traffic conditions in the network. The values of different criteria included in Table 3 also suggest a sustained congestion from 4th to 7th hour of the evacuation. Though case 3 and 4 show similar kinds of traffic congestion, 98% of the evacuees are moved in case 3 against the 83% in case 4 since the arrival rate of the evacuees at destinations in case 3 is higher than that of case 4. The total distance traveled is the lowest in case 4 as only 83% of the evacuees make trips from the Peninsula to shelters.

Conclusion and Future Direction

This paper presented a dynamic traffic microsimulation modeling of an emergency evacuation of the Halifax Peninsula due to floods of different water levels informed by a GIS-based flood risk model. The DEM and DTA-based microsimulation modeling techniques are used to evaluate flood-related damages to the network and the subsequent interruptions to evacuation traffic flow.

The simulation results suggest that it would take 15 hours to evacuate 98% of evacuees from the Peninsula with a flood water level of 3.9 m. A flood of water level 7.9 m reduces evacuation paths from the Peninsula to shelters by 31.2%. This creates a bottleneck traffic condition which allows only 83% of evacuees to reach shelters, against the 98% in case of 3.9 m. The simulation results show that traffic accumulates increasingly as the evacuation time progresses, and flood water level rises in different evacuation cases. The number of vehicles peaks at 13,000 in case of 7.9 m flood, indicating that the arrival rate of the evacuees is much lower than that of the other flood scenarios. Traffic congestion is found to peak within 4–7 hours from the beginning of the evacuation.

This study has some limitations. For instance, it has not considered the temporal aspect of flooding which could be useful for future work for determining timing of evacuation decisions. In addition, this study has not considered modeling of evacuees' decision processes. Future work should include developing behavioral modeling

Table 3. Hourly Network Performance Results during Evacuation of the Halifax Peninsula

Case	Criteria	1st hour	2nd hour	3rd hour	4th hour	5th hour	6th hour	7th hour	8th hour	9th hour	10th hour	11th hour	12th hour	13th hour	14th hour	15th hour
Case-1 (No-links flooded)	Avg. delay per vehicle	17.79	22.70	25.81	32.14	33.85	27.91	22.64	14.64	11.55	17.41	13.87				
	Avg. number of stops per vehicle	117.4	94.1	80.9	89.1	93.1	98.8	117.4	120.3	140.9	203.3	149.1				
	Total delay time	2382.78	4014.22	5151.85	6743.69	6258.01	4469.65	2720.20	1097.88	446.87	393.50	187.27				
	Total travel time	3871.75	5685.24	6886.53	8245.07	7907.34	6203.40	4258.53	2177.23	945.17	669.97	435.32				
	Total distance traveled	65933.3	74293.0	77201.8	66800.8	72596.7	74912.0	66214.6	46022.1	21782.9	12213.8	10968.4				
	Delay stop avg.	11.98	18.43	21.97	28.19	29.51	23.09	17.38	9.83	6.55	10.31	8.26				
Case-2 (2.9 m)	Avg. delay per vehicle	20.27	29.94	35.09	36.01	37.17	35.79	33.47	31.83	32.67	29.11	27.06	20.06	8.18		
	Avg. number of stops per vehicle	107.8	105.7	61.5	72.9	75.4	74.7	76.9	79.5	108.3	140.3	193.0	170.0	71.8		
	Total delay time	2418.71	5010.94	6950.63	8547.55	8403.50	7396.99	6103.95	4221.44	2702.76	1647.89	1007.10	422.67	54.78		
	Total travel time	3489.31	5972.21	7956.98	9905.80	9880.28	8865.92	7584.86	5370.36	3480.34	2177.49	1429.75	660.09	150.12		
	Total distance traveled	47438.7	43043.0	44780.6	60673.2	65781.2	64925.2	64847.3	49621.7	32911.8	22411.2	17911.7	10330.6	4201.9		
	Delay stop avg.	15.18	25.71	32.37	32.92	33.72	32.24	29.74	28.52	28.44	24.05	20.12	14.20	5.15		
Case-3 (3.9 m)	Avg. delay per vehicle	21.23	32.29	35.95	40.23	41.19	41.15	41.98	40.37	43.68	46.31	45.75	43.12	40.87	40.80	29.70
	Avg. number of stops per vehicle	71.15	60.79	64.24	64.01	75.61	74.64	72.62	73.43	73.02	60.60	83.70	101.99	101.70	94.03	77.69
	Total delay time	2581.84	4990.67	7205.36	9323.51	9363.47	8523.52	7900.22	7298.95	6183.01	4814.80	3978.87	3162.50	2235.51	1434.21	664.79
	Total travel time	3697.02	5831.13	8178.67	10352.28	10617.22	9657.15	8839.48	8401.02	6944.08	5261.69	4406.77	3609.78	2653.01	1711.65	874.33
	Total distance traveled	47817.3	36363.8	42156.9	44338.6	54437.7	49016.6	40331.9	47034.5	32380.4	18904.6	17768.2	18388.7	17032.6	11400.5	8531.4
	Delay stop avg.	17.23	29.46	33.35	37.70	38.07	38.04	39.10	36.95	40.68	44.17	42.81	39.71	37.44	37.65	26.93
Case-4 (7.9 m)	Avg. delay per vehicle	25.49	34.69	39.14	45.10	47.09	44.46	42.32	43.65	40.76	42.37	37.53	42.26	39.40	26.24	2.72
	Avg. number of stops per vehicle	122.6	96.2	65.3	54.2	67.4	86.3	101.1	66.3	100.7	86.2	37.5	35.6	52.3	93.6	24.6
	Total delay time	3147.99	6158.10	8275.63	10032.45	9898.87	8971.30	8142.71	7458.32	6397.76	5062.32	2996.80	1942.01	1204.39	484.15	18.05
	Total travel time	3955.89	7076.69	9159.57	10783.18	10694.39	9923.45	9119.81	8309.69	7319.49	5881.02	3662.42	2236.51	1481.46	750.80	118.74
	Total distance traveled	35634.0	40869.6	39524.7	33746.0	35573.3	42131.9	43541.1	37281.0	40301.7	35435.6	28475.8	12587.7	11626.5	10930.5	4064.9
	Delay stop avg.	20.05	30.62	36.37	42.89	44.47	41.14	38.16	40.90	37.02	39.08	35.43	40.16	36.35	21.08	1.46

components, specifically regarding decision to move, destination choice, mode choice, and departure time choice to develop a behaviorally representative traffic microsimulation model. Based on network analysis, modeling departure time choice should be prioritized which would enable testing different scenarios of communication and reaction time of evacuees.

Nevertheless, this research contributes to the literature by developing a city-scale evacuation microsimulation model that considers the network interruptions informed by a flood risk model. The findings of this research are relevant for other coastal cities, particularly for densely populated urban cores with limited access points. The study found that a sudden increase in traffic demand exceeds the capacity of the densely populated transport network. For instance, 17% of residents could not be evacuated for an extreme flood risk scenario in Halifax. Other coastal cities should prepare contingency plans for such circumstances. Moreover, evacuating all citizens at once could take a longer time since spillback could gridlock the narrow roads of historical towns. Therefore, similar types of cities might need staged evacuation planning. This study will guide emergency professionals to tackle the anticipated challenges of a mass evacuation due to flood risk scenarios.

Acknowledgments

The authors would like to thank the Marine Environmental Observation Prediction and Response Network (MEOPAR), the Natural Sciences and Engineering Research Council (NSERC), and the Nova Scotia Department of Energy for their contribution in supporting the research. The authors would also like to thank Sara Campbell and Stephen McCarthy for their valuable time in proof reading.

Author Contributions

The authors confirm contribution to the paper as follows: study conception and design: MAH, MJA; data collection: MJA, MAH; analysis and interpretation of results: MJA, MAH, TW, KQ; draft manuscript preparation: MJA, MAH, TW. All authors reviewed the results and approved the final version of the manuscript.

References

1. Rahman, M., and M. A. Habib. *Evaluating Alternate Transit Options in Halifax using Travel Demand Forecasting Model Informed by a Stated Preference Survey*. Canadian Transport Research Forum, Montreal, QC, Canada, 2015.
2. California Coastal Commission. *The Tohoku Tsunami of March 11, 2011: A Preliminary Report on Effects to the California Coast and Planning Implications*. State of California Natural Resources Agency, San Francisco, CA, 2011, pp. 1–40.
3. Iddri, R. H. *Disaster Evacuation from Japan's 2011 Tsunami Disaster and the Fukushima Nuclear Accident*. IDDRI, Paris, France, 2013.
4. Lohr, U. *Laser Scanning for DEM Generation*. Computational Mechanics Publications, Southampton, UK, 1998, pp. 243–249.
5. Forbes, D. L., G. K. Manson, J. Charles, K. R. Thompson, and R. B. Taylor. *Halifax Harbour Extreme Water Levels in the Context of Climate Change: Scenarios for a 100-Year Planning Horizon*. Geological Survey of Natural Resource Canada, Open File 6346, Ottawa, ON, Canada, 2009.
6. Ge, Q., and M. Menendez. An Efficient Sensitivity Analysis Approach for Computationally Expensive Microscopic Traffic Simulation Models. *International Journal of Transportation*, Vol. 2, No. 2, 2014, pp. 49–64.
7. Florian, M., M. Mahut, and N. Tremblay. A Hybrid Optimization-Mesoscopic Simulation Dynamic Traffic Assignment Model. In *Intelligent Transportation Systems*, IEEE, Piscataway, NJ, 2001, pp. 118–121.
8. Sheffi, Y., H. Mahmassani, and W. B. Powell. A Transportation Network Evacuation Model. *Transportation Research Part A: Policy and Practice*, Vol. 16, No. 3, 1982, pp. 209–218.
9. KLD Associates Inc. *Formulations of the DYNEV and I-DYNEV Traffic Simulation Models Used in ESF*. Washington, D.C., 1984.
10. Lieberman, E., and W. Xin. *Macroscopic Traffic Modeling for Large-Scale Evacuation Planning*. Transportation Research Board of the National Academies, Washington, D.C., 2012.
11. Chunfu, S. H. A. O., Z. H. A. O. Yi, Y. U. Lei, and X. U. Juan. Case Analysis on Vehicle Parking and Departure Plans for the National Stadium. *Journal of Transportation Systems Engineering and Information Technology*, Vol. 8, No. 6, 2008, pp. 52–55.
12. Bae, J. W., S. Lee, J. H. Hong, and I. C. Moon. Simulation-Based Analyses of an Evacuation from a Metropolis during a Bombardment. *Simulation*, Vol. 90, No. 11, 2014, pp. 1244–1267.
13. Jha, M., K. Moore, and B. Pashaie. Emergency Evacuation Planning with Microscopic Traffic Simulation. *Transportation Research Record: Journal of the Transportation Research Board*, 2004. 1886: 40–48.
14. Yang, H., E. F. Morgul, K. Ozbay, and K. Xie. Modeling Evacuation Behavior under Hurricane Conditions. *Transportation Research Record: Journal of the Transportation Research Board*, 2016. 2599: 63–69.
15. Sadri, A. M., S. V. Ukkusuri, P. Murray-Tuite, and H. Gladwin. Hurricane Evacuation Route Choice of Major Bridges in Miami Beach, Florida. *Transportation Research Record: Journal of the Transportation Research Board*, 2015. 2532: 164–173.
16. Cheng, G., C. Wilmot, and E. Baker. Dynamic Gravity Model for Hurricane Evacuation Planning. *Transportation Research Record: Journal of the Transportation Research Board*, 2011. 2234: 125–134.
17. Park, B., and J. Schneeberger. Microscopic Simulation Model Calibration and Validation: Case Study of VISSIM

Simulation Model for a Coordinated Actuated Signal System. *Transportation Research Record: Journal of the Transportation Research Board*, 2003. 1856: 185–192.
18. Wiedemann, R. *Simulation des straßenverkehrsflusses, Schriftenreihe Heft 8*. Institute for Transportation Science, Karlsruhe, Germany, 1974.
19. *PTV 6.0. User Manual*. PTV AG, Karlsruhe, Germany, 2014.
20. Olstam, J. J., and A. Tapani. *Comparison of Car-Following Models*. VTI Report 960A. Swedish National Road and Transport Research Institute (VTI), Stockholm, Sweden, 2004.
21. Miller, D. M. *Developing a Procedure to Identify Parameters for Calibration of a VISSIM Model*. Master's thesis. School of Civil and Environmental Engineering, Georgia Institute of Technology, Atlanta, GA, 2005.

The Standing Committee on Emergency Evacuations (ABR30) peer-reviewed this paper (18-06677).

Article

TRR
JOURNAL OF THE TRANSPORTATION RESEARCH BOARD

Transportation Research Record
2018, Vol. 2672(1) 161–170

DOI: 10.1177/0361198118784404
journals.sagepub.com/home/trr
SAGE

Estimating Choice Models to Quantify the Effect of Herding on the Decision to Evacuate: Application of a Serious Gaming Experimental Setup

Mignon Van den Berg[1], **Rob van Nes**[1], **and Serge Hoogendoorn**[1]

Abstract

Insight into factors influencing the choices people make in case of an evacuation from a natural disaster can help governments and emergency management personnel to manage people in case of such a situation. One of the aspects that influences the choices that people make in such a situation is herding. Since herding has not been quantified, this paper focuses on quantifying the effect of herding on the decision to evacuate by using an experimental setup that is based on the serious game *Everscape*. Around 400 people participated in 13 experiments with this setup. Choice models were estimated with the data from these experiments by including observable characteristics of herding as an attribute into the utility function. It is concluded that an important step is made in quantifying herding. It is shown that the more people someone sees leaving, the more inclined this person is to leave. Seeing people leave has more impact than seeing people stay. When people have no information from official sources, they tend to use other people as a source of information. In case of a disaster, this might result in following people who make the situation even more dangerous (for themselves and possibly for others as well). The information provided by official sources is therefore essential in managing people in the best possible way in case of a natural disaster.

It is known from research that many different aspects influence the decision to evacuate or not in case of a natural disaster, for example, disaster characteristics, instructions, and a person's social situation. In the case of a hurricane, for instance, people are more likely to evacuate when the storm is more severe (*1*, *2*). When people receive evacuation instructions that are directed to them personally, they are more likely to evacuate than when these instructions are not personally directed (*3*). With respect to a person's social situation, the household especially has an impact on the mobility patterns (*4*).

In case of an evacuation, people may also be influenced by the behavior of other people, and copy this. Ariely considers this to be herding behavior, and defines it as seeing other people doing something and believing that what they are doing is a good alternative, resulting in doing the same thing (*5*). With animals, herding leads to having anti-predator or foraging benefits (*6–10*). With people, individualism allows people to detect a good solution; for example, in the case of an evacuation from a building, people are able to detect the exits. Herding might lead to the imitation of successful alternatives (*11*). To the authors' knowledge, the effect of herding on evacuation choices has not been quantified empirically.

Doirado et al. and Van den Berg et al. described the development of an experimental setup using the 3D multi-user virtual environment (or serious game) *Everscape*, in which a group of people participates simultaneously (*12*, *13*). A serious game is defined as a game which is used for other purposes than pure entertainment. In this case it is used for research, but it could also be used for education or training purposes (*14–16*).

Everscape consists of a virtual island, where an avatar is appointed to each participant. With this avatar, each participant can walk and drive around on this island. He or she can also see the other avatars and what they are doing. Beforehand, participants know they will go to this island to see a concert but during this concert there is an earthquake and they have to evacuate for a tsunami. The authors believe that the experimental setup using *Everscape* is a suitable technique to quantify the effect of herding on the decision to evacuate.

[1]Delft University of Technology, Faculty of Civil Engineering and Geosciences, Delft, the Netherlands

Corresponding Author:
Address correspondence to Mignon Van den Berg: mignon_berg@hotmail.com

Figure 1. Impression of the concert area in *Everscape*.

The main objective of this paper is therefore to quantify the effect of herding on the decision to evacuate with the use of the experimental setup with *Everscape*. More specifically, the study aims to answer two specific research questions:

1. Does herding behavior influence the decision to evacuate?
2. If herding influences the decision to evacuate, how does it work?

To answer the research questions, the research approach consists of several steps. The experimental setup using *Everscape* is developed to focus on quantifying the effect of herding. Choice models are subsequently estimated with the data from the *Everscape* experiments to answer the research questions. Finally, the conclusions and recommendations for further research are presented.

Data Collection with *Everscape*

The experimental setup with *Everscape* at its core was designed to focus on quantifying the effect of herding on the decision to evacuate. The setup consisted of three parts: an introduction, the experiment with *Everscape*, and filling in a questionnaire.

Everscape *Tool*

In *Everscape*, which was created in cooperation with the National Institute of Informatics (in Tokyo, Japan), the considered scenario is as follows: people arrive on an island by helicopter and upon arrival they all take a car and drive to a concert themselves. For an impression of the concert area, see Figure 1. When all people are at the concert, an earthquake is triggered.

The earthquake consists of shaking of the ground and a rumbling sound. As a result of the earthquake, the concert stage partly collapses. After the earthquake, there is a moment of silence, followed by a news item, which informs people they have to evacuate the island because of a tsunami that will soon arrive. In the news item, they are instructed to travel back to the helicopter and they are informed about their travel options: a direct train connection and two options by car. By car, they can drive via a short route along the beach and a longer route through the mountains. The news item abruptly ends. Those who are not able to make it to the helicopter on time do "not survive." After the news item, the expected arrival time of the tsunami is shown throughout the rest of the scenario. For an impression of the *Everscape* scenario, see the movie of one of the first experiments online (*17*).

Setup of Everscape *Experiments*

The full experimental setup consists of three parts. These are explained below.

Part 1: Introduction to Experiment. Upon arrival at the experiment, each participant sat down behind a computer. At their computer they found a set of documents. The first document included their participant id, which they had to use throughout the entire experiment. They also had to sign this document before the start of the experiment, stating they participated voluntarily and the researchers could use the data anonymously. Parents of participants younger than 18 had to sign this form for them. The second document included a brief description on how to use the controls during the virtual part of the experiment with *Everscape*.

When every participant had signed his or her informed consent form, an introduction was given to the participants. In this introduction, it was explained that the experiment consisted of a virtual part and a questionnaire. In the virtual part, they would visit an island to see a concert. They would arrive on the island by helicopter and then each of them would have to drive to the concert by car. The introduction included a short demonstration on how to use the controls.

Part 2: Virtual Experiment with Everscape. The second part of the experiment was the actual experiment with *Everscape*. During this part of the experiment, one person was managing the process. This same person gave the introduction. A second person was behind the admin computer to see if everything went well and to give a sign when everyone was at the concert area. When this was the case, the earthquake was triggered. Two extra people were in the room to help if there were any questions.

Part 3: Questionnaire. After the virtual experiment with *Everscape*, each participant filled in a questionnaire, which included questions on their socio-economic characteristics, their computer and gaming experience, their experience with (evacuations from) disasters, and questions on what they did during the virtual experiment and why.

The Experiments and the Participants. In total, 13 experiments were conducted with on average 27 people per experiment. The first two experiments were with students only, who were recruited via a student employment agency. Because of technical problems, it was decided to first do experiments with students only. When this worked, other people were invited for the rest of the experiments.

To recruit the other participants, announcements were placed at the website of Delft University of Technology, at the website of the Faculty of Civil Engineering and Geosciences, via the news paper of Delft University of Technology (Delta), via a local newspaper (*Delft op Zondag*), via a local radio station (Omroep West), and on social media (LinkedIn, Facebook, and Twitter). The announcements stated that the study was going to test a new type of experiment for research on travel choice behavior. Nothing was said about disasters or evacuations. Participants could register via a website and choose their preferred day and timeslot. They all received a €25 gift check after participation. Except for the first two experiments in which only students participated, in the other experiments a mixed population took part, with ages ranging from 12 to 78 and around an equal share of men and women. Because of technical limitations the maximum number of participants was fixed to 38. This resulted in five more experiments than originally planned but with fewer people per experiment.

Available Data

After the experiments, the available data consisted of two datasets. The first dataset included data from the virtual experiment with *Everscape*. The second dataset included data from the questionnaire. In *Everscape*, the behavior of each participant was closely monitored during the virtual experiment (*13*): "the data were collected at a 1.0 (s) time and at a 0.1 (m) position resolution along with viewing directions (resolution of 1). Furthermore, the following data from the events that occurred were logged:

- starting time of the earthquake
- timing of the tsunami
- departure time of the train
- departure time of the helicopter
- when a participant got in / out of a car, the train, or the helicopter"

To link the *Everscape* data to the questionnaire, each participant received a participant id. They filled this in before both parts of the experiment.

Approach to Quantify Effect of Herding on Evacuation Decisions

Before discussing the approach to quantify the effect of herding, the mindset of the participants during the experiments was considered. As data were collected with a serious game, the question could be if people participated seriously. During the experiments nothing was said about a game; the participants were informed that they took part in an experiment with a virtual part. To find out how their mindset was, the authors asked about participants' goals during the *Everscape* part of the experiment and also looked at the trajectory data. Of every participant their goal was to, for example, "survive" or "get back to the helicopter." The trajectory data showed that all participants at least tried to evacuate. From this, it was concluded that the participants were taking the experiments seriously.

Every participant evacuated the concert area. The differences were the characteristics of the situation they were in before they evacuated. For example, did people leave before they had received the information and instructions via the news item, or did they leave after they had received this? How many people did they see and what were these people doing?

To quantify the effect of herding on the decision to evacuate, (discrete) choice models were estimated. As discussed earlier, with herding, it is the behavior of other people that is copied. It was therefore decided to quantify the effect of herding by including the actions of other people. This makes it possible to use observable characteristics of herding as an attribute into the utility function.

The herding phenomenon involves the behavior of other people, but it is not known how this exactly works: what do people actually see? Do they see other people leaving or staying (or both)? Do they see this in a split second or over a longer period? And when do they make the decision to evacuate? Is this a continuous decision process or not? The

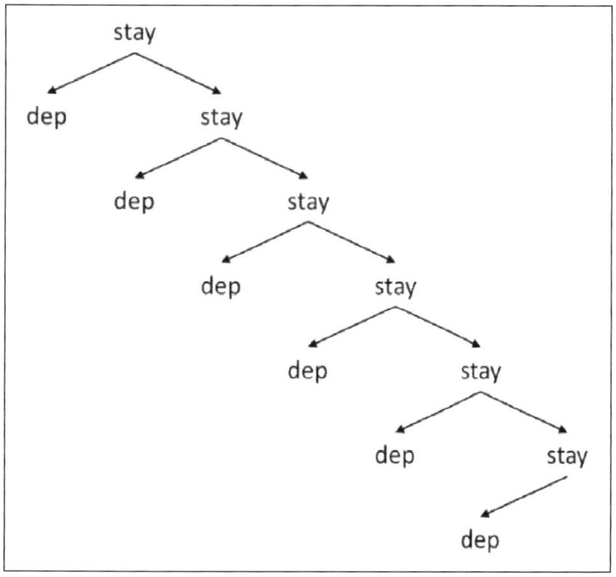

Figure 2. Decision tree on staying at and leaving the concert area.

moment that each participant left the concert area was known. This was used as the basis, and it was decided to use a systematic approach to consider herding and cover a range of possibilities of the herding phenomenon.

To include observable characteristics of herding as an attribute into the utility function, the *Everscape* data had to be processed. After this process, it was known, per participant, when he or she started to leave toward the exit of the concert area. Also known per participant and per second was the numbers of people that were seen while they were staying at and leaving the concert area.

To make it possible to estimate choice models, it was assumed that people also decided to stay. Figure 2 shows the decision tree that has been used.

Figure 3 shows a sequential choice process with one departure choice (D) and six assumed stay choices (S). Intervals of 5 seconds were added between the choices to make sure there were intervals with different characteristics, for example, different numbers of people that were seen but also differences in what was happening in the *Everscape* scenario itself.

To include the actions of other people in a systematic way, the authors wanted to cover a range of options on what participants might have observed. Therefore, different observation times were included. Note that Figure 3 only shows the observation times before the departure choice moment; data were also calculated for the stay choice moments.

An observation time of 5 seconds, for example, means that within the 5 seconds prior to the decision to leave, the numbers of unique people that were seen within that time frame were calculated. As with the interval times, the observation times were also chosen in a systematic way to cover a range of options. Because of the duration of the *Everscape* scenario and the duration of the different phases in *Everscape* (before the earthquake, during the earthquake, during the news item, after the news item), a maximum of 30 seconds of observation time was used.

Choice models were estimated with Biogeme (*18*) in two steps. Step 1 of the choice modeling process involved estimating binary choice models, which included herding as an attribute in the utility function. See Equation 1 for the utility function representing the choice to leave the concert area and Equation 2 for the utility function representing the choice to stay at the concert area. Both utility functions consist of an alternative specific constant (ASC), which was fixed in U_{stay} to easily compare both utilities and find out if people were inclined to leave or stay. The herding attribute is β. The analysis was focused on the *ASCs, βs, adjusted* ρ^2, and *final log-likelihood*:

$$U_{\text{dep}} = ASC_{\text{dep}} + \beta_n \times attribute_n \quad (1)$$

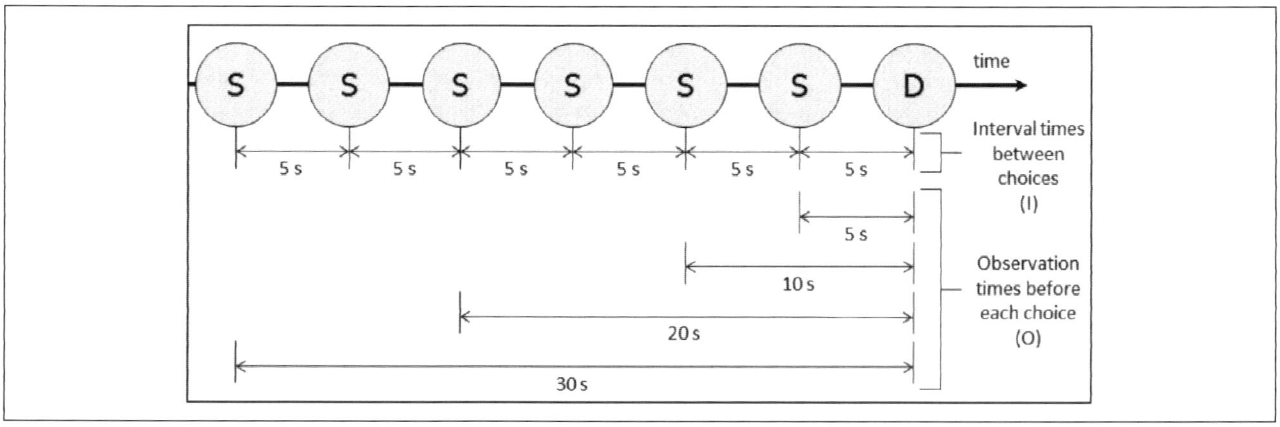

Figure 3. Herding data used to estimate the choice models.

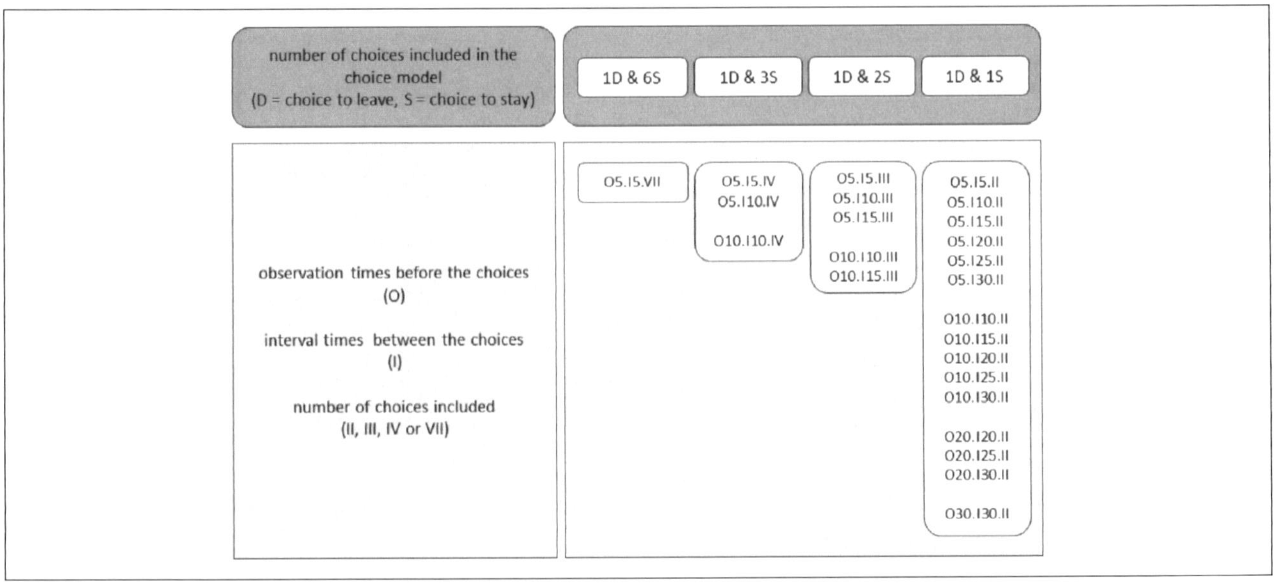

Figure 4. Step 1 of choice modeling process, structured approach to determine herding.

$$U_{stay} = ASC_{stay} \tag{2}$$

Figure 4 shows all different choice set combinations of staying at (S) and leaving the concert area (D). Depending on the choice set, different combinations of observation and interval times were considered. For example, O5.I5.II means that two choices are included (one departure choice and one stay choice) for an observation time (O) of 5 seconds before each choice and an interval time (I) of 5 seconds between the choices. No overlapping observation times were included to consider every different action of each person once.

Step 2 of the choice modeling process involved estimating latent class models to find out if a segmentation on herding can be made based on other characteristics. To find out if personal characteristics influenced herding, age and gender were considered. To consider the effect of the situation, the news item was used. This meant determining if people had received information through the news item or not before their decisions to stay and leave. Finally, to find out if gaming experience caused a bias in the results, this is considered as well.

With latent class models it is assumed that different classes exist that each have homogeneous preferences. The classes cannot be observed, and emerge from the estimation process. A separate set of parameters (of multinomial logit [MNL] type) is estimated for each class.

Before discussing the results in detail, it should be stressed that the panel effect was considered to find out if the choices from the same person were correlated. Based on too high p-values, it was concluded that there was no panel effect. Therefore, the panel effect is not considered later in the choice modeling process.

Estimation Results on Herding Effect on Decision to Evacuate

This paper discusses a selection of the results from the estimated choice models, for a complete overview, see Van den Berg (Chapter 5) (*19*). The results of the two steps introduced in Chapter 3 are here.

Step 1 focuses on determining if and how herding works. How often do people make the decision to stay or leave? Do they consider other people that stay or leave (or both)? What is the effect of observation time before the decision to stay or leave, and what is the effect of different interval times between choices? Step 2 will discuss the results of latent class models to find out if a segmentation on herding can be made based on age, gender, and information via the news item. Gaming experience is considered to find out if it caused a bias in the results.

Results of Step 1: Structured Approach to Herding Phenomenon Modeling

In the first step of the choice modeling process, the goal was to find out if herding exists and, if so, how it works. Tables 1–4 show the results for (Part A) the effect of the number of included choices, (Part B) the effect of seeing people leave versus seeing people stay, (Part C) the effect of observation times before each choice, and (Part D) the effect of interval times between choices. The p-values are included between parentheses.

With respect to the number of included stay choices, Part A (Table 1) shows that when more choices are included, the adjusted ρ^2 and the *final log-likelihood* increase. Because the sizes of the datasets that are used to estimate these four

Table 1. Results of Step 1, Structured Approach to Determine Herding: Part A

A. The effect of including a different number of stay choices

	O5.I5.II T_{obs} = 5 T_{int} = 5 2 choices	O5.I5.III T_{obs} = 5 T_{int} = 5 3 choices	O5.I5.IV T_{obs} = 5 T_{int} = 5 4 choices	O5.I5.VII T_{obs} = 5 T_{int} = 5 7 choices
Adjusted ρ^2	0.011	0.093	0.201	0.417
Final log-likelihood	-561,770	-773,305	-908,094	-1,161,422
ASC_{NSTAY}	0 (0.00)	0 (0.00)	0 (0.00)	0 (0.00)
ASC_{NDEP}	-0.159 (0.05)	-0.856 (0.00)	-1.28 (0.00)	-1.96 (0.00)
β_{NDEP5}	0.261 (0.00)	0.269 (0.00)	0.307 (0.00)	0.289 (0.00)

Note: The *p*-values can be found in parentheses.

Table 2. Results of Step 1, Structured Approach to Determine Herding: Part B

B. The effect of seeing people stay versus seeing people leave

	O5.I5.II NDEP5 T_{obs} = 5 T_{int} = 5 2 choices	O5.I5.II NSTAY5 T_{obs} = 5 T_{int} = 5 2 choices	O5.I5.VII NDEP5 T_{obs} = 5 T_{int} = 5 7 choices	O5.I5.VII NSTAY5 T_{obs} = 5 T_{int} = 5 7 choices
Adjusted ρ^2	0.011	0.003	0.417	0.407
Final log-likelihood	-561,770	-569,740	-1,161,422	-1,179,891
ASC_{NSTAY}	0 (0.00)	0 (0.00)	0 (0.00)	0 (0.00)
ASC_{NDEP}	-0.159 (0.05)	-0.0128 (0.89)	-1.96 (0.00)	-1.80 (0.00)
β_{NSTAY5}	n.a.	0.00466 (0.82)	n.a.	0.00242 (0.87)
β_{NDEP5}	0.261 (0.00)	n.a.	0.289 (0.00)	n.a.

Note: The *p*-values can be found in parentheses.

Table 3. Results of Step 1, Structured Approach to Determine Herding: Part C

C. The effect of different observation times

	O5.I30.II T_{obs} = 5 T_{int} = 30 2 choices	O10.I30.II T_{obs} = 10 T_{int} = 30 2 choices	O20.I30.II T_{obs} = 20 T_{int} = 30 2 choices	O30.I30.II T_{obs} = 30 T_{int} = 30 2 choices
Adjusted ρ^2	0.014	0.008	0.005	0.006
Final log-likelihood	-559,543	-563,457	-564,896	-564,432
ASC_{NSTAY}	0 (0.00)	0 (0.00)	0 (0.00)	0 (0.00)
ASC_{NDEP}	-0.176 (0.03)	-0.143 (0.08)	-0.141 (0.09)	-0.157 (0.06)
β_{NDEP5}	0.300 (0.00)	0.201 (0.00)	0.154 (0.00)	0.142 (0.00)

Note: The *p*-values can be found in parentheses.

models are different, the only option to compare the models is to find out if the results are consistent. The most important result is that even though people were inclined to stay at the concert area (ASC_{NDEP} is smaller than ASC_{NSTAY}), the positive value for β_{NDEP5} shows that the more people they saw leaving, the more inclined they were to leave themselves. The results of these models are consistent in confirming herding.

Part B (Table 2) presents the results for determining the effect of seeing people leave versus seeing people stay. The results are shown when including two choices in the dataset (columns 1 and 2) and when including seven choices (columns 3 and 4) in the dataset. Columns 1 and 3 show the results when only the number of people seen that were leaving are included; columns 2 and 4 show the results when only the number of people seen when they were staying are

Table 4. Results of Step 1, Structured Approach to Determine Herding: Part D

D. The effect of different interval times				
	O5.110.II T_{obs} = 5 T_{int} = 10 2 choices	O5.115.II T_{obs} = 5 T_{int} = 15 2 choices	O5.120.II T_{obs} = 5 T_{int} = 20 2 choices	O5.125.II T_{obs} = 5 T_{int} = 25 2 choices
Adjusted ρ^2	0.012	0.024	0.014	0,017
Final log-likelihood	-560,761	-554,375	-559,990	-558,279
ASC_{NSTAY}	0 (0.00)	0 (0.00)	0 (0.00)	0 (0.00)
ASC_{NDEP}	-0.174 (0.03)	-0.225 (0.01)	-0.178 (0.03)	-0.183 (0.02)
β_{NDEP5}	0.288 (0.00)	0.405 (0.00)	0.299 (0.00)	0.321 (0.00)
$ASC_{NDEP} / \beta_{NDEP5}$	0.604	0.556	0.595	0.570

Note: The *p*-values can be found in parentheses.

included. Because of differences in sizes of the datasets, columns 1 and 2 can be compared, and columns 3 and 4 can be compared. From comparing columns 1 and 2, seeing people leave has more effect than seeing people stay. This is confirmed by comparing the results of column 3 with column 4 (β_{NDEP5} is larger than β_{NSTAY5}). Throughout the rest of the choice modeling proces of Step 1, the focus is on including the people that were seen while they were leaving.

To find out how much observation time is considered, Part C (Table 3) shows four modeling results when different observation times are included. The interval times and number of included stay choices are consistent over these models. Based on the adjusted ρ^2 and the *final log-likelihood*, the model that performs best is the model with an observation time of 5 seconds. This suggests that the last few seconds before the decision is made influences this decision the most.

Part D (Table 4) presents the final aspect of Step 1 of the choice modeling process, which is the effect of interval times. When considering different interval times between the choices to stay and leave, the situation that influenced these choices was more likely to be different before these choices. When the choices were closer together, the situation of the *Everscape* scenario was more similar and it was more likely that only herding was different as an influencing factor. Differences in adjusted ρ^2, ASC_{NDEP}, and β_{NDEP5} can be found for the different models. No behavioral conclusions can be drawn from this. It might be a coincidence, but it might also be related to the result that all participants left. To determine if a segmentation can be made based on herding, latent class models were estimated.

Results of Step 2: Estimated Latent Class Models

To find out if a segmentation could be made based on herding, latent class models were estimated. A selection of the results is presented in Table 5. For more results, see Van den Berg (Chapter 5) (*19*). The assumption with latent class models is that different classes exist that each have homogeneous preferences. The classes, with each a separate set of MNL-type parameters, emerge from the estimation process and cannot be observed.

In all cases both NDEP5 and NSTAY5 are included, so the number of people seen while they were leaving and staying within the last 5 seconds before the decision to stay or leave was made. The different columns show the results for estimated latent class models with a segmentation on age, gender, information via the news item, and gaming experience. Age and gender are considered to find out if certain types of people respond differently, and information via the news item is considered to find out what the effect of information is. Gaming experience is considered to find out if it caused a bias in the results, which is undesirable.

According to Akaike's Information Criterion (AIC), the lowest *AIC* (0.807) is for the model NEWS. This means that the effect of the news item determines class membership more than age, gender, and gaming experience. There are two classes in the model. The people in class 1 are influenced by the behavior of other people, while people in class 2 are not influenced by this behavior. These results show the effect of the news item. Before the news item, participants were more likely to follow than after the news item. This makes sense, because before the news item, the participants did not have any information and "used" other people as a source of information.

The shares are about 25% and 75%, meaning 25% of the participants are part of class 1 and could be defined as the followers. This model shows (and confirms models in Step 1) that seeing people leave has more impact than seeing people stay.

The results of Step 1 on the effect of interval times, suggested that herding occurred because everyone left. However, the latent class results of Step 2 showed the effect of information on herding. It more specifically showed that herding exists when there is no information. In that case other people are a source of information. The other participants who left did not leave because they were following; they were more likely to leave as a result of the information and instructions

Table 5. Results of Step 2, Latent Class Results on Herding with Class Membership

Membership attribute	Age	Gender	News	Gaming exp.
MNL – Log-likelihood function	-1,160,002	-1,160,002	-1,160,002	-1,160,002
MNL – Inf.Cr.AIC	.808	.808	.808	.808
LC – Log-likelihood function	-1,152,762	-1,154,278	-1,152,213	-1,154,293
LC – restricted log-likelihood	-1,160,002	-1,160,002	-1,160,002	-1,160,002
LC – Significance level	.399	1.000	.029	.121
LC – McFadden Pseudo R-squared	.002	.000	.007	.005
LC – Inf.Cr.AIC	.809	.824	.807	.808
Constant	1.90 (.00)	1.90 (.00)	1.90 (.00)	1.90 (.00)
NDEP5	-.32 (.00)	-.32 (.00)	-.32 (.00)	-.32 (.00)
NSTAY5	0.03 (.10)	0.03 (.10)	0.03 (.10)	0.03 (.10)
Constant class 1	13.90 (.29)	.06 (.95)	2.69 (.19)	7.37 (.28)
NDEP5 class 1	-17.03 (.29)	-1.20 (.04)	-4.30 (.06)	27.38 (1.00)
NSTAY5 class 1	1.04 (.33)	.13 (.20)	.22 (.00)	-.34 (.41)
Constant class 2	1.83 (.00)	7.34 (.29)	1.74 (.00)	.05 (.95)
NDEP5 class 2	-.08 (.56)	34.90 (1.00)	2.66 (.56)	-1.21 (.03)
NSTAY5 class 2	.02(.23)	-.33(.42)	.02 (.50)	.13 (.16)
ONE1	-1.08 (.06)	-.94 (.24)	-.85 (.00)	.98 (.12)
AGE1, GENDER1, NEWS1, GAME1	-.04 (.12)	-.02 (.86)	-.46 (.02)	-11d-04 (.96)
ONE2	Fixed	Fixed	Fixed	Fixed
AGE2, GENDER2, NEWS2, GAME2	Fixed	Fixed	Fixed	Fixed
Class1Pr	.10	.27	.25	.73
Class2Pr	.90	.93	.75	.27

from the news item. The modeling results show that a segmentation on age, gender, and gaming experience is not possible (see especially the p-values in parentheses).

Conclusions and Recommendations

The main objective of this paper was to quantify the effect of herding behavior with the use of the experimental setup with *Everscape*. Based on the results, it is concluded that an important step is made in quantifying herding with empirical research.

The first question was if herding behavior influences the decision to evacuate; and if so, the second question was: how does this herding work? To answer these research questions, several conclusions are drawn in this paper. From the estimated choice models, it can be concluded that herding influenced the decision to leave the concert area and therefore the decision to evacuate. The models showed that people were inclined to stay at the concert area but when they saw others leaving, they were inclined to leave as well. The results seem to imply that herding is impulsive because, especially within the last few seconds before the decision to leave, herding influenced this decision.

Latent class models were estimated to find out if a segmentation on herding could be made. The results showed that the effect of information is especially important. When people have no information, they tend to use other people as a source of information and copy their behavior. This might be positive when "good" behavior is copied. In case of a disaster, it might also happen that people are followed who move toward a disaster instead of away from it. It is important in practice to provide people (if possible) with information and instructions to control or manage a disaster as much as possible.

As an important step is made in quantifying the effect of herding, several recommendations for further research are to be made.

Experiments in Known Environments

In real life, when a natural disaster occurs, people are usually in known environments. With *Everscape*, everyone had the same prior knowledge, which is useful for excluding the effect of this prior knowledge. The effect of herding behavior might be different when people are in a known environment. When people are in a known environment, they probably have more knowledge on their evacuation options, which might have a different effect on herding. Collecting data in known environments could provide insights on this.

More People Taking Part in the Same Experiment

In the *Everscape* environment, the maximum number of people participating at the same time was fixed at 38 because of technical limitations. Increasing the number of people could provide extra insights. In real life, especially in large-scale events such as concerts, emergency management personnel need to know how they can influence and manage people in case of a stressful situation. Knowledge on how people influence each other will provide insights on which information to provide to whom, and how.

Measuring Stress Levels

A natural disaster is a stressful situation. The effect of the stress levels that people experience might also influence their behavior. Knowing how much stress people experience, for example by measuring heart rates, could show differences in the effect of herding on evacuation choices.

Measuring Viewing Directions

The number of people that were seen, including if they were staying at or leaving the concert area, was based on the viewing directions. It remains unknown if participants were actually considering this. Adding eye trackers could help in finding out what people were actually looking at and considering for their choices.

Different Setup of the Choices

For the choice modeling process a structured approach was used. It could be useful to find out what the effect would be of a different setup, for example, when including the situation of the *Everscape* scenario in the choice itself. The choices could then be to leave before the news item or to leave as a result of the news item. This was tested, but the available dataset was too small. It would be recommended to collect more data to use this setup of the choices.

Different Type of Model

The herding phenomenon involves the behavior of other people, and as it is not known how this exactly works, it was decided to start with basic choice models and use a systematic approach to cover a range of possibilities of the herding phenomenon. A useful next step and different type of approach with the data could be to use a continuous duration model in which a hazard function could capture the probability of leaving after a certain time. Characteristics of the participants as well as the situation can then be included.

Acknowledgments

This paper is based on edited parts of the dissertation of Van den Berg (*19*). It is part of the research program "Traffic and Travel Behavior in case of Exceptional Events" sponsored by the Dutch Foundation of Scientific Research MaGW-NWO.

Author Contributions

The authors confirm contribution to the paper as follows: study conception and design: Mignon Van den Berg, Rob van Nes, Serge Hoogendoorn; data collection: Mignon Van den Berg; analysis and interpretation of results: Mignon Van den Berg, Rob van Nes, Serge Hoogendoorn; draft manuscript preparation: Mignon Van den Berg. All authors reviewed the results and approved the final version of the manuscript.

References

1. Dow, K., and S. L. Cutter. Public Orders and Personal Opinions: Household Strategies for Hurricane Risk Assessment. *Environmental Hazards*, Vol. 2, 2000, pp. 143–155.
2. Whitehead, J. C., B. Edwards, M. Van Willigen, J. R. Maiolo, K. Wilson, and K. T. Smith. Heading for Higher Ground: Factors Affecting Real and Hypothetical Hurricane Evacuation Behavior. *Global Environmental Change, Part B: Environmental Hazards*, Vol. 2, 2000, pp. 133–142.
3. Baker, E. Hurricane Evacuation Behavior. *International Journal of Mass Emergencies and Disasters*, Vol. 9, No. 2, 1991, pp. 287–310.
4. Murray-Tuite, P., and H. S. Mahmassani. Model of Household Trip Chain Sequencing in Emergency Evacuation. *Transportation Research Record: Journal of the Transportation Research Board,* 2003. 1831: 21–29.
5. Ariely, D. *Predictably Irrational: The Hidden Forces that Shape our Decisions*. HarperCollins, New York, 2008.
6. Hamilton, W. D. The Genetical Evolution of Social Behavior. *Journal of Theoretical Biology*, Vol. 31, 1970, pp. 295–311.
7. Bertram, B. Living in Groups: Predators and Prey. In *Behavioral Ecology*, 1st ed. (J. R. Krebs and N. B. Davies, eds.), Blackwell Scientific, Oxford, 1978, pp. 64–96.
8. Alexander, R. The Evolution of Social Behavior. *Annual Review of Ecology, and Systematics*, Vol. 5, 1974, pp. 325–383.
9. Rubenstein, D. I. On Predation, Competition, and the Advantages of Group Living. *Perspectives in Ethology*, Vol. 3, 1978, pp. 205–231.
10. Gueron, S., S. A. Levin, and D. I. Rubenstein. The Dynamics of Herds: From Individual to Aggregations. *Journal of Theoretical Biology*, Vol. 182, 1996, pp. 85–98.
11. Helbing, D., I. Farkas, and T. Vicsek. Simulating Dynamical Features of Escape Panic. *Macmillan Magazines Ltd, Nature*, Vol. 47, 2000, pp. 487–490.
12. Doirado, E., M. Van den Berg, J. W. C. Van Lint, S. P. Hoogendoorn, and H. Prendinger. Everscape: The Making of a Disaster Evacuation Experience. Presented at CHI12, Austin, Tex., 2012.
13. Van den Berg, M., E. Doirado, J. W. C. Van Lint, A. Nakasone, S. P. Hoogendoorn, H. Prendinger, and R. Van Nes. Application of a 3D Multi-User Virtual Environment for Research on Travel Choice Behavior. The Case of a Tsunami. In *International Conference on Travel Behavior Research* (M. Roorda and E. Miller, eds.), Lulu Publishers, 2012, pp. 103–125.
14. Wideman, H. H., R. D. Owston, C. Brown, A. Kushniruk, F. Ho, and K. C. Pitts. Unpacking the Potential of Educational Gaming: A New Tool for Gaming Research. *Simulation & Gaming*, Vol. 38, No. 1, 2007, pp. 1–21.
15. Smith, R. (2009). The Long History of Gaming in Military Training. *Simulation & Gaming*, Vol. 41, No. 6, pp. 6–19.
16. Thompson, D., T. Baranowski, R. Buday, J. Baranowski, V. Thompson, R. Jago, and M. J. Griffith. Serious Video Games for Health: How Behavioural Science Guided the Development of a Serious Video Game. *Simulation & Gaming*, Vol. 41, No. 4, 2010, pp. 587–606.

17. *Everscape* movie. https://www.youtube.com/watch?v=udISaRMWl-c. Accessed on June 18, 2018.
18. Bierlaire, M. Biogeme: A Free Package for the Estimation of Discrete Choice Models. *Proc., 3rd Swiss Transportation Research Conference*, Ascona, Switzerland, 2003.
19. Van den Berg, M. *The Influence of Herding on Departure Choice in Case of Evacuation: Design and Analysis of a Serious Gaming Experimental Setup*. T2016/22, TRAIL Thesis Series, the Netherlands, 2016. doi:10.4233/uuid:dc3bb80c-781f-4eb6-b331-356a0165bdef.

The Standing Committee on Emergency Evacuations (ABR30) peer-reviewed this paper (18-01460).

Article

TRR
JOURNAL OF THE TRANSPORTATION RESEARCH BOARD

Transportation Research Record
2018, Vol. 2672(1) 171–182

DOI: 10.1177/0361198118796351
journals.sagepub.com/home/trr
SAGE

Simulating Indoor Evacuation of Pedestrians: The Sensitivity of Predictions to Directional-Choice Calibration Parameters

Milad Haghani[1,2], Majid Sarvi[1], and Abbas Rajabifard[1]

Abstract

The increasing occurrence of safety-related incidents like fire and terror attacks in crowded public facilities and mass gatherings has heightened the importance of planning for efficient evacuations through optimizing evacuation routes and architectural designs. This calls for the development of simulation and analytical tools that can replicate occupants' responses and thereby their most likely movement patterns. Such models must be accurate to prevent inappropriate design and planning. One major factor connected to prediction accuracy is the sensitivity of modeling outputs to the value of their various parameters. We report on implementation of a calibrated model of directional choices in a microscopic simulation model of pedestrians' evacuation. We show how estimates of the aggregate measures of prediction are sensitive to the parameters of this tactical level (i.e., directional choice) model. Results demonstrate that the prediction of the total evacuation time and average individual evacuation times are closely correlated with one another in terms of their variation, and are both very sensitive to the specification of each directional choice parameter. Simulated evacuation time could vary up to nearly 30% depending on parameter values. The observed sensitivity highlighted the significance of importing well-calibrated parameters into such simulation models and practicing consistent degrees of accuracy for all levels of decision modeling. We also inferred that the two aggregate measures (i.e., total evacuation time and average individual evacuation times) can be used interchangeably as the basis for evacuation optimization or sensitivity analysis practices.

The design of efficient evacuation plans and accessing accurate estimates of evacuation times have become major concerns related to the safety of crowded public facilities, high-rise buildings, and mass gatherings (*1*, *2*). Attention to this problem has heightened as a result of the increasing occurrence of emergency incidents in such environments within recent years (*3*). Research on this problem has grown at a fast pace aiming at developing simulation and optimization tools for supporting fast discharge of occupants (*4*, *5*). Such simulation models are meant to allow planners to evaluate the efficiency of their evacuation plans and the architectural design of facilities, perform reliable risk analysis, and identify problematic locations and critical crowding levels.

The accuracy of such models has always been a major issue given the potential ramifications of inappropriate design or poor planning in this context. Clearly, the accuracy of such models relies heavily on how accurately they replicate the element of human behavior. This has particularly been a significant modeling aspect for the microscopic simulation models that attempt to replicate evacuation patterns by simulating individuals' behavior (as happens in reality) (*6*).

The paucity of data and observations in this particular context has been regarded as a major obstacle towards calibrating and validating many of the existing methodologies and heuristic approaches (*7*, *8*). To address this challenge, one has to resort either to the available field observations from previous emergency incidents (e.g., from CCTV footage) or to collect data in experimental settings. Both approaches, however, pose their own challenges.

While observations from naturally-occurring emergencies are valued for their realism (*9*, *10*), they are rarely

[1]Department of Infrastructure Engineering, Centre for Disaster Management and Public Safety, The University of Melbourne, Melbourne, Australia
[2]Institute of Transport and Logistics Studies, Business School, The University of Sydney, Australia

Corresponding Author:
Address correspondence to Milad Haghani: milad.haghani@sydney.edu.au

accessible to researchers or do not provide sufficient material for modeling or model calibration. Such observations are often one-off and do not offer replications, and the analyst does not have any influence on the observations and thus cannot manipulate any factor for a systematic analysis of a specific phenomenon. They also may be subject to high levels of measurement error (e.g., due to low quality or limited coverage of recorded images).

On the other hand, experimental data provides greater flexibility in that the analyst is in charge of all control factors and can vary them through efficient designs and observe the behavior systematically (*11–15*). Measurement errors can be kept to a minimum level, and the analyst can replicate the experiments multiple times and draw robust inferences. The major issue is how realistically the context of an experiment can represent a case of an emergency (*16*). There are certainly limits to how realistic an experiment of evacuation behavior could be, given the ethical and safety restrictions.

A number of authors in this field have suggested that the aspects of the human behavior that need to be investigated and modeled could be classified into three levels of decision making (*17–20*): (a) occupants' reaction times and pre-evacuation activities, known as "strategic" decisions, (b) their directional route choices, known as "tactical" decisions, and (c) their momentary step choices to avoid collision, known as "operational" decisions. It is clear that accurate modeling in this context necessitates consistent modeling accuracy at all three levels.

In a recent review of the empirical studies in this field, Haghani and Sarvi (*21*) quantitatively showed that studies related to operational (also known as "walking behavior") models greatly outnumber studies that have addressed the two other aspects of evacuation modeling. Well-functioning and data-supported walking models have been proposed that can adequately replicate the step choices of evacuees (given their directional preferences) (*22–24*). How occupants make their directional choices has been much less understood, however (*25–27*). A number of studies have resorted to pure theoretical or modeler-defined assumptions for this aspect (*28–31*). In the absence of empirical observations, however, one may question the accuracy of such assumptions which could often be quite arbitrary. This could be partly attributed to the greater challenges involved in experimenting with the strategic and tactical levels of decision making as opposed to observing and analyzing pedestrians' walking.

In a recent literature survey, Haghani and Sarvi (*21*) reviewed the possible relevant data collection methods in this context and identified seven major empirical methods in their categorization: (1) animal crowd experiments (*32, 33*), (2) virtual reality experiments, (3) controlled laboratory experiments with human crowds, (4) evacuation drills in real buildings, (5) video analysis of naturally-occurring emergencies, (6) video analysis of natural walking, and (7) interview surveys with survivors of mass emergencies. They discussed the advantages and disadvantages offered by each class of data collection and also determined that each class only provides data for particular levels of escape decision making that we categorized earlier.

For the particular problem of directional choices, laboratory experiments with human crowds (*34*), virtual reality experiments (*27, 35*), and evacuation drills (*36*) have proven most popular in former studies. Experiments with human crowds in artificial environments, in particular, have been regarded as a reasonable compromise between the level of contextual realism and the level of controllability. Despite the great efforts that have been made in recent studies, a missing element has been experimental design and analysis at the level of individuals, as previous studies have predominantly been based on the analysis of aggregate (or macroscopic) measures of behavior (*26, 37, 38*). Although useful insight can still be inferred from such aggregate measures, this common feature of the past studies has partly hindered the derivation of behavioral models suitable for microscopic simulation models.

In this study, we report on a series of controlled laboratory experiments with human crowds dedicated to wayfinding and directional choice behavior. The experimental observations have been extracted at disaggregate (i.e., individual) level from which we derived a multivariable model of directional choices. The model probabilistically replicates the directional choices by quantifying evacuees' tradeoffs between a number of factors related to the physical architecture of the escape environment (i.e., physical factors) and a number of factors related to peer influence (i.e., social factors). This tactical level model is then integrated with a conventional operational level model of walking to form a bi-layer simulation model of evacuation. We report on sensitivity analyses on the directional choice parameters of this model to examine the sensitivity of system-wise aggregate measures (e.g., total evacuation times) to these calibration parameters.

Methods

Operational Model Specification

The simulation model of pedestrian evacuations employed in this analysis consists of two layers of decision making. At the operational level, the movement of pedestrians is simulated based on the notion of social force model originally proposed by Helbing, Farkas, and Vicsek (*39*). At the tactical level, the directional route choices of pedestrians are simulated using a probabilistic

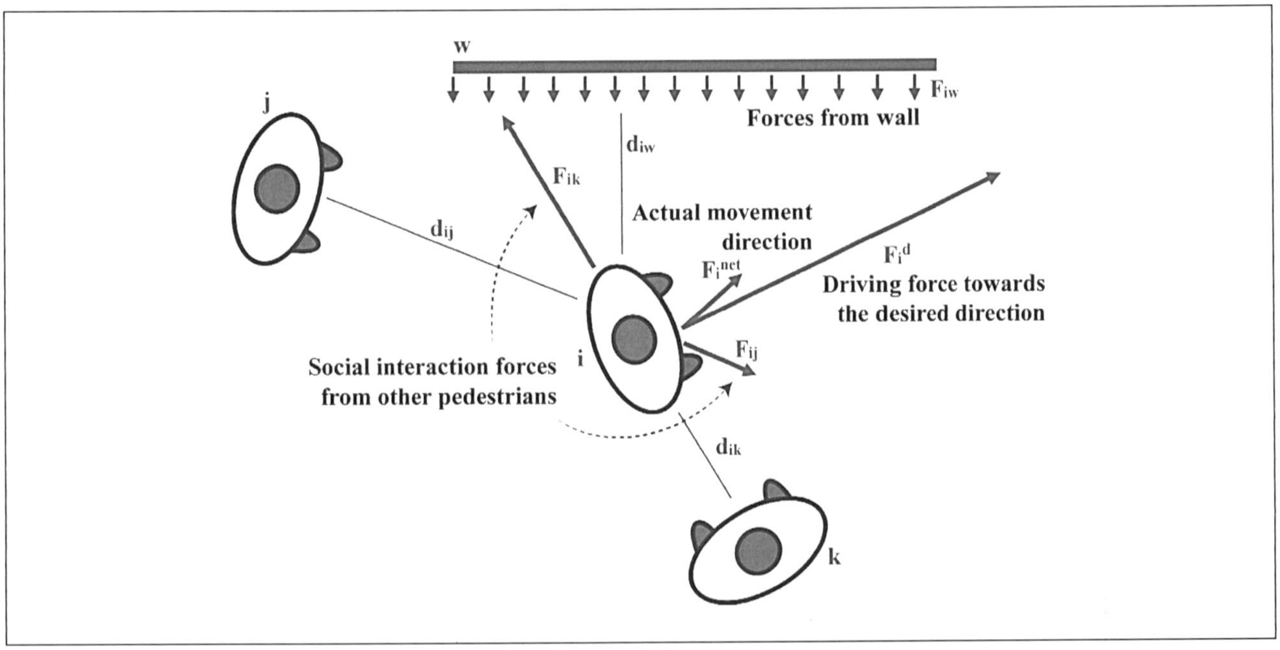

Figure 1. Conceptual illustration of the social force based model for pedestrians' walking, the operational layer of the simulation model employed in this study.

econometric model of discrete choice. The outcome generated from the tactical level model is used as input by the lower level (i.e., the operational level) to generate the simulated movements.

At the operational level model, for a particular pedestrian i, each step motion is determined by the net outcome ($\vec{F}_i^{net}$) of three types of virtual forces: (1) the desired force of pedestrian i driving him/her towards his/her preferred direction $\vec{F}_i^d$, (2) the social interaction forces from other pedestrians $j \neq i$, $\vec{F}_{ij}$; and (3) the forces from walls (w) (or any architectural barrier that the pedestrian would avoid collision with) $\vec{F}_{iw}$ (Equation 1). These three components are defined by Equations 2, 3, and 4 respectively. In these equations, m_i is the body mass of individual i, $\vec{v}_i$ is the vector of velocity for individual i, $\vec{v}_i = d\vec{r}_i/dt$, with $d(.)$ denoting the derivation operator and $\vec{r}_i$ signifying the vector of position for individual i. Also, t represents time and dt signifies the time increment. The unit vector $\vec{e}_i^d$ denotes the desired direction of pedestrian i and v_i^d is the desired (magnitude of) velocity that pedestrian i aims to achieve. The body radius of pedestrian i is denoted by r_i, and d_{ij} denotes the inter-individual Euclidian distance between pedestrians i and j; and e signifies the (natural) exponential function. The function $G(r_i + r_j - d_{ij})$ equals $r_i + r_j - d_{ij}$ when $r_i + r_j - d_{ij} > 0$ and equals 0 otherwise. Also $\vec{n}_{ij}$ is defined as $\vec{n}_{ij} = \frac{1}{d_{ij}}\overrightarrow{(r_i r_j)} = \left(n_{ij}^1, n_{ij}^2\right)$ indicating the unit vector that connects pedestrian j to i. The term Δv_{ji}^t is defined as $\Delta v_{ji}^t = \left(\vec{v}_j \vec{v}_i\right).\vec{t}_{ij}$ in which $\vec{t}_{ij} = \left(-n_{ij}^2, n_{ij}^1\right)$ represents the unit vector for tangential direction. Similarly, d_{iw} signifies the distance of pedestrian i to wall w, and $\vec{n}_{iw}$ denotes the unit vector on the direction perpendicular to the wall w from the position of pedestrian i, and $\vec{t}_{iw}$ is the unit vector on the direction perpendicular to $\vec{n}_{iw}$ (similar to the relation between $\vec{n}_{ij}$ and $\vec{t}_{ij}$). Also, τ_i, A_i, B_i, k_1 and k_2 are all calibration parameters. We specified these values according to the article that originally proposed the social force method (*39*). Figure 1 visually conceptualizes the force based operational layer of the model as described in the previous lines.

$$\vec{F}_i^{net} = m_i \frac{d\vec{v}_i}{dt} = \vec{F}_i^d + \vec{F}_{ij} + \vec{F}_{iw} \tag{1}$$

$$\vec{F}_i^d = m_i \frac{v_i^d \vec{e}_i^d - \vec{v}_i}{\tau_i} \tag{2}$$

$$\begin{aligned}\vec{F}_{ij} = &\left(A_i e^{(r_i + r_j - d_{ij})/B_i} + k_1 G(r_i + r_j - d_{ij})\right) \\ &\vec{n}_{ij} + k_2 G(r_i + r_j - d_{ij}) \Delta v_{ji}^t \vec{t}_{ij}\end{aligned} \tag{3}$$

$$\begin{aligned}\vec{F}_{iw} = &\left(A_i e^{(r_i - d_{iw})/B_i} + k_1 G(r_i - d_{iw})\right) \\ &\vec{n}_{iw} - k_2 G(r_i - d_{iw}) (\vec{v}_i \vec{t}_{iw}) \vec{t}_{iw}\end{aligned} \tag{4}$$

Tactical Model Specification

The operational level layer of the model obtains the component of the desired direction $\vec{e}_i^d$ for each pedestrian i, from the tactical layer (as given input). The tactical model assumes that each pedestrian i perceives a certain level of desirability (or utility) from each exit direction n, U_{ni}, when given a finite and discrete set of N possible exit directions. This utility is composed of a systematic linear-in-parameters component, V_{ni}, and a random error term component, ε_{ni} (Equation 5) added together. In our specification, the systematic part consists of a number of factors related to pedestrian's social interaction as well as a number of factors related to the physical structure of the environment (Equation 6). This term accommodates the pedestrian's tradeoffs between these set of factors. In Equation 6, β_k denotes the utility coefficient (or the relative weight) associated with the kth attribute and x_{nik} is the kth attribute of the alternative exit direction n perceived by (or measured for) pedestrian i, and K is the number of attributes. The error term is assumed to follow an extreme value type I distribution and this distribution is assumed to be identical and independent over all possible exit directions and all simulated pedestrians. Equation 7 determines the probability density function, $f(\varepsilon_{ni})$, of this distribution. This term accommodates measurement errors, perception errors, and model specification errors (e.g., reflecting any variable that is not captured by our systematic part of the model). Based on the premise that the pedestrian would choose the direction with greatest perceived utility, the probability of i choosing any alternative exit direction n will be given by Equation 8.

$$U_{ni} = V_{ni} + \varepsilon_{ni} \quad n = 1, 2, \ldots, N \tag{5}$$

$$V_{ni} = \sum_{k=1}^{K} \beta_k x_{nik} \tag{6}$$

$$f(\varepsilon_{ni}) = e^{-\varepsilon_{ni}} e^{-e^{-\varepsilon_{ni}}} \tag{7}$$

$$P_{ni} = \frac{e^{V_{ni}}}{\sum_{m=1}^{N} e^{V_{mi}}} \tag{8}$$

The systematic part of the utility is specified in our model as Equation 9. The main attributes that are recorded include: $(DIST)_{ni}$, the Euclidian distance of the individual n to exit i measured in meters, $(CONG)_{ni}$ the congestion of exit n for individual i measured in terms of the number of pedestrians (who are present at the vicinity of the exit n, defined as a pre-specified region near the exit), $(FLTOEX)_{ni}$ the size of flow observed by individual i moving towards exit n, measured in terms of the number of pedestrians that are closer to exit n than individual i, are moving towards exit n and are not yet in the vicinity of exit n (i.e., are not counted as congestion), $(VIS)_{ni}$ the visibility status of exit n, for individual i that equals 1 if there is no obstacle or architectural barrier between individual i and the center point of exit n and equals 0 otherwise.

By multiplying $(FLTOEX)_{ni}$ by $(VIS)_{ni}$ and $(1 - VIS)_{ni}$, we basically differentiate between the coefficients of the flows that are moving to visible exits (denoted as $(FLTOVIS)_{ni}$) from the flows moving to the invisible exits (denoted as $(FLTOINVIS)_{ni}$) according to individual i. When exit n is visible to individual i, then $(FLTOVIS)_{ni}$ equals $(FLTOEX)_{ni}$ and $(FLTOINVIS)_{ni}$ becomes zero. Similarly, when exit n is invisible to individual i, then $(FLTOINVIS)_{ni}$ equals $(FLTOEX)_{ni}$ and $(FLTOVIS)_{ni}$ becomes zero. So for each alternative exit n available to each individual i, only one of $(FLTOVIS)_{ni}$ and $(FLTOINVIS)_{ni}$ takes a non-zero value (and that value simply equals $(FLTOEX)_{ni}$). Figure 2 conceptualizes in a visual way the measurements of the social and physical factors related to the tactical layer of the model described above.

$$\begin{aligned} V_{ni} = {} & \beta_1 (DIST)_{ni} + \beta_2 (CONG)_{ni} \\ & + \beta_3 \left(\underbrace{(VIS)_{ni} \times (FLTOEX)_{ni}}_{= (FLTOVIS)_{ni}} \right) \\ & + \beta_4 \left(\underbrace{(1 - VIS)_{ni} * (FLTOEX)_{ni}}_{= (FLTOINVIS)_{ni}} \right) \\ & + \beta_5 (VIS)_{ni} \end{aligned} \tag{9}$$

For each pedestrian i, once the probabilities of all available exit directions are generated based on the tactical model, the choice of the exit direction is simulated and the outcome, $\vec{e}_i^{\,d}$, is then introduced to the operational level of the model. Currently, in the analysis we reported in this study, the exit direction choice is a one-off decision and is simulated as soon as the pedestrian steps in an environment (i.e., room). The directional choice parameters are also assumed to be generic and also constant over the population of the simulated pedestrians for this study.

Tactical Model Calibration

The parameters of the tactical layer model described in the previous section were estimated using a dataset of individual directional choice observations collected from a set of (field-type) controlled laboratory experiments. Nearly 150 individuals were recruited and invited to participate in the experiments which took place at a basketball court in Melbourne, Australia. We made a number of temporary built environments using panels designed for the experiments and asked the participants to escape from that environment as if they were escaping from an imminent danger. They were allowed to run and were asked to compete with other participants and choose the best escape direction (using the set of available alternatives) as they would do in an actual case of emergency.

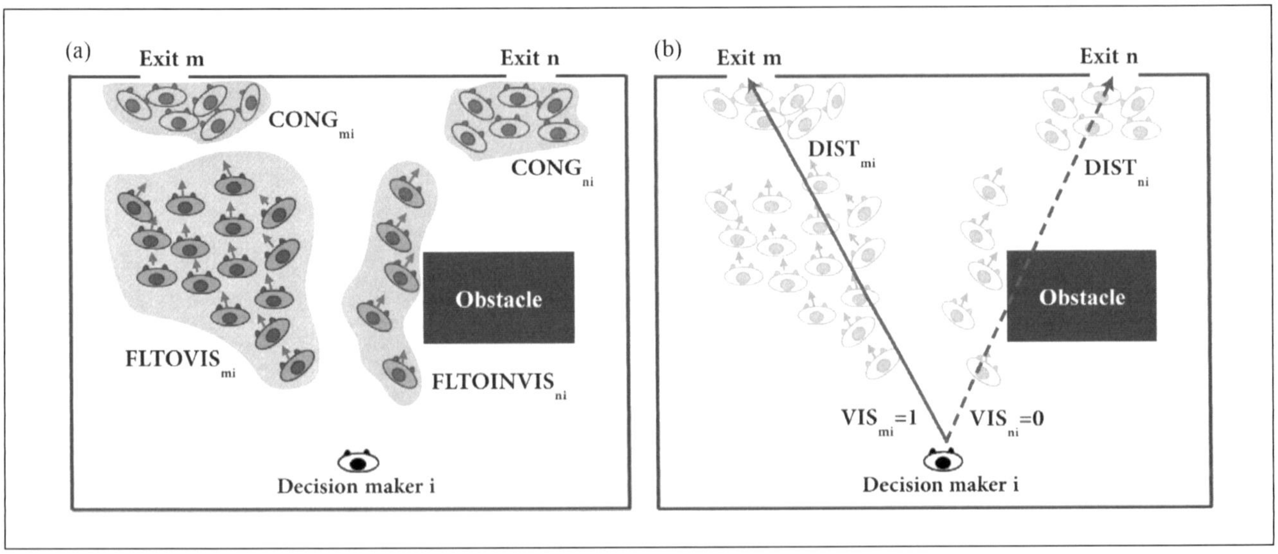

Figure 2. Conceptual illustration of the variable measurements for the directional discrete exit-choice model (the tactical layer model) in the simulation method employed in this study: (*a*) visualization of the measurement of social interaction factors, (*b*) visualization of the measurement of physical factors.

The experiment in total was composed of 29 runs of simulated escape as described above, over which we varied the levels of certain control factors rather randomly in order to create maximum variation in the individual-level observations (*40*). These control factors include the number of evacuees we let in for each scenario, the number and widths of exit doors, and the positioning of available exits. The participants were not informed of these factors (except for the number of evacuees, which they noticed themselves) prior to each run in order to mitigate the effect of habituation to a fixed architectural setting. Figure 3 depicts the blueprint of the environment that was built and manipulated during the experimental scenarios. Figure 4*a* and *b* presents two still images from two scenarios (one with 150 and one with 75 evacuees).

The experiments were videorecorded using a GoPro camera mounted 8 meters above the floor. The legs of the camera tripod were positioned inside the obstacle in the room. The footage was analyzed using PeTrack software to extract the trajectories and the choices of participants. An individual-based data extraction approach was performed in order to provide a modeling dataset from which the parameters of the described tactical model could be estimated. The behavior of every single participant in these scenarios was examined individually, and their trajectories, their choices, and the characteristics of their exit directional choice were extracted. A total of 3,338 observations were extracted. For each observation we recorded three essential elements of the directional choice: (1) the choice set of available exit directions (treated as unlabeled alternatives), (2) the attributes of each exit direction (including the social and physical

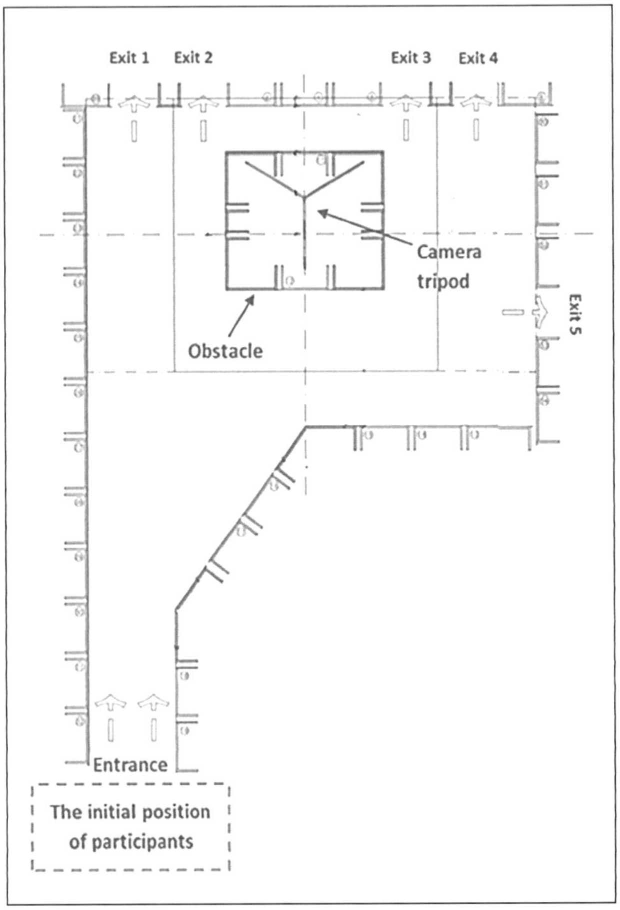

Figure 3. Blueprint of the experimental setup. In each scenario, only a subset of exits was available.

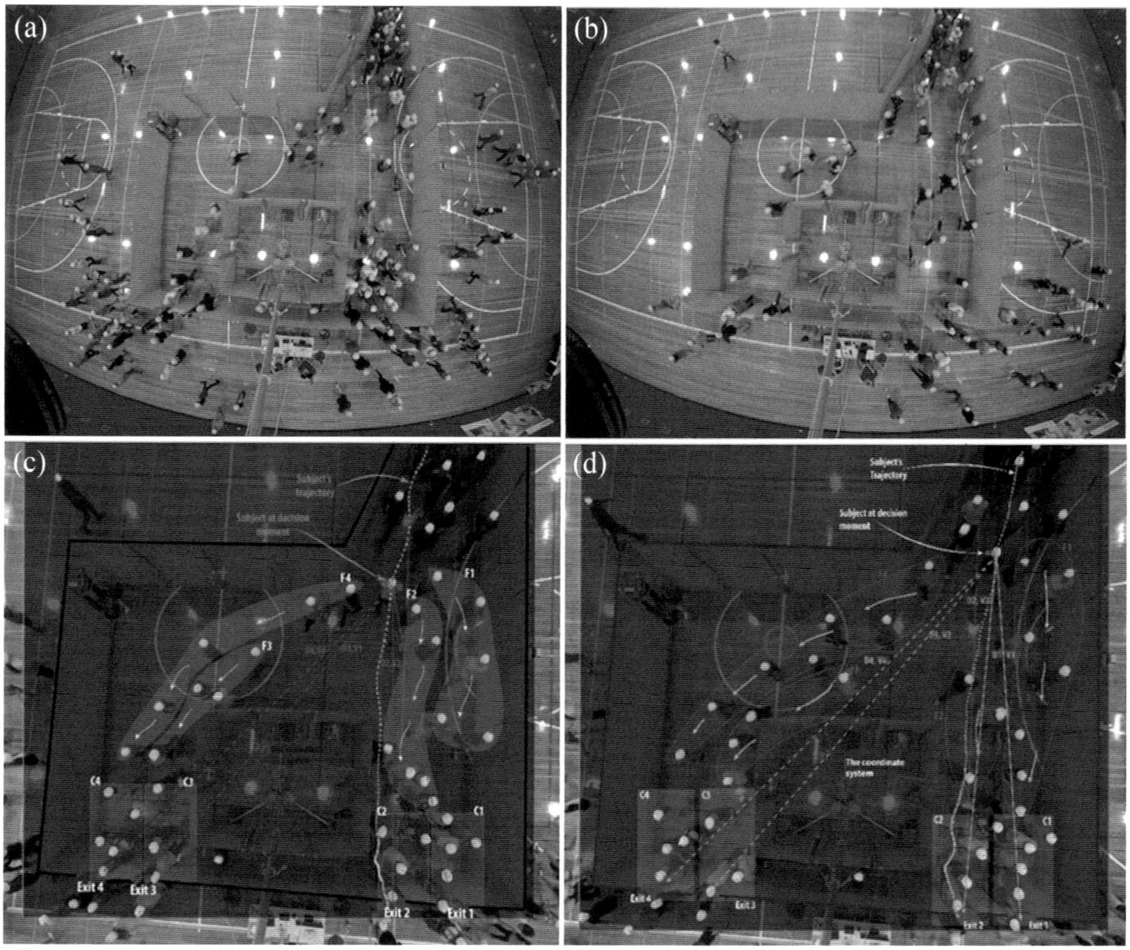

Figure 4. Still images from footage of the simulated evacuation experiments. (*a*) and (*b*) show snapshots from the raw footage of scenarios with 150 and 75 evacuees respectively. (*c*) and (*d*) illustrate the data extraction process at the level of individuals. The process of recording of choice observations is demonstrated by superimposing the visualization of the choice components (consistent with Figure 2) on the interface of the image-processing software. The trajectories show the "choice." The "choice set" and the "alternative attributes" (including the social and physical factors) are illustrated on the pictures. The collection of these three elements qualifies as one disaggregate choice observation.

factors described earlier), and (3) the choice (as the dependent variable). Figure 4*c* and *d* visually demonstrate how these observations were recorded by exemplifying two pedestrians and the characteristics of their choices.

Analyses and Results

Based on the dataset described in the previous section, the mathematical formulation presented in Equation 9 was calibrated using a maximum likelihood estimation process. The calibrated model is presented in Equation 10 where the numbers in subscripts of the estimates represent the *t*-statistic values associated with those estimates. All estimates were statistically significant at conventional levels.

$$V_{ni} = -0.256_{(-11.40)}(DIST)_{ni} - 0.138_{(-12.90)}(CONG)_{ni} - 0.024_{(-2.07)}\left(\underbrace{(VIS)_{ni} * (FLTOEX)_{ni}}_{=(FLTOVIS)_{ni}}\right) + 0.093_{(9.52)}\left(\underbrace{(1-VIS)_{ni} * (FLTOEX)_{ni}}_{=(FLTOINVIS)_{ni}}\right) + 0.710_{(2.85)}(VIS)_{ni} \qquad (10)$$

The combined microscopic simulation model was then investigated by performing a sensitivity analysis on the values of these parameters. The architectural setup presented in Figure 5 was used as the basis of this analysis.

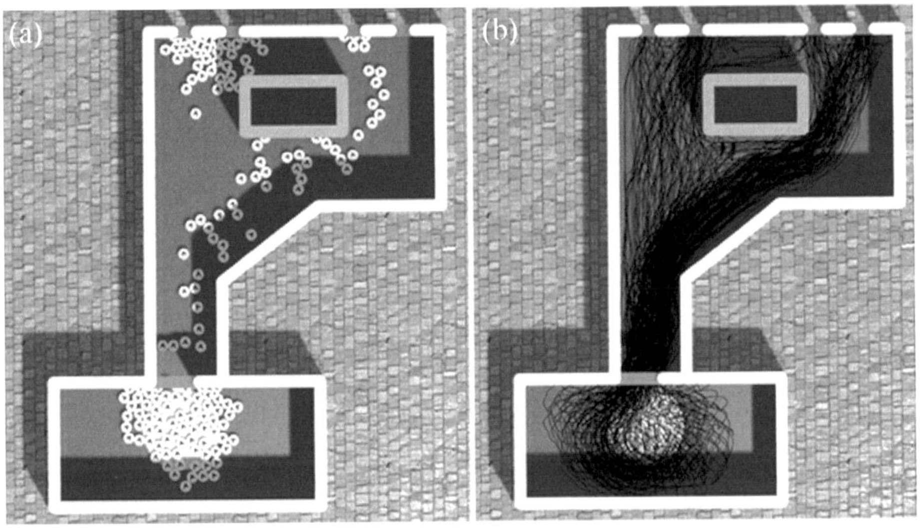

Figure 5. Still images from the visualization of the simulation model employed in this study and the architectural setup used for the sensitivity analyses. (*a*) simulation process of 150 pedestrians while calculations are in progress, (*b*) trajectories of all simulated pedestrians once a simulation run is over.

Figure 5 illustrates still images from the visualization of a simulated scenario while the simulation calculation is in progress as well as an image depicting the trajectories of all simulated pedestrians at the end of the simulation calculations for that scenario. For the sensitivity analysis, 150 simulated pedestrians were generated all at the same time in the rectangular area shown in the pictures. Two aggregate measures of performance were recorded for each run of the simulation: (1) the total evacuation time and (2) the average of individuals' evacuation times. Sensitivity analyses were performed by taking each parameter of the tactical model and departing from its estimated value in both increasing and decreasing directions to the level that no further sensitivity was observed. Given the fact that the variables of the tactical model have different units of measurement, the ranges of sensitivity for different parameters were different. Therefore, no fixed common range for changing the level of parameters could be set a priori.

The simulated experiments associated with the sensitivity analysis of each parameter were performed while the values of all other parameters were kept constant at their estimated values. For the parameter under investigation, the values were gradually increased and decreased while recording the two aforesaid measures. For each individual parameter value, 50 repetitions of the simulation were performed. A total of nearly 7,500 simulation repetitions was practiced for these analyses.

The results of these analyses are summarized in Figures 6–10, where the values of "total evacuation times" (left vertical axis) and "average individual evacuation times" (right vertical axis) have been averaged over 50 simulation repetitions and have been visualized for each parameter value as line charts. The error envelopes represent the standard deviation of the measurements. The graphs are respectively related to the analysis on the values of the parameters β_1 to β_5 in Equation 9. Also, each graph visualizes the estimated value for that parameter along with the 95% confidence interval of that estimate as vertical lines superimposed on each plot.

A common observation based on these sensitivity analyses reflected in Figures 6–10 is the consistency between the two aggregate measures of performance that we applied, total evacuation time, and average of individual evacuation times. As can be seen in these plots, these two measures always invariably display the same type of "sensitivity patterns" to the values of the behavioral parameters. The choice of the most relevant aggregate measure of performance for evacuations is important when system-wise optimization (of architectural design or evacuation routes) becomes the analyst's concern. This high level of correlation between these two measures of performance could be used as evidence that either of these measures could be alternatively applied as the objective function of such optimization analyses.

Another common observation in our analyses was that, the outcome of the simulation is only sensitive to the values of each of the directional parameters within a certain range that varies from parameter to parameter. Also, the values of the estimates for these parameters fell invariably within this sensitive range for all five attributes. In our experience, for some attributes, particularly for the physical attributes whose values are not as time dependent as the social attributes and do not change much by

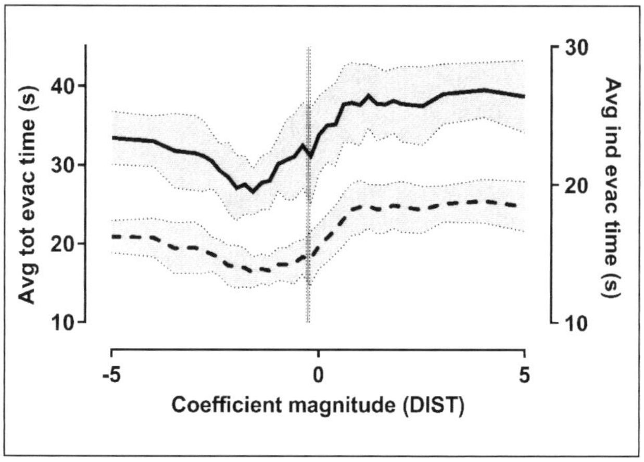

Figure 6. The variation of total evacuation times (solid line, represented by left vertical axis) and average individual evacuation times (dashed line, represented by right vertical axis) in response to the change of the parameter for DIST in the tactical level of the model. Each point on the charts has been obtained by averaging 50 simulation repetitions. The error bands represent the standard deviation of the simulated observations. The vertical line shows the calibrated value of the parameter and the dashed vertical lines specify the 95% confidence interval for that estimate.

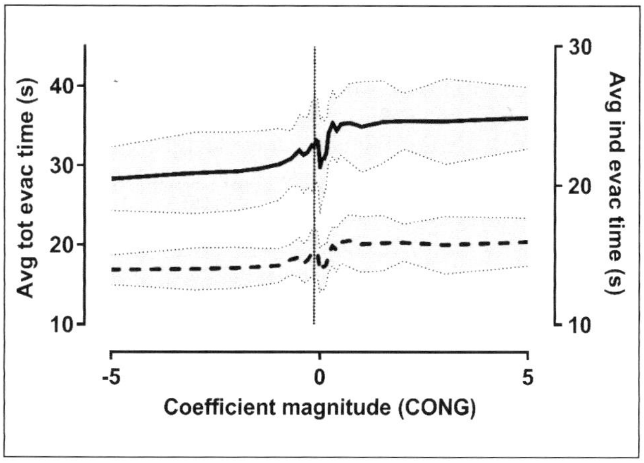

Figure 7. The variation of total evacuation times (solid line, represented by left vertical axis) and average individual evacuation times (dashed line, represented by right vertical axis) in response to the change of the parameter for CONG in the tactical level of the model. Each point on the charts has been obtained by averaging 50 simulation repetitions. The error bands represent the standard deviation of the simulated observations. The vertical line shows the calibrated value of the parameter and the dashed vertical lines specify the 95% confidence interval for that estimate.

the dynamics of the system (take VIS for example), once the value of the parameter associated with an attribute is taken to an "extreme" level for that attribute, that attribute becomes dominant. Once this happens, a very similar pattern of movement is observed at each repetition, as if the directional choices are virtually deterministic.

We observed in our simulated experiments that the level of sensitivity of the aggregate performance measures to the parameter values varies from attribute to attribute. For instance, depending on how we specify the value of DIST parameter, the prediction of total evacuation time for our simulated setup can vary from nearly 27 s to nearly 40 s. Among the attributes included in our tactical model, the least degree of sensitivity of evacuation performance was observed for CONG. Also, it is observable that, except for CONG, the pattern of the sensitivity of the evacuation time to the parameter values is not monotonic and they each show a minimum at an intermediate level for that parameter.

Another aspect of this analyses is connected to the fact that each of the parameters investigated here conveys behavioral interpretation and is reflective of how a "typical" evacuee prioritizes these factors when choosing the direction of movement. Similarly the values of our parameter estimates inferred from our experimental observations are reflective of the dominant prioritization that subjects displayed in the experimental setting. The question one may raise is whether the observed "natural" prioritization of evacuees was optimal from a system optimization perspective. What if the evacuees had shown a different type of prioritization (or relative decision weights)? Could that have benefited the system in terms of the total evacuation times? Also, another relevant question is whether a multi-attribute tradeoff is more beneficial than a single-attribute decision. In other words, would it be beneficial from a system perspective if all evacuees make their decisions based on choosing the closest exit direction, or make their decisions based on only choosing among the visible exit directions, or make their decisions based on following the majority?

The answers to the above questions are case specific, as shown in Figures 6–10. According to Figure 7, avoiding congestion at exits is always beneficial from the system perspective. Decreasing the value of this parameter (equivalent of reinforcing a congestion-avoidance tendency) resulted almost consistently in shorter evacuation times. For the other four parameters, however, the minimum evacuation time was never achieved at extreme values. According to Figure 6, making decisions dominantly based on minimizing distance (associated with an extremely negative parameter value for DIST) would benefit the system more than making decisions dominantly based on choosing the furthest exit directions (associated with an extremely positive parameter value for DIST). The minimum evacuation time, however, occurred at neither of these extreme values, but rather at an intermediate level. According to Figure 8, for visible exits, complete avoidance or complete imitation of the behavior of moving peers choosing that direction are almost equally non-beneficial, and the optimum is again

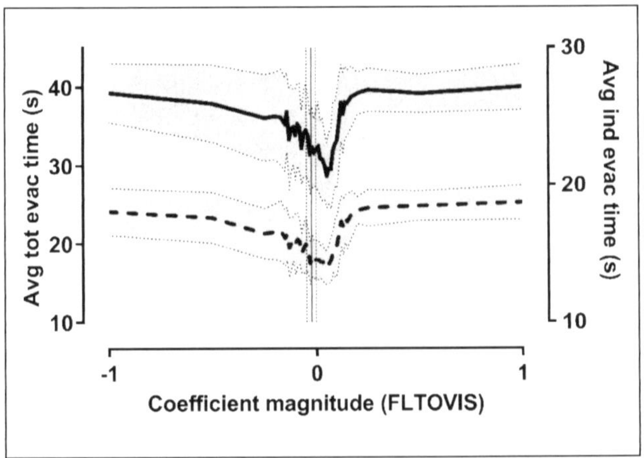

Figure 8. The variation of total evacuation times (solid line, represented by left vertical axis) and average individual evacuation times (dashed line, represented by right vertical axis) in response to the change of the parameter for FLTOVIS in the tactical level of the model. Each point on the charts has been obtained by averaging 50 simulation repetitions. The error bands represent the standard deviation of the simulated observations. The vertical line shows the calibrated value of the parameter and the dashed vertical lines specify the 95% confidence interval for that estimate.

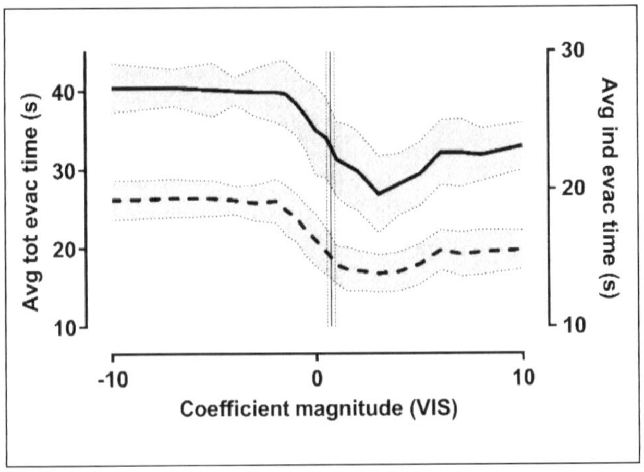

Figure 10. The variation of total evacuation times (solid line, represented by left vertical axis) and average individual evacuation times (dashed line, represented by right vertical axis) in response to the change of the parameter for VIS in the tactical level of the model. Each point on the charts has been obtained by averaging 50 simulation repetitions. The error bands represent the standard deviation of the simulated observations. The vertical line shows the calibrated value of the parameter and the dashed vertical lines specify the 95% confidence interval for that estimate.

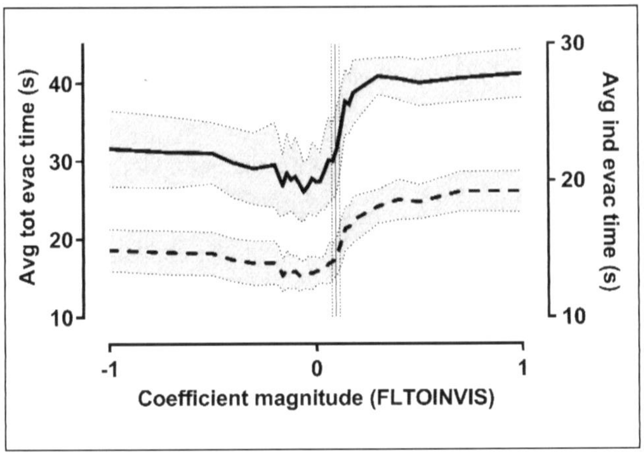

Figure 9. The variation of total evacuation times (solid line, represented by left vertical axis) and average individual evacuation times (dashed line, represented by right vertical axis) in response to the change of the parameter for FLTOINVIS in the tactical level of the model. Each point on the charts has been obtained by averaging 50 simulation repetitions. The error bands represent the standard deviation of the simulated observations. The vertical line shows the calibrated value of the parameter and the dashed vertical lines specify the 95% confidence interval for that estimate.

at an intermediate level. Whereas, according to Figure 9, for invisible exit directions, complete avoidance (extreme negative value) is less suboptimum than complete imitation of the moving peer evacuees choosing that direction. According to Figure 10, making directional decisions purely based on choosing between visible exits (extreme positive values) is also less suboptimal than making decisions purely based on choosing between invisible exits and yet again, an optimum tendency is observable at intermediate levels.

It is also clear that, in terms of the behavior (i.e., set of directional parameters) that is mathematically optimum and results in minimum evacuation times in our setup, there may not exist any unique solution and our analysis was not meant to find such a set of parameters either. It was, however, reflected indirectly in our analysis that for each of the behavioral parameters at tactical level, assuming all other parameters fixed at their estimated values, there could be certain behavioral tendencies (i.e., certain set of attribute prioritization) that could result in better system performance (in terms of the aggregate measures of evacuation).

Summary and Conclusions

This paper reports on the calibration of the directional choice parameters of a microscopic simulation model for pedestrian evacuations using a large data set of disaggregate experimental observations. An analysis was performed on the sensitivity of the simulated evacuation time estimates to the value of these parameters.

The results showed a great degree of sensitivity for the evacuation time estimates to the specification of these behavioral parameters, highlighting the importance of accurate calibration for such parameters. The results also showed that the variation of total evacuation time and average individual evacuation time estimates are closely correlated, suggesting they can be almost indifferently used as the objective function for optimization purposes.

From the behavioral perspective, the simulated results indicated that directional choice-making behavior based on single attributes is less beneficial to the system than multi-attribute tradeoffs (which was the type of behavior people actually displayed in our experimental observations). Extreme values for the parameters proved almost invariably suboptimal from a system perspective, and for each parameter, the value corresponding with the best system performance (minimum evacuation time) occurred at an intermediate level. In other words, and as a practical example, the theoretical microscopic models that replicate the directional route choices purely based on the shortest distance criterion (*29*) may actually result in overestimation of the evacuation time. The same argument can be applied to the models that use "follow-the-crowd" behavior (*41*, *42*) as the directional choice criterion.

The results emphasize the importance of accurate calibration of simulation models for evacuation as opposed to the less costly approach of applying intuitive or pure heuristic rules of behavior. The type of sensitivity analysis reported in this work is relatively time consuming given the very large number of simulation repetitions needed to collect the required data. However, there is potential for this type of analysis to lead to the identification of the optimum behavioral tendencies, which could also have important practical applications. As a simple question, one may ask whether evacuation managers should advise people to follow each other in case of an emergency or whether it is more beneficial for the occupants to avoid heavily used paths (the same question could also be raised for choosing the nearest exit). Answering these types of question was not the primary purpose of this study and this aspect is pointed out as a side outcome of this analysis as preliminary insight. We assume that there is no short universal answer to the problem of "optimum behavior" and that it may depend on many factors, such as the architectural and geometric features of the environment as well as the level of crowding in the environment (and possibly other factors). We intend to provide more comprehensive analysis on these aspects in subsequent studies. Finally, as pertinently pointed out by a referee of this paper, it is crucial to have an understanding of the universality of the parameters inferred from the laboratory experiments that we proposed. The extent to which these parameter estimates are the artefact of the physical setup of the experiment is a valid question that is worth investigation through further experimentation. The authors will report on their findings about the parameter transferability in a future work.

Author Contributions

The authors confirm contribution to the paper as follows: study conception and design: M. Haghani, M. Sarvi; data collection: M. Haghani, M. Sarvi; analysis and interpretation of results: M. Haghani; draft manuscript preparation: M. Haghani, M. Sarvi, A. Rajabifard. All authors reviewed the results and approved the final version of the manuscript.

References

1. Illiyas, F. T., S. K. Mani, A. P. Pradeepkumar, and K. Mohan. Human Stampedes During Religious Festivals: A Comparative Review of Mass Gathering Emergencies in India. *International Journal of Disaster Risk Reduction*, Vol. 5, 2013, pp. 10–18.
2. Yogameena, B., and C. Nagananthini. Computer Vision Based Crowd Disaster Avoidance System: A Survey. *International Journal of Disaster Risk Reduction*, Vol. 22, 2017, pp. 95–129.
3. Kobes, M., I. Helsloot, B. de Vries, and J. G. Post. Building Safety and Human Behaviour in Fire: A Literature Review. *Fire Safety Journal*, Vol. 45, No. 1, 2010, pp. 1–11.
4. Vermuyten, H., J. Beliën, L. De Boeck, G. Reniers, and T. Wauters. A Review of Optimisation Models for Pedestrian Evacuation and Design Problems. *Safety Science*, Vol. 87, 2016, pp. 167–178.
5. Yu, Z., D. Li, S. Zhu, W. Luo, Y. Hu, and L. Yuan. Multisource Multisink Optimal Evacuation Routing with Dynamic Network Changes: A Geometric Algebra Approach. *Mathematical Methods in the Applied Sciences*, Vol. 41, No. 11, 2018, pp. 4179–4194.
6. Duives, D. C., W. Daamen, and S. P. Hoogendoorn. State-of-the-art Crowd Motion Simulation Models. *Transportation Research Part C: Emerging Technologies*, Vol. 37, 2013, pp. 193–209.
7. Shi, L., Q. Xie, X. Cheng, L. Chen, Y. Zhou, and R. Zhang. Developing a Database for Emergency Evacuation Model. *Building and Environment*, Vol. 44, No. 8, 2009, pp. 1724–1729.
8. Haghani, M., and M. Sarvi. Stated and Revealed Exit Choices of Pedestrian Crowd Evacuees. *Transportation Research Part B: Methodological*, Vol. 95, 2017, pp. 238–259.
9. Yang, X., Z. Wu, and Y. Li. Difference Between Real-life Escape Panic and Mimic Exercises in Simulated Situation with Implications to the Statistical Physics Models of Emergency Evacuation: The 2008 Wenchuan Earthquake. *Physica A: Statistical Mechanics and Its Applications*, Vol. 390, No. 12, 2011, pp. 2375–2380.
10. Bernardini, G., E. Quagliarini, and M. D'Orazio. Towards Creating a Combined Database for Earthquake Pedestrians' Evacuation Models. *Safety Science*, Vol. 82, 2016, pp. 77–94.
11. Haghani, M., and M. Sarvi. Social Dynamics in Emergency Evacuations: Disentangling Crowd's Attraction and Repulsion Effects. *Physica A: Statistical Mechanics and Its Applications*, Vol. 475, 2017, pp. 24–34.
12. Bode, N. W. F., S. Holl, W. Mehner, and A. Seyfried. Disentangling the Impact of Social Groups on Response Times

and Movement Dynamics in Evacuations. *PLoS One*, Vol. 10, No. 3, 2015, p. e0121227.
13. Wagoum, A. K., A. Tordeux, and W. Liao. Understanding Human Queuing Behaviour at Exits: an Empirical Study. *Royal Society Open Science*, Vol. 4, No. 1, 2017, p. 160896.
14. Shahhoseini, Z., M. Sarvi, and M. Saberi. Pedestrian Crowd Dynamics in Merging Sections: Revisiting the "Faster-is-slower" Phenomenon. *Physica A: Statistical Mechanics and its Applications*, Vol. 491, 2018, pp. 101–111.
15. Shahhoseini, Z., M. Sarvi, M. Saberi, and M. Haghani. Pedestrian Crowd Dynamics Observed at Merging Sections. *Transportation Research Record: Journal of the Transportation Research Board*, 2017. 2622: 48–57.
16. Haghani, M., and M. Sarvi. Human Exit Choice in Crowded Built Environments: Investigating Underlying Behavioural Differences Between Normal Egress and Emergency Evacuations. *Fire Safety Journal*, Vol. 85, 2016, pp. 1–9.
17. Asano, M., T. Iryo, and M. Kuwahara. Microscopic Pedestrian Simulation Model Combined with a Tactical Model for Route Choice Behaviour. *Transportation Research Part C: Emerging Technologies*, Vol. 18, No. 6, 2010, pp. 842–855.
18. Haghani, M., and M. Sarvi. Pedestrian Crowd Tactical-level Decision Making During Emergency Evacuations. *Journal of Advanced Transportation*, Vol. 50, No. 8, 2016, pp. 1870–1895.
19. Duives, D., and H. Mahmassani. Exit Choice Decisions During Pedestrian Evacuations of Buildings. *Transportation Research Record: Journal of the Transportation Research Board*, 2012. 2316: 84–94.
20. Haghani, M., M. Sarvi, and Z. Shahhoseini. Accommodating Taste Heterogeneity and Desired Substitution Pattern in Exit Choices of Pedestrian Crowd Evacuees Using a Mixed Nested Logit Model. *Journal of Choice Modelling*, Vol. 16, 2015, pp. 58–68.
21. Haghani, M., and M. Sarvi. Crowd Behaviour and Motion: Empirical Methods. *Transportation Research Part B: Methodological*, Vol. 107, 2018, pp. 253–294.
22. Moussaïd, M., D. Helbing, and G. Theraulaz. How Simple Rules Determine Pedestrian Behavior and Crowd Disasters. *Proceedings of the National Academy of Sciences of the United States of America*, Vol. 108, 2011, pp. 6884–6888.
23. Robin, T., G. Antonini, M. Bierlaire, and J. Cruz. Specification, Estimation and Validation of a Pedestrian Walking Behavior Model. *Transportation Research Part B: Methodological*, Vol. 43, No. 1, 2009, pp. 36–56.
24. Seitz, M. J., N. W. F. Bode, and G. Köster. How Cognitive Heuristics Can Explain Social Interactions in Spatial Movement. *Journal of the Royal Society Interface*, Vol. 13, No. 121, 2016, p. 20160439.
25. Haghani, M., and M. Sarvi. Identifying Latent Classes of Pedestrian Crowd Evacuees. *Transportation Research Record: Journal of the Transportation Research Board*, 2016. 2560: 67–74.
26. Bode, N. W. F., and E. A. Codling. Human Exit Route Choice in Virtual Crowd Evacuations. *Animal Behaviour*, Vol. 86, No. 2, 2013, pp. 347–358.
27. Lovreglio, R., A. Fonzone, and L. dell'Olio. A Mixed Logit Model for Predicting Exit Choice During Building Evacuations. *Transportation Research Part A: Policy and Practice*, Vol. 92, 2016, pp. 59–75.
28. Lo, S. M., H. C. Huang, P. Wang, and K. K. Yuen. A Game Theory Based Exit Selection Model for Evacuation. *Fire Safety Journal*, Vol. 41, No. 5, 2006, pp. 364–369.
29. Kemloh Wagoum, A. U., A. Seyfried, and S. Holl. Modeling the Dynamic Route Choice of Pedestrians to Assess the Criticality of Building Evacuation. *Advances in Complex Systems*, Vol. 15, No. 07, 2012, p. 1250029.
30. Fu, L., W. Song, W. Lv, and S. Lo. Simulation of Exit Selection Behavior Using Least Effort Algorithm. *Transportation Research Procedia*, Vol. 2, 2014, pp. 533–540.
31. Abdelghany, A., K. Abdelghany, H. Mahmassani, H. Al-Ahmadi, and W. Alhalabi. Modeling the Evacuation of Large-Scale Crowded Pedestrian Facilities. *Transportation Research Record: Journal of the Transportation Research Board*, 2010. 2198: 152–160.
32. Shahhoseini, Z., and M. Sarvi. Collective Movements of Pedestrians: How We Can Learn from Simple Experiments with Non-human (Ant) Crowds. *PLoS One*, Vol. 12, No. 8, 2017, p. e0182913.
33. Shahhoseini, Z., M. Sarvi, and M. Saberi. Insights Toward Characteristics of Merging Streams of Pedestrian Crowds Based on Experiments with Panicked Ants. *Transportation Research Record: Journal of the Transportation Research Board*, 2016. 2561: 81–88.
34. Haghani, M., and M. Sarvi. Following the Crowd or Avoiding It? Empirical Investigation of Imitative Behaviour in Emergency Escape of Human Crowds. *Animal Behaviour*, Vol. 124, 2017, pp. 47–56.
35. Moussaïd, M., M. Kapadia, T. Thrash, R. W. Sumner, M. Gross, D. Helbing, and C. Hölscher. Crowd Behaviour During High-stress Evacuations in an Immersive Virtual Environment. *Journal of the Royal Society Interface*, Vol. 13, No. 122, 2016, p. 20160414.
36. Kobes, M., I. Helsloot, B. de Vries, J. G. Post, N. Oberijé, and K. Groenewegen. Way Finding During Fire Evacuation; An Analysis of Unannounced Fire Drills in a Hotel at Night. *Building and Environment*, Vol. 45, No. 3, 2010, pp. 537–548.
37. Haghani, M., and M. Sarvi. How Perception of Peer Behaviour Influences Escape Decision Making: The Role of Individual Differences. *Journal of Environmental Psychology*, Vol. 51, 2017, pp. 141–157.
38. Heliövaara, S., J.-M. Kuusinen, T. Rinne, T. Korhonen, and H. Ehtamo. Pedestrian Behavior and Exit Selection in Evacuation of a Corridor – An Experimental Study. *Safety Science*, Vol. 50, No. 2, 2012, pp. 221–227.
39. Helbing, D., I. Farkas, and T. Vicsek. Simulating Dynamical Features of Escape Panic. *Nature*, Vol. 407, No. 6803, 2000, p. 487.
40. Haghani, M., M. Sarvi, Z. Shahhoseini, and M. Boltes. How Simple Hypothetical-Choice Experiments Can Be Utilized To Learn Humans' Navigational Escape Decisions in Emergencies. *PLoS One*, Vol. 11, No. 11, 2016, p. e0166908.

41. Low, D. J. Statistical Physics: Following the Crowd. *Nature*, Vol. 407, No. 6803, 2000, pp. 465–466.
42. Haghani, M., M. Sarvi, and A. Rajabifard. Common Misconceptions about Herd-type Behavior in Emergency Evacuations of Pedestrian Crowds. Presented at 96th Annual Meeting of the Transportation Research Board, Washington, D.C., 2017.

The Standing Committee on Emergency Evacuations (ABR30) peer-reviewed this paper (18-02165).

Article

TRR
JOURNAL OF THE TRANSPORTATION RESEARCH BOARD

Transportation Research Record
2018, Vol. 2672(1) 183–197

DOI: 10.1177/0361198118787343
journals.sagepub.com/home/trr
SAGE

Fear in Humans: A Glimpse into the Crowd-Modeling Perspective

Alastair Shipman[1] and Arnab Majumdar[2]

Abstract

There have been a number of high-profile incidents in recent years in which appropriate evacuation has been necessary due to the extreme nature of the incident (e.g., terrorist attacks). Evacuation under extreme situations causes specific behavioral responses and understanding these can significantly affect the efficacy of an evacuation, potentially saving lives. This paper offers a view of the current understanding of this area of research by performing a literature review into the required areas (sociological, psychological and computer modeling). It finds that although there has been significant progress in sociological models of human behavior, there is a lack of appropriate software models for extreme emergencies that can accurately model states of fear. This point is independently established in interviews with key stakeholders in the area of human behavior and emergencies. The interviewees identify that a better model is required and provide their desired areas of progress, specifically that more data is required to develop new theories and calibrate software models.

In recent years numerous high-profile incidents have required appropriate evacuation (e.g., terrorist attacks). Such evacuations, and accompanying stress or "fear," can cause unique behavioral responses and understanding these can significantly affect the efficacy of the evacuation, potentially saving lives.

To assess evacuation strategies, currently quantitative human movement modeling is performed using computer-based "pedestrian dynamics" simulation models. The use and availability of these models has greatly increased since the late 1990s given vast improvements in computing and simulation capabilities. Incorporating human behavior into these models is key to their predictive capabilities (*1*, *2*). Typically, these models investigate quotidian events, such as movement through a train station. There is guidance available on how to mimic certain behavioral characteristics (*3*), which can be used as an indicator of how predictive the models are.

However, available simulation models lack validated predictive capability in many areas, including the modeling of how human behavior and decision-making changes under stress, or fear, as experienced in extreme emergencies such as terrorist attacks.

This deficiency is partly because of the difficulties of obtaining reliable data—for both developing an experimental procedure, and in recording data during an actual emergency. This is further compounded by a poor understanding of the difference between individual and crowd reactions in these scenarios. Without this data or understanding, the models used to plan evacuations in these scenarios will not provide the required level of reliability.

This paper examines how human behavioral responses to stressful (or dangerous) scenarios have been analyzed and modeled previously. This will allow better-directed future development of theoretical models of human behavior, and improve the predictive capabilities of computer models. Specifically, it offers an overview of the theoretical approaches and an examination of experimental and empirical studies which aim at describing the nature of individual and collective reactions to fear-inducing situations. The literature review concludes by examining various relevant pedestrian dynamics models, including models that have explicitly included these theoretical or experimental elements. Notably, there is only slight overlap between these two areas.

Furthermore, several interviews with the relevant stakeholders were undertaken. The interviewees come from areas of pedestrian dynamics, emergencies, and resilience to assess the current state of these industries and how they use behavioral reactions to fear. Finally, the interviews examine the demand for improvements in this area.

[1]Centre for Transport Studies, Department of Civil and Environmental Engineering, Imperial College London, London, UK
[2]Lloyds Register Foundation Transport Risk Management Centre, Centre for Transport Studies, Department of Civil and Environmental Engineering, Imperial College London, London, UK

Corresponding Author:
Address correspondence to Alastair Shipman:
a.shipman16@imperial.ac.uk

Literature Review

This literature review tackles two separate fronts: firstly, by understanding the individualistic and sociological theories behind human behavior under stress, and secondly, by assessing the current state-of-the-art computer modeling techniques.

To understand current computer modeling techniques, the focal starting point was reference books (*4–8*). Further searches of online repositories, including code words "pedestrian modeling" and "fear" provided the remaining references. This literature also guided the research on the theoretical background to human behavior. This knowledge was subsequently extended by semi-structured interviews with subject matter experts in the field, and a search of online sources, e.g., Google Scholar, including code words "human behavior" and "stress" or "fear."

The main findings of this literature review are presented below, summarizing the key findings of each paper in Tables 1, 2 and 3, together with a commentary on key themes under the relevant subcategory.

Theoretical Approaches

Numerous theories (both individualistic and crowd-based) have attempted to detail how human behavior changes after the onset of an emergency, all attempting to provide a framework with which to explain and eventually predict empirically observed phenomena.

The major crowd theories are Panic Theory, Emergent Norm Theory, Social Attachment Theory and Self-Categorization Theory. This section will also outline an individualistic theory, the Protective Action Decision Model.

Panic Theory (*7*, *9*, *10*) was the primary forerunner theory in understanding how human beings react to physical threats, arguing that within a group acting in the presence of a threat, human beings lose their individuality and become an anonymous part of a whole. Therefore, any action within the group is immediately copied unquestioningly by every other individual, causing "mob mentality." However, this theory is now considered outdated and does not account for empirically observed effects.

Quarantelli (*11–13*) described the required features for a panic to develop, before concluding that "compared with other reactions panic is a relatively uncommon phenomenon." This has also been argued by others (*14*, *15*) and seen empirically (*16*). Interestingly, despite this academic research, Panic Theory is still prevalent, especially in official guidance (*17–19*).

This is not to say that panic does not occur, but rather that it occurs in very specific scenarios. Schultz (*20*) has offered compelling evidence of "panic" in military settings, and Johnson (*21*) summarizes several different incidents involving some form of emergency crowd movement. The existence of "panicking" individuals might be explained by Entrapment Theory, where people begin to ignore social rules, and act in irrational, non-adaptive, and competitive ways in very specific scenarios. These are when there is

1) An immediate, pressing threat to the individual's life.
2) An opportunity to escape, which is diminishing quickly.
3) "A feeling of helplessness in otherwise dealing with the threat" (*22*).

This is mirrored by Chertkoff and Kushigian (*8*), who go on to state that large-scale emergencies and fatalities occur when several preconditions are met in combination. These are when there exists a

1) Severe limitation on space.
2) Large number of people present.
3) Lack of knowledge about the local layout.
4) Lack of emergency management and guidance.

Elsewhere, the Emergent Norm Theory (ENT) (*23*) suggests that, instead of "panicking," people group together during an emergency and, out of an ambiguous scenario, define what is occurring, and the best course of action is developed from within the group, by the generation of new social norms. This theory permits a heterogeneous group to start acting in a homogeneous way and has been used to inform conceptual models (*25*).

The Social Attachment model (*26*) attempts to explain the empirically observed actions of individuals who seek out "familiar" others in an evacuating crowd ("affiliative behavior"), and has been examined against various scenarios, including combat and fires. This theory provides an avenue into modeling the likely choices of people moving within a group, especially in an evacuation scenario.

The Self-Categorization Theory (SCT) allows people to define themselves dynamically within an emergency, showing how "emergent properties of group processes can be explained in terms of a shift in self-perception, from personal to social identity" (*27*, p.2). Crucially, SCT provides a way of explaining why strangers tend to help each other in emergencies (*28*). SCT has been examined against several real-world incidents, see for example Stott and Drury (*29*).

The Protective Action Decision Model (PADM) (*30*) focuses on individual behavior. PADM describes the internal processes of an agent through an emergency scenario and can be readily used to explain the "milling" and "keynoting" processes that occur in ENT. However, it is not easily applied to the social aspects of emergency evacuation, detailed by the SCT and the Social Attachment model.

Empirical Approaches

Empirical research gathers and uses data from previous emergencies to inform theoretical analysis. A key example includes the investigation by Feinberg and Johnson (*31*) examining the

Table 1. Summary of Selected Theoretical Approaches

Author	Area of research	Method	Key findings	Comments
Le Bon (*10*)	Crowd behaviors Emergencies	Holistic	Panic Theory: The loss of individual identity in favor of crowd anonymity in emergencies leads to mob mentality, riots and panic.	The initial major theory of human behavior and has informed most of the guidance on crowd management. Superseded by more recent theories and empirical evidence.
Chertkoff and Kushigan (*8*)	Crowd behavior Emergencies	Review and case study analysis	Provides an overview of theories of panic and human behavior in emergencies. Provides an analysis of several examples of emergencies (largely based on fire), both successful and unsuccessful examples of evacuation.	Exceptional review of previous panic theories and case studies.
Turner and Killian (*23*)	Crowd behavior Emergencies	Holistic	ENT: the process of information gathering in ambiguous, threatening scenarios, provides a way for groups to generate new social norms.	A primary theory of human crowd behavior during emergencies. Does not account for social attachments, or the presence of multiple groups, or the reasoning behind norm selection.
Mawson (*26*)	Crowd behavior Emergencies	Incident reports Interviews Theoretical approach	Agrees with the argument that people generally don't panic in emergencies. Instead develops the Social Attachment (SA) model, stating that during evacuation people look for familiar others/places while moving, or "affiliative behavior."	Powerful theory explaining why people re-enter dangerous situations. Fails to account for stranger-interactions seen empirically.
Turner et al. (1987)	Crowd behavior	Holistic	Argues that people form groups based on SCT into a shared identity, leading to a distinction of "us" and "them." This can be readily extended to emergencies.	Powerful theory that can also explain why strangers help each other in emergencies.
Lindell and Perry (*30*)	Individual behavior Emergencies	Theoretical	Develops the PADM, providing an approach to understanding individual decision-making during the change from normal to emergency scenarios.	Good individual approach, incorporating behavioral considerations. Has some testable conclusions for real-world application.
Quarantelli (*13*)	Crowd behaviors Emergencies Panic	Incident report analysis	States that panic is rare, and is an unhelpful way of analyzing behavioral response to emergencies	Major text saying that panic is rare. Quarantelli is therefore considered a referential author on the subject.
Johnson (*21*)	Crowd behaviors Review of emergencies	Incident report analysis Interviews	Corroborates Quarantelli's arguments providing multiple scenarios for comparison of behavioral reactions.	Introduces analogues of emergency scenarios, including a bank run.
Schultz (*20*)	Review	Interviews Incident reports	Assesses occurrences of panic in the military, specifically the requirements for mass flight.	Considers "predisposing" and "precipitating" factors that can affect the likelihood of a flight reaction in trained individuals. Another precursor to SA theory.
Pastel (*16*)	Panic Military	Interviews Incident reports	Examines the occurrence of panic in military contexts. Introduces the idea of an outbreak of mass unexplained symptoms (OMUS).	Highlights the fact that there are many different types of reaction to fear.
Stott and Drury (*29*)	Crowd behavior Emergencies	Interviews Video footage Incident reports	Analyzes self-categorization processes in an incident where a civilian march turned to violence. Found that the police charges against the crowd formed an "us" versus "them" mentality within the crowd, that did not exist before.	Strong evidence shown in favour of self-categorization processes within a stressful situation.

(Continued)

Table 1. (Continued)

Author	Area of research	Method	Key findings	Comments
Fritz and Marks (*18*)	Emergencies	Interviews Incident reports	Examines five separate emergencies, in 1951 and 1952, all of which included loss of life. Reaffirms Quarantelli's arguments and provides a statistical description of various reactions. Argues for improved communication techniques.	Provides inconclusive evidence about traumatic scenarios (including separation from family members) but is one of the bases for SA theory.
Gantt and Gantt (*22*)	Panic	Case studies	Provides an explanation of the requirements for panic to set in, as well as an investigation of previous case studies.	A brief overview, with useful references.
Schultz (*42*)	Review	Holistic	Provides an overview of theories of individual and crowd panic.	Touches on multiple theories of panic but does not answer any apparent contradictions from other areas of the literature.

behavior of occupants of the Beverly Hills Supper Club fire, aiming to correlate their ultimate fate with one of several models of the breakdown of social order. Similarly, Baddeley (*32*) examines various papers on human performance in a range of extreme scenarios, including combat, parachuting and scuba diving. He observes significant attention-narrowing, i.e., peripheral activities are given less attention and consequently are performed worse. Schultz (*20*) provides examples of panic within military contexts, which is referenced by Mawson (*26*) to help form the basis of the Social Attachment model. More recently, similar analysis has been performed on the evacuation procedures during the World Trade Center Attacks, 2001. For details see Kuligowski (*33*).

Experiments that allow measurement of reactions to a stressful situation are inherently difficult to perform, given both practical and increasing ethical considerations. Thus, there have been far fewer experiments in recent years than in previous decades. Mintz (*34*) formulates a thought experiment into the conditions necessary for panic to occur and provides some of the first experimental evidence of how actions change in the presence of a threat or fear stimulus. He shows that people can engage in non-adaptive group behavior when their needs conflict with others. Kelley et al. (*35*) comment on Mintz's research, decrying the lack of a sufficient stressor to properly investigate the demonstrated effect. They further Mintz's research by threatening an electric shock to investigate how social cues affect people's actions. Keinan et al. (*36*) subsequently timed several different effects seen when subjects were provided with cognitive tests, while again exposed to the threat of electric shocks. These effects include: "temporal narrowing" (assessing each option for less time than previously), "premature closure" (choosing an answer before having assessed all options), and "non-systematic scanning" (not checking the answers in order). They provide a potential basis for modeling individual panic responses in a computer simulation, for a microscopic-level, agent-based, cognitive model.

More recently, experimental research has been performed by Drury et al. (*37*), using a virtual reality scenario to provide ecological validity. This experiment examines how likely people are to push past or help each other during the evacuation of a train station, depending on the level of agreement between the perceived category of the simulated agent and the self-categorization of the volunteer (*37*). Bar-Haim et al. (*38*) examined how anxiety might affect how people experience time during an emergency scenario, which could inform various elements of a cognitive model, but would more likely serve to support Baddeley's proposition of attention narrowing.

Field experiments are a key element of this research but can be difficult to study, as any ecological validity needs to be controlled, and extraneous effects are often not repeatable. For example, Proulx and Sime (*39*) examine how people react in an underground train station to different types of fire alarm, including different public-address (PA) system announcements and verbal cues from the emergency services. Zinke et al. (*40*) performed a similar study, in which commuters in underground stations were asked to imagine an emergency scenario, and then evacuate the underground station. Their results indicated strong sociological effects, where cohesion was maintained within certain groups, and then showed altruistic behavior in all groups, providing solid supporting evidence for the SCT.

Quantitative measurements have been performed by Li et al (*41*), that provide an example experimental design for the investigation of bladed attacks.

Finally, two relevant reviews were written by Schultz (*20*) and Starcke and Brand (*42*). In 1965, Schultz (*43*) discussed various theories of "panic", analyzing the required elements for this response to occur, ignoring Quarantelli's 1954 arguments, instead focusing on Le Bon's crowd theories. In contrast, Starcke and Brand (*42*) provide a review of papers exploring decision-making under stress, examining a range of concepts such as the neuro-biological and psycho-physiological concepts of stress, before going on to critically

Table 2. Summary of Selected Empirical Approaches

Author	Area of research	Method	Key findings	Comments
Mintz (*34*)	Laboratory experiment	Coordinative tests under incentive.	Non-adaptive* behavior can occur with competing priorities between agents. High emotional states of fear are not required.	Useful view at how competitive behavior can occur but does not account for any differences in a situation with a "fear" variable.
Kelley et al. (*35*)	Laboratory experiment	Coordinative tests under threat of electric shock.	As threat level increases, co-ordination decreases. If people can communicate confidence, then co-ordination increases. If communication allows people to observe waiting behavior, then the outcome is dependent on "optimism" within the group.	Interesting view into how communication can affect outcomes within a stressful scenario. Difficult to repeat owing to ethical considerations.
Keinan and Friedland (*36*)	Laboratory experiment	Cognitive tests under threat of electric shock.	Found evidence for stress induced non-systematic scanning and premature closing. Found no evidence for temporal narrowing.	An early attempt to quantitatively characterize specific reactions to fear. Difficult to reproduce owing to ethical considerations.
Baddeley (*32*)	Review	Selective review.	The presence of a threat will narrow the attention and increase the arousal of individuals in the scenario. Experience of a particular threat tends to inhibit this reaction.	A selective review that provides a concise way of stating some behavioral reactions under threat.
Proulx and Sime (*39*)	Field experiment Evacuation efficiency	Measurement of real evacuation efficiency given different PA announcements.	When provided with more specific information, passengers on the train system take more effective action at evacuating. Passengers must trust the PA system if they are to take action.	Confirmation of the idea that panic doesn't occur, and that people are rational when under threat. Requires that people trust the information, so false alarms and regular drills may be counterproductive; also, that management should be fully informed.
Bar-Haim et al. (*38*)	Laboratory experiment Perception of time	Anxious and non-anxious subjects were asked to estimate the length of time a picture was shown. The picture was of a person with a "fearful" or "calm" expression.	The results indicate that anxious individuals subjectively perceive time passing more slowly in the presence of a threat.	An introductory way to test individual reactions to stress, but it is difficult to envisage how this data will be used for predictions of reactions.
Feinberg and Johnson (*31*)	Emergencies Breakdown of social structure	A comparison of casualty and injury levels obtained through incident reports and interviews.	According to a statistical review of casualties in the Beverly Hills Supper Club fire, and predicted correlations, social order did not break down during the fire.	Relies on an intuitive prediction of casualties given various social norms.
Drury et al. (*28*)	Laboratory experiment and model	Virtual reality. Visualization and immersion.	Places volunteers into a virtual scenario where they have to evacuate a train station. Compares how the volunteers identify with strangers, and how various theories can explain the results. Argues that SCT is the most appropriate.	Provides a good example of how mass emergency reactions can be studied in the laboratory. An improvement would be to use fully immersive headsets, rather than computer screens.
Starcke and Brand (*42*)	Review	Systematic searches of papers between 1985 and 2011.	Provides an in-depth review and discussion of the neurophysiology of stress, decision-making under stress and stress related disorders.	Useful starting point to examine different areas of stress/fear related research.

(Continued)

Table 2. (Continued)

Author	Area of research	Method	Key findings	Comments
Kuligowski (*33*)	PhD Thesis Incident review Interviews	World Trade Center human movement analysis.	Provides a detailed view of the thought processes and evacuation procedures of the occupants of the World Trade Center.	Detailed predictive model is developed for behavioral responses in emergencies.
Zinke et al. (*40*)	Data generation of emergency behavior in underground stations	A review of previous laboratory and field tests. The study then asked participants (the demography was analyzed prior to the test) from three separate underground stations to imagine themselves in an emergency and evacuate the station as fast as possible.	Altruistic behavior was seen in all 3 tests, even though post-experiment questionnaires suggested a moderate level of stress. Group cohesion was also seen, with people remaining as family or social groups. Whereas evacuation time and efficiency depended on group size, helping behavior was independent of size.	A good provisional assessment of human behavior in emergencies, with respect to sociological theories. However, it does not include quantitative results that can be used to input into computer models.
Li et al. (*41*)	Data generation of movement behaviour and model	A novel investigation into how pedestrians react to a bladed attacker, and a useful example of how this could be simulated.	Reactions are asymmetric, and strongly dependent on the starting position of the attacker.	Exceptional first step towards understanding reactions to bladed attackers. Further work will provide much needed validation.

Note: *Non-adaptive behavior can be defined in many ways. Mintz uses it solely to highlight where rational individualistic behavior causes a sub-optimal group outcome (for example in crowding an exit during a fire). In this field, the term has also described individual behavior that does not react to present and clear information.

examine papers investigating decision-making while exposed to stressful scenarios.

Computer Modeling Approaches

The major aim of a computer model of pedestrian dynamics is to provide predictive capability, allowing a scenario to be instantiated and subsequently modeled to completion. A standard test of whether a model is predictive relates to whether it reproduces various phenomena observed in relevant scenarios. Thus, to model emergencies, the types of behavior and cognition seen in emergencies must be used to calibrate the models. A significant amount of specific data is needed for statistical certainty. As noted later in the interviews, there is a high demand for a model that can accurately reproduce how human beings behave in emergencies (such as a fire or a terrorist attack).

This section examines examples of how previous studies have explored how to introduce human behavioral and emotional considerations to pedestrian dynamics models, in light of some of the considerations examined earlier in the literature review.

Several review papers present the ability of various software packages to represent certain scenarios, and a comparison of them with respect to certain metrics. For example, Duives et al. (*44*) provide a review of the different types of models available, comparing them with respect to computational efficiency and phenomena reproduction. Further to this, Kuligowski (*45*) provides a review of the available egress models, for use with fire planning and engineering, building on the research by Gwynne and Galea (*46*).

Finally, there are reviews that incorporate social science concepts with computer modeling. Templeton et al. (*2*) provide an analysis from a sociology viewpoint, classifying computer models in terms of their ability to represent human behavior (from "homogeneous mass" to "cognitive groups"). They argue for the inclusion of behavioral and group effects in pedestrian dynamics models to achieve a satisfactory level of realism. Aguirre et al. (*47*) detail how agent-based modeling can be used to replicate concepts from social science while modeling emergencies. They conclude with seven different recommendations to computer modelers, finally arguing for the use of video footage to design and calibrate models.

There are several ways of modeling behavioral responses under stress. For instance, Hollmann et al. (*48*) describe a method of understanding emotional behavior in travelling pedestrians. The emotional state is inferred as agents in a train station alter their itinerary and route choice depending on the remaining available time. This is an initial step toward being able to model a fear response, as it allows for variation of the agent's plans within the model simulation itself, as opposed to prescribing a reaction ahead of time.

Pan et al. (*49*) provide an algorithm-based approach for modeling behaviors under stress, including social theory within the model, focusing on modeling panic theory.

Table 3. Summaries of Selected Computer Modeling Approaches

Author	Area of research	Method	Key findings	Comments
Templeton et al. (*2*)	Review	Systematic review of available models with respect to human behavioral representation.	Most models approximate humans to a homogeneous mass, with cognitive groups being the least represented.	Homogeneous mass is the least individualistic way of modeling human behavior, but the most computationally efficient. Cognitive groups allow for individual behavior to be modeled but require calibration.
Kuligowski (*45*)	Review	Review of available models with respect to egress.	Models have a range of applications which they can model, with some models being better suited than others to certain scenarios.	Useful list and review of available models.
Duives et al. (*44*)	Review	Review of types of modeling approaches (Cellular Automata, etc.) with respect to various metrics.	Modeling approaches have a range of benefits and drawbacks. Some approaches are better suited to design questions than others.	Specifically focuses on high density pedestrian dynamics.
Aguirre et al. (*47*)	Review	Review of available models with respect to human behavioral representation.	Seven major recommendations on future modeling of human behavior in emergencies.	Useful breakdown of the Rhode Island night club fire (2003).
Lawson (*70*)	PhD Thesis Review	Comparison and analysis of methods of obtaining data to predict human behavior in emergencies.	An exhaustive view of the validity of data gathering approaches, including laboratory experiments, field experiments and hypothetical approaches.	Excellent reference literature but should be updated to account for improvements in virtual reality environments.
Helbing, Farkas and Vicsek (*50*)	Emotional modeling	Theoretical software approach.	An extra force in the Social Force Model allows "herding" behavior to be approximated. Competitive behavior also modeled as clogging of exits.	An interesting step toward adapting the SFM to account for different behavioral states. It asks for video data at the end for calibration. However, it approximated humans to a homogeneous mass, which is argued against by Templeton et al.
Franca et al. (*25*)	Conceptual model	A theoretical model, showing how ENT could be incorporated into a computer model.	n/a	Interesting approach but does assume ENT is the correct approach for emergency scenarios.
Zou et al. (*51*)	Emotional modeling	Theoretical software approach, modeling competitive behavior.	Competitive behavior is modeled by allowing lattice-gas agents to "overtake" each other. Provides analysis.	Useful analysis into the mathematics behind competitive movement but does not account for calibration with real world data.
Tsai et al. (*52*)	Emotional modeling	Theoretical software approach applied to LAX airport.	By incorporating an integer fear variable, and allowing "Spread of Knowledge" concepts, this model attempts to understand the reaction dynamics to a threat, and the best crowd management techniques to counter it. This model also approximates family movement and emotional contagion.	A very useful step toward an accurate behavioral model, allowing for various group level concepts, as well as stress considerations. This model does not mention calibration against real-world data but was discussed with LAX professionals.
Haghani and Sarvi (*66*)	Route choice and decision analysis	Interviews and questionnaires.	Shows a statistical breakdown of route and exit choices during a hypothetical emergency. Compares the desirability of crowding against proximity of an exit or threat.	Provides useful insight into the various priorities of evacuating pedestrians. Suffers from a lack of ecological validity.

(Continued)

Table 3. (Continued)

Author	Area of research	Method	Key findings	Comments
Hollmann et al. (*48*)	Emotional modeling	Theoretical software approach.	Itinerary based approaches can approximate the decision process of an individual in a dynamic scenario.	Effective first step toward modeling fear, however, will need to be calibrated with respect to options available and probability of selection.
Shiwakoti et al. (*60*)	Movement data	Experiments on ants to replicate human behavior. Development of a model to approximate this effect.	There is replication of the tested phenomena, namely that there is a faster flow of ants if there is an obstruction immediately in front of the exit.	Using ant repellent to induce movement is a method of replicating stressed behavior in later papers. This result may only be valid for behavior of ants in emergencies.
Dias et al. (*64*)	Data generation to compare normal and panic scenarios.	People are asked to walk along a corridor, then jog, then repeat for various angle turns.	Provides data on flow/density relationships and interpersonal distance at walking and jogging speed.	Very useful as a proxy for a panic or emergency state. Will be accurate for a steady-state fast moving simulation.
Altshuler et al. (*57*)	"Panic" movement data	Experiments on ants to replicate herding effects.	Directly observes herding by placing ant repellent in the center of a dish containing ants, with two diametrically opposed exits. Herding is observed to produce up to 80:20 split in door selection.	Provides useful insight into how behavioral differences under stress can be observed. This would form the basis of a calibration for computer models.
Saloma et al. (*63*)	"Panic" movement data	Mice are placed within a deep pool of water and their movement is recorded.	Provides data about how mice move when placed in deep water. Builds a model based on this data.	By altering the escape door width, this model states an optimal sampling rate for measuring escape panic. It is unclear how relevant this data is for human behavioral considerations.
Tsai et al. (*67*)	Empirical validation of emotional contagion models	By comparing the output of two emotional contagion models (within the ESCAPES pedestrian dynamics software) with CCTV footage of two separate emergency scenarios, this study begins calibrating models with real data.	The ASCRIBE model (72) performs best in both datasets. The comparison with CCTV footage is done by analyzing positional difference in pixels over short periods of time (~10s).	One of the first attempts to calibrate and validate pedestrian dynamics models in emergencies with real data.

The use of proxy "panic" variables is another approach, used by many researchers. Helbing et al. (*50*) include herding behavior, attempting to explain arching and clogging at exits during an emergency evacuation by introducing a p factor that dictates an agent's likelihood of following their neighbors. Zou et al. (*51*) describe a lattice-gas model that shows competitive behavior, examining the various outcomes when agents are either "gentle" or "flustered," with the latter able to overtake the agent in front of them. Finally, Tsai et al. (*52*) incorporate a "fear" characteristic that takes an integer value between 0 (no fear) and 2 (full fear). This characteristic affects the agent's actions, causing them to move radially away from a "threat."

There are two common features within these papers. The first is the concept of emotional contagion, where the emotional state of an individual agent can affect that of another. This has a biological basis (*53*) and has been examined on a theoretical level on multiple occasions. For instance, Elzie et al. (*54*) studied group dynamics under various emotional and cognitive conditions, allowing the spread of information to vary between simulations. The other common feature in these papers is the difficulty in comparing the results with real world data, given the lack of readily available data and the difficulty of obtaining experimental evidence. This is the case both for the underlying theory of pedestrian dynamics (i.e., where do people go when they're scared), and for the abstract theory of emotional contagion (i.e., how does panic spread). For an introduction to these, Helbing et al. (*55*) provide a useful examination of the different phenomena seen during "normal" and "panic" situations.

To calibrate these models, it is first necessary to obtain relevant data, from scenarios that closely resemble the one being modeled. Given the extreme nature of the scenario, animal behavior is the most widespread of data sources available. Although several datasets of human behavior exist, questions remain as to their applicability for scenarios that elicit genuine fear.

Animal data can be used as a way of approximating human responses, with ants being a common example of experimental data sources. For example, Shiwakoti et al. (*56*) achieved this by placing ants within an area and introducing ant repellent (citronella), causing the ants to move toward the exit. Shiwakoti et al. (*56*) also develop an emergency evacuation simulation model (EmSim), based on the Social Force Model. Altshuler et al. (*57*) also attempted to understand herding behavior in humans by observing ants in an acrylic drum with two, diametrically opposed exits, after also introducing ant repellent (citronella) to the system. Altshuler et al. (*57*) observed a statistically significant effect of symmetry braking (i.e., herding behavior) in ants, as they followed other ants toward a presumed exit rather than choosing another, potentially more beneficial, exit. This type of model has been used to analyze design of turning on emergency evacuations (*58–60*) and multiple stream merging behavior (*61*, *62*). Other animals have been studied, for example Saloma et al. (*63*), who performed experiments by putting mice into a pool of deep water.

A problem with using animal data as a proxy for human responses is that it requires validation with real human data before it can be reliably used as a predictor. This human data must be obtained either experimentally or from previous records. Obtaining experimental data with human subjects requires a higher level of ethical review, and it is difficult to simulate an emergency with the sense of realism required to validate models. However, there exist several datasets of human movement, including that produced by Dias et al. (*64*), where groups of subjects were asked to walk and then jog through corridors and round corners. This is an approximation of the movement likely to be seen in emergencies and serves to partially calibrate models.

Survey data has been used as a method of calibrating computational models, including looking at route choice and symmetry-breaking behavior (*65*), or the impact of pedestrian density and flow direction on exit choice (*66*).

Tsai et al. (*67*) examine calibration by comparing a pedestrian dynamics model, combined with an emotional transmission model, with CCTV footage of two separate incidents. By analyzing the difference in pixel location of several agents within the footage and within the model, they evaluated the performance of both the behavioral and the behavioral transmission models. Similar approaches have been performed looking at movement during earthquakes (*68*, *69*).

Finally, there are several review papers (*70*, *71*) that summarize the quantitative methods of gathering data.

This review has described the major concepts and papers that currently dominate how the pedestrian modeling and behavioral psychology industries understand human reactions during emergencies. It is seen that, although there has been considerable research on both theoretical and empirical fronts, there is still much to be done before individual and group behavior in emergencies is sufficiently well modeled in computer software.

Interviews

Interviews were conducted to fully understand the current use of, and demand for, research into this domain. These interviews took the form of a semi-structured interview, allowing the interviewees to elaborate on their own area, as well as comment on other areas of this review. The interviews (mean: 33.4 minutes, range: 23.7–48.9 minutes) were partially transcribed and thematically grouped.

Prompt Questions

The prompts aimed to instigate discussion on human behavior. They comprised of three sections, initially establishing the interviewee's role and responsibilities, followed by investigating their work within pedestrian and group dynamics. Finally, the interviewees were asked about how they incorporate the concept of fear into their work, and how they would like this area to improve.

Section A: Job Description

What is your current main sector of research/industry?

How does your work relate to crowd/pedestrian movement and emergency evacuations?

Section B: Human Behaviors Involvement

How do you incorporate human behavior into your work?

More specifically, pedestrian and pedestrian group dynamics?

Section C: Use of Fear and Desired Progress

How do you incorporate fear in your work?

If your work is related to emergencies, how do you incorporate behavioral responses with respect to this?

If you think that you could make use of an improved model of behavioral response to fear, what would be the most useful tool you would use?

Responses

Section A: Interviewees. The interviewees were chosen according to their work in the area of human reactions during emergencies, and their position as stakeholders in the area of resilience and pedestrian dynamics. Their job titles and descriptions (as part of their responses to Section A) are detailed below:

Interviewee 1: Resilience Practice Leader, engineering consultancy

Interviewee 2: Resilience Planning Manager for a UK city council.

Interviewee 3: Social psychologist: "my work primarily focuses on risk perceptions and behavioral responses to extreme events (such as terrorism)."
Interviewee 4: Assistant Professor in "psycho-social aspects of the management of emergencies and disasters, in particular looking at human behavior in mass emergencies."
Interviewee 5: Industry pedestrian dynamics practitioner and part time researcher.
Interviewee 6: Professor in social psychology: "Behavior in emergencies and disasters is one of my research areas and so is behavior in pedestrian flow. I specialize in crowd behavior."
Interviewee 7: Pedestrian dynamics modeler and consultant on industrial projects: "also involved in research related to crowd modeling."
Interviewee 8: Associate Professor in evacuation simulation and human behavior in fire.
Interviewee 9: Senior Adviser, UK Security Services.

Section B: Human Behaviors Involvement. Section B aimed to understand the intricacies of each interviewee's area of work, outside of their job title and single line description. There was almost unanimous general agreement between all interviewees that their work had to consider human behavior when performing their job. The exception is that Interviewee (Int.) 1 did not routinely use human behavioral concerns in their counterterrorism work, focusing instead on structural design. Int. 1 continued to say that behavioral considerations were important when working in resilience in general.

Int. 5, 7, 8 use pedestrian dynamics models daily, contributing to their development in a research capacity. They use these models for a variety of purposes, including validation and crowd modeling, noting that each type of model has its benefits and drawbacks. Int. 3, 4, 6, 8 regularly perform research into different human behaviors, either by investigating previous emergency scenarios, or by developing experiments to test and develop models of behavior. Int. 3 and 4 commented that their work is aimed at providing insight into the types of behavioral response seen during extreme emergencies and therefore informing policy decisions and guidance regarding these scenarios. Int. 2 uses behavioral concepts in their work in resilience alongside the operational and structural responses, as this is a required element of the planning (such as communication methods, cordon zones). As such, they are also required to train in behavioral science as part of their continuous professional development (CPD). Int. 9 routinely considers human behavior in their work within counterterrorism, attempting to improve the resilience and preparedness of the UK, with multiple approaches (including event management, building and infrastructure design).

Group dynamics are implicit in the work of Int. 3 and 4 while Int. 5, 6, 7, 8, 9 noted that their work can specifically include the concept of groups (or crowds). Int. 5 notes the difference in capability between the types of macroscopic and microscopic models in replicating group behavior, with microscopic (agent-based) models able to incorporate the most intricate concepts, and therefore being the most computationally expensive. They then comment on the limitations with these models as "You don't have models that have a default behavioral system that predicts what people will do. You as the user have [to decide] what proportion of things happen." Int. 5 also comments on the current limitations of modeling group behavioral theories in emergencies with models such as ENT and PADM, saying that they could be replicated in computer models, but "you'd have to make big assumptions about what people would do". Int. 6 comments on how the different theories have treated group behavior in emergencies, citing ENT, the Social Attachment (SA) model and SCT. They note that ENT and SA "seem less good [than SCT] at explaining collective co-ordinated behavior amongst strangers... There are studies showing...that even when there aren't people that we know, people still behave in [a] socially structured way."

Section C: Use of "Fear" and Desired Progress. This section aimed to understand the interviewee's work on behavioral reactions in a state of fear, and if there is demand for a particular improvement in this area.

The primary common theme was the unanimous agreement that a better model of human behavior during fear-inducing scenarios would be beneficial, either for mitigating the effects of an emergency (Int. 1, 3, 5, 7, 8, 9), or for informing guidance for the emergency services (Int. 2, 4, 9). From a computer-modeling viewpoint, Int. 5, 7, 8 confirmed the lack of an accurate representation of "fear" in the various software models but commented that there are workarounds for implementing what might be reasonably assumed in a real scenario. Int. 5 referenced Hollmann et al. (*48*) on utility optimization and itinerary adjustment "...which presumably could then be used in terms of modeling other emotions such as fear...".

Another fundamental common theme in desired areas of progress was the requirement to produce high-quality data for future work. This varied between the interviewees, e.g., Int. 2 specifically requests more 'case studies', for a lessons-learned style of analysis. In contrast, Int. 3, 5, 7, 8 want more basic, quantitative data to improve modeling capabilities; both by producing theoretical approaches and improving software models. Int. 1 and 9 stated that a better understanding of how people react would be beneficial, to provide guidance on design processes and test cases, specifically within the context of terrorism. Finally, Int. 4 and 6 preferred an improvement to the methods of gathering data, as a way of improving the ecological validity of the data and the opportunity to examine a wider range of scenarios.

Finally, a common theme in several interviews (Int. 3, 4, 6, 8) stated that an appropriate level of fear was sometimes a prerequisite for adaptive and preservative behavior. For example,

Table 4. Summary of Interviewee Responses

Interviewee	Section A key points	Section B key points	Section C key points	Key area of desired progress
1	Works on resilience, including on counterterrorism "bomb proofing" and "invacuation*." Other areas of the company work on people flow modeling, including in hospitals and security queues. "Consulting, specifically in the area of resilience for businesses and cities."	Routinely uses human behavioral considerations in resilience work. Does not use human behavior in counterterrorism work.	Suggests that providing guidance will allow for an improvement in the design of buildings. Also, the development of a predictive model will allow for more in-depth analysis of critical infrastructure. An example would be design case guidance to protect against a Marauding Terrorist Firearms Attack (MTFA).	Predictive model and guidance for architects.
2	Planning emergency procedures and preparedness for UK city, including cross-organizational responses. "I am responsible for planning major emergencies that can impact the city, and ensuring the council is able to respond to them."	Does not use pedestrian dynamics models but needs to include likely human behavioral responses in plans. Training involves CPD in latest behavioral considerations.	Further to answers in Section B, an improvement would be further "case studies" of emergency planning and reactions; more data to help predict potential issues.	More data ("case studies")
3	Studies human behavior in emergency scenarios as a research area.	Aims to produce studies on human behavior that can inform policy and guidance.	Fear is something that needs to be managed, as it is often needed for a behavioral response, but too much can produce unexpected results.	Improved modeling capability. More data.
4	Studies human behavior and management of emergencies as a research area.	Has worked on operational considerations in mass decontamination events, including behavioral considerations. Aims to produce studies on management of emergencies and human behavior that can inform policy and guidance.	The difficulty is that when doing experiments, you cannot be sure that the subjects are showing the same behavior as in a real scenario, and you can only go so far to increase the reality of the experiment. Qualitative research is carried out to enable data to be triangulated across different research methods. However, there is still a need to develop experimental studies with high ecological validity.	More data/ improved research capabilities.
5	Works as a senior consultant in modeling pedestrian dynamics and flows. "Senior consultant, in pedestrian and evacuation dynamics … As an academic role I research with various international institutions in this same area."	All their models (macroscopic and microscopic) have some form of behavioral assumptions, so all work involves human behavior. Also works as a researcher, aiming to improve the quality of models.	There is a way of including emotional modeling in the software, by way of variable itineraries. However, this is not a true fear response. By collating responses, and then statistically representing them as likelihoods, you can edit models to mimic fear responses. However, improving the theory of how people respond will also be helpful.	Theory development and improved data.
6	Works as a researcher in social psychology, looking at crowd behavior in emergencies.	Works at verifying and modifying crowd psychology theories of crowd behavior during emergencies. This includes laboratory and field experiments and interviews.	Any computer models need to take into account the sociological and crowd psychological theories of behavioral responses, or risk being inaccurate. Fear is a huge variable in response to these scenarios, so any test needs to include it as much as practicable.	Improved testing capabilities

(Continued)

Table 4. (Continued)

Interviewee	Section A key points	Section B key points	Section C key points	Key area of desired progress
7	Works with commercial and in-house pedestrian dynamics software, analyzing large scale events, including for real time prediction of safety concerns and evacuations. Also works with researchers developing models.	The behavioral inputs are user specific. The models themselves can be altered to replicate a certain type of behavior ("any behavioral element is done by the modeler").	The use of fear is unquantified in current models and it can have a large effect on what people do. There is speculation on how fear transmits from one person to another, but without validation, the work is still based on modeler experience. If this information were available, a large impact would be the ability to improve crowd management ("the sort of information you need to provide the crowd … how you need to guide them").	"What effect does fear actually have?" from a quantitative point of view, including movement speeds, directions and emotional interdependence.
8	Works in academia as a researcher in human behavior during emergencies and simulating this behavior. Works with international agencies to validate current models.	Has conducted data collection for human behavior during emergencies, both in fire and other emergencies. Understanding human behavior is integral to approaching this field, and to modeling evacuations. There are several considerations when gathering data on human behavior in emergencies- key being ecological validity.	Many models do not explicitly incorporate fear, "not because it can't be done, but because there is a limited understanding of the psychological basis of the consequences of fear." The specifics of modeling fear might include spatial navigation algorithms, for example, avoidance behavior.	"My answer is very simple: data." Basic research will provide calibration for models.
9	Works as part of a counterterrorism unit within the UK Security Services.	A large part of their work revolves around classifying the types of behavior to expect within emergency scenarios, specifically terrorism. This is the behavior of both an aggressor and a civilian.	Use fear in their work, but "not in a scientific way"; the aim is to improve the design of scenarios (e.g., buildings, events) such that their use is intuitive when people are scared. This understanding of how people behave is integral to the planning for emergencies.	"A better understanding of fear." This will enable testing of threat scenarios and assess different designs.

Note: *The term "invacuation" refers to the use of buildings and structures, such as a stadium, as temporary accommodation within extreme scenarios (for instance a terrorist attack). This poses a complex issue, as there may be a requirement for quick ingress or egress in buildings that are often designed to improve flow in only one direction.

it is commonly seen that, during floods and fires, people will move slowly, if at all, until the threat is confirmed (e.g., visually). However, in the context of mass decontamination, there is a level of reaction that is unhelpful (e.g., running through the decontamination unit without being properly processed). These are two examples of behavioral reactions that lead to suboptimal outcomes. Int. 8 went further by stating "The problem is to get people out of the [burning] building ... Panic ... that's [a] false myth ... in the majority of cases we have the opposite problem." This is aptly summarized by Int. 3, "Fear isn't necessarily always something that is negative, but in the context of emergency planning, we have to think of fear as something that needs to be managed."

Conclusion and Discussion

In the literature review section, this paper examined the understanding and use of behavioral models and panic behavior, as well as their utilization to inform pedestrian dynamics software. Although there are many sociological models of human behavior, there is an apparent lack of appropriate software models for extreme emergencies that can accurately model human responses during states of fear. This could be for several reasons, including the ethical and practical barriers to such research, implying that, although this is an open area for research, any future experimental approach should necessarily consider the ethical implications of inducing this type of reaction.

This conclusion was corroborated by analysis of interviews with key stakeholders, examining both areas where behavioral response is currently considered, and where models need to progress before reliable use. The stakeholders, who are all involved in the various areas of human behavior, emergency planning and pedestrian dynamics, agreed that a better model is required. These stakeholders then envisioned progress in several different forms, including a potential computer model that could be used on the MTFA design case, allowing "qualitative guidance for architects."

All interviewees commented on how more data is needed for this research to progress, and that for a model to be relevant, it needs to be calibrated against data of previous events and case studies. This calibrated data could take multiple forms, including incident reports, sociological theory and CCTV data. Without this calibration, these models remain theoretical.

To conclude, although this paper finds that the area of pedestrian dynamics has made huge leaps in recent decades, there still remain unanswered questions about behavioral characteristics during fear-inducing scenarios. This largely unexplored area of research can have far-reaching consequences, including saving lives.

Author Contributions

The authors confirm contribution to the paper as follows: study conception and design: A Shipman, A Majumdar; data collection: A Shipman; analysis and interpretation of results: A Shipman; draft manuscript preparation: A Shipman, A Majumdar. All authors reviewed the results and approved the final version of the manuscript.

References

1. Kuligowski, E. D., and S. M. V. Gwynne. The Need for Behavioral Theory in Evacuation Modeling. In *Pedestrian and Evacuation Dynamics 2008* (W. W. F. Klingsch, C. Rogsch, A. Schadschneider, and M. Schreckenberg, eds.), Springer, Berlin, Heidelberg, 2010, pp. 721–732.
2. Templeton, A., J. Drury, and A. Philippides. From Mindless Masses to Small Groups: Conceptualizing Collective Behavior in Crowd Modeling. *Review of General Psychology*, Vol. 19, No. 3, 2015, pp. 215–229.
3. Kuligowski, E. D., S. M. Gwynne, M. J. Kinsey, and L. Hulse. Guidance for the Model User on Representing Human Behavior in Egress Models. *Fire Technology*, Vol. 53, No. 2, 2017, pp. 649–672.
4. Cuesta, A., O. Abreu, and D. Alvear. *Evacuation Modeling Trends*. Springer Publishing Company, Inc., 2015, 133 p.
5. Cristiani, E., B. Piccoli, and A. Tosin. *Multiscale Modeling of Pedestrian Dynamics*. Springer, 2014.
6. Timmermans, H. J. P. *Pedestrian Behavior: Models, Data Collection and Applications*. Emerald Group Publishing Limited, 2009.
7. McPhail, C. *The Myth of the Madding Crowd*. Walter de Gruyter, Inc., 1991.
8. Chertkoff, J. M., and R. H. Kushigian, *Don't Panic: The Psychology of Emergency Egress and Ingress*. Praeger, Westport, Conn, 1999, 145 p.
9. Smelser, N. J. *Theory of Collective Behavior*. Collier-MacMillan Canada, Ltd., 1962.
10. Le Bon, G. *The Crowd. A Study of the Popular Mind*. T. F. Unwin, London, 1896.
11. Quarantelli, E. Panic Behavior Some Empirical Observations. In *American Institute of Architects Conference*, 1975.
12. Quarantelli, E. Images of Withdrawal Behavior in Disasters: Some Basic Misconceptions. *Social Problems*, Vol. 8, 1960, pp. 68–79.
13. Quarantelli, E. L. The Nature and Conditions of Panic. *American Journal of Sociology*, Vol. 60, No. 3, 1954, pp. 267–275.
14. Keating, J. The Myth of Panic. *Fire*, 1982, 147 p.
15. Fahy, R. F., G. Proulx, and L. Aiman. Panic or Not in Fire: Clarifying the Misconception. *Fire and Materials*, Vol. 36, No. 5–6, 2012, pp. 328–338.
16. Pastel, L. R. H. *Collective Behaviors: Mass Panic and Outbreaks of Multiple Unexplained Symptons*. U.S. Army Medical Research Institute of Infectious Diseases, 2002.
17. Drury, J., D. Novelli, and C. Stott. Representing Crowd Behaviour in Emergency Planning Guidance: 'Mass Panic' or Collective Resilience? *Resilience*, Vol. 1, No. 1, 2013, pp. 18–37.
18. Fritz, C. E., and E. S. Marks. The NORC Studies of Human Behavior in Disaster. *Journal of Social Issues*, Vol. 10, No. 3, 1954, pp. 26–41.
19. Drury, J., D. Novelli, and C. Stott. Psychological Disaster Myths in the Perception and Management of Mass Emergencies. *Journal of Applied Social Psychology*, Vol. 43, No. 11, 2013, pp. 2259–2270.

20. Schultz, D. P. *Panic in the Military*. Department of Psychology, University of North Carolina, 1971.
21. Johnson, N. R. Panic and the Breakdown of Social Order: Popular Myth, Social Theory, Empirical Evidence. *Sociological Focus*, Vol. 20, No. 3, 1987, pp. 171–183.
22. Gantt, P., and R. Gantt. *Disaster Psychology: Dispelling the Myths of Panic*, 2012.
23. Turner, R. H. E., and L. M. E. Killian. *Collective Behaviour*. Eaglewood Cliffs, NJ, 1957.
24. Turner, J. C., Hogg, M. A., Oakes, P. J., Reicher, S. D., & Wetherell, M. S. (1987). Rediscovering the Social Group: A Self-Categorization Theory. Cambridge, MA, US: Basil Blackwell.
25. Franca, R. D. S., M. D. G. B. Marietto, and M. B. Steinberger. Proposing a Cognitive Multi-agent Model for the Panic in Crowds Phenomenon. In *2009 2nd International Conference on the Applications of Digital Information and Web Technologies*, 2009.
26. Mawson, A. R. Understanding Mass Panic and Other Collective Responses to Threat and Disaster. *Psychiatry: Interpersonal and Biological Processes*, Vol. 68, No. 2, 2005, pp. 95–113.
27. Turner, J. C., P. J. Oakes, S. A. Haslam, and C. McGarty. Self and Collective: Cognition and Social Context. *Personality and Social Psychology Bulletin*, Vol. 20, No. 5, 1994, pp. 454–463.
28. Drury, J., C. Cocking, and S. D. Reicher. The Nature of Collective Resilience: Survivor Reactions to the 2005 London Bombings. *International Journal of Mass Emergencies and Disasters*, Vol. 27, 2009, pp. 66–95.
29. Stott, C., and J. Drury. Crowds, Context and Identity: Dynamic Categorization Processes in the 'Poll Tax Riot'. *Human Relations*, Vol. 53, No. 2, 2000, pp. 247–273.
30. Lindell, M. K., and R. W. Perry. The Protective Action Decision Model: Theoretical Modifications and Additional Evidence. *Risk Analysis*, Vol. 32, No. 4, 2012, pp. 616–632.
31. Feinberg, W. E., and N. R. Johnson. The Ties that Bind: A Macro-Level Approach to Panic. *International Journal of Mass Emergencies and Disasters*, Vol. 19, No. 3, 2001, pp. 269–295.
32. Baddeley, A. D. Selective Attention and Performance in Dangerous Environments. *Journal of Human Performance in Extreme Environments*, Vol. 5, No. 1, 2000.
33. Kuligowski, E. *Terror Defeated: Occupant Sensemaking, Decision-making and Protective Action in the 2001 World Trade Center Disaster* (K. Tierney, L. Downey, and W. Grosshandler, eds.), University of Colorado at Boulder, 2011.
34. Mintz, A. Non-Adaptive Group Behavior. *Journal of Anormal Psychology*, Vol. 46, No. 2, 1951, pp. 150–159.
35. Kelley, H., J. C. Condry, Jr, A. E. Dahlke, and A. H. Hill. Collective Behavior in a Simulated Panic Situation. *Journal of Experimental Social Psychology*, Vol. 1, No. 1, 1965, pp. 20–54.
36. Keinan, G., N. Friedland, and Y. Ben-Porath. Decision Making Under Stress: Scanning of Alternatives Under Physical Threat. *Acta Psychologica*, Vol. 64, No. 3, 1987, pp. 219–228.
37. Drury, J., C. Cocking, S. Reicher, A. Burton, D. Schofield, A. Hardwick, D. Graham, and P. Langston. Cooperation Versus Competition in a Mass Emergency Evacuation: A New Laboratory Simulation and a New Theoretical Model. *Behavior Research Methods*, Vol. 41, No. 3, 2009, pp. 957–970.
38. Bar-Haim, Y., A. Kerem, D. Lamy, and D. Zakay. When Time Slows Down: The Influence of Threat on Time Perception in Anxiety. *Cognition & Emotion*, Vol. 24, No. 2, 2010, pp. 255–263.
39. Proulx, G., and J. D. Sime. To Prevent 'Panic' in an Underground Emergency: Why Not Tell People the Truth? In *Fire Safety Science- Proceedings of the Third International Symposium*, 1991.
40. Zinke, R., G. Hofinger, and L. Künzer. Psychological Aspects of Human Dynamics in Underground Evacuation: Field Experiments. In *Pedestrian and Evacuation Dynamics 2012* (U. Weidmann, U. Kirsch, and M. Schreckenberg, eds.), Springer International Publishing, Cham, 2014, pp. 1149–1162.
41. Li, S., et al., *Driving-Forces Model on Individual Behavior in Scenarios Considering Moving Threat Agents*. Physica A: Statistical Mechanics and its Applications, 2017. 481: p. 127–140.
42. Starcke, K., and M. Brand. Decision Making Under Stress: A Selective Review. *Neuroscience & Biobehavioral Reviews*, Vol. 36, No. 4, 2012, pp. 1228–1248.
43. Schultz, D. P. Theories of Panic Behavior: A Review. *The Journal of Social Psychology*, Vol. 66, No. 1, 1965, pp. 31–40.
44. Duives, D. C., W. Daamen, and S. P. Hoogendoorn. State-of-the-art Crowd Motion Simulation Models. *Transportation Research Part C-Emerging Technologies*, Vol. 37, 2013, pp. 193–209.
45. Kuligowski, E. Review of 28 Egress Models. In *Workshop on Building Occupant Movement During Fire Emergencies*. NIST SP 1032, 2005.
46. Gwynne, S., E. R. Galea, M. Owen, P. J. Lawrence, and L. Filippidis. A Review of the Methodologies Used in the Computer Simulation of Evacuation from the Built Environment. *Building and Environment*, Vol. 34, No. 6, 1999, pp. 741–749.
47. Aguirre, B. E., S. El-Tawil, E. Best, K. B. Gill, and V. Fedorov. Contributions of Social Science to Agent-based Models of Building Evacuation. *Contemporary Social Science*, Vol. 6, No. 3, 2011, pp. 415–432.
48. Hollmann, C., P. J. Lawrence, and E. R. Galea. Introducing Emotion Modelling to Agent-Based Pedestrian Circulation Simulation, in Pedestrian and Evacuation Dynamics. In (R. D. Peacock, E. D. Kuligowski, and J. D. Averill, eds.), Boston, MA, Springer, 2011, pp. 799–803.
49. Pan, X., C. S. Han, K. Dauber, and K. H. Law. Human and Social Behavior in Computational Modeling and Analysis of Egress. *Automation in Construction*, Vol. 15, No. 4, 2006, pp. 448–461.
50. Helbing, D., I. Farkas, and T. Vicsek. Simulating Dynamical Features of Escape Panic. *Nature*, Vol. 407, No. 6803, 2000, pp. 487–490.
51. Zou, Y., J. Xie, and B. Wang. Evacuation of Pedestrians with Two Motion Modes for Panic System. *PLoS One*, Vol. 11, No. 4, 2016, p. e0153388.
52. Tsai, J., N. Fridman, E. Bowring, M. Brown, S. Epstein, G. Kaminka, S. Marsella, A. Ogden, I. Rika, A. Sheel, and M. E. Taylor. ESCAPES: Evacuation Simulation with Children, Authorities, Parents, Emotions, and Social Comparison. In *The 10th International Conference on Autonomous Agents and Multiagent Systems - Volume 2*. International Foundation for Autonomous Agents and Multiagent Systems, Taipei, Taiwan, 2011, pp. 457–464.

53. Huis in 't Veld, E. M. J., and B. de Gelder. From Personal Fear to Mass Panic: The Neurological Basis of Crowd Perception. *Human Brain Mapping*, Vol. 36, No. 6, 2015, pp. 2338–2351.
54. Elzie, T., E. Frydenlund, A. J. Collins, and R. M. Robinson. Panic that Spreads Sociobehavioral Contagion in Pedestrian Evacuations. *Transportation Research Record: Journal of the Transportation Research Board*, 2016. 2586: 1–8.
55. Helbing, D., I. J. Farkas, P. Molnar, and T. Vicsek. Simulation of Pedestrian Crowds in Normal and Evacuation Situations. In *Pedestrian and Evacuation Dynamics*, Duisburg, Germany, 4–6 April 2001, (M. Schreckenberg and S. D. Sharma, eds.), Springer, 2002, p. 58.
56. Shiwakoti, N., M. Sarvi, G. Rose, and M. Burd. Biologically Inspired Modeling Approach for Collective Pedestrian Dynamics Under Emergency Conditions. *Transportation Research Record: Journal of the Transportation Research Board*, 2010. 2196: 176–184.
57. Altshuler, E., O. Ramos, Y. Núñez, J. Fernández, A. J. Batista-Leyva, and C. Noda. Symmetry Breaking in Escaping Ants. *The American Naturalist*, Vol. 166, No. 6, 2005, pp. 643–649.
58. Dias, C., M. Sarvi, N. Shiwakoti, and M. Burd. Turning Angle Effect on Emergency Egress. *Transportation Research Record: Journal of the Transportation Research Board*, 2012. 2312: 120–127.
59. Dias, C., M. Sarvi, N. Shiwakoti, O. Ejtemai, and M. Burd. Investigating Collective Escape Behaviours in Complex Situations. *Safety Science*, Vol. 60, 2013, pp. 87–94.
60. Shiwakoti, N., M. Sarvi, G. Rose, and M. Burd. Consequence of Turning Movements in Pedestrian Crowds During Emergency Egress. *Transportation Research Record: Journal of the Transportation Research Board*, 2011. 2234: 97–104.
61. Shahhoseini, Z., M. Sarvi, and M. Saberi. Insights Toward Characteristics of Merging Streams of Pedestrian Crowds Based on Experiments with Panicked Ants. *Transportation Research Record: Journal of the Transportation Research Board*, 2016. 2561: 81–88.
62. Shahhoseini, Z., M. Sarvi, M. Saberi, and M. Haghani. Pedestrian Crowd Dynamics Observed at Merging Sections. *Transportation Research Record: Journal of the Transportation Research Board*, 2017. 2622: 48–57.
63. Saloma, C., G. J. Perez, G. Tapang, M. Lim, and C. Palmes-Saloma. Self-organized Queuing and Scale-free Behavior in Real Escape Panic. *Proceedings of the National Academy of Sciences*, Vol. 100, No. 21, 2003, 11947–11952.
64. Dias, C., M. Sarvi, O. Ejtemai, and M. Burd. Elevated Desired Speed and Change in Desired Direction. *Transportation Research Record: Journal of the Transportation Research Board*, 2015. 2490: 65-75.
65. Shiwakoti, N., R. Tay, P. Stasinopoulos, and P. J. Woolley. Likely Behaviours of Passengers Under Emergency Evacuation in Train Station. *Safety Science*, Vol. 91, 2017, pp. 40–48.
66. Haghani, M., and M. Sarvi. Human Exit Choice in Crowded Built Environments: Investigating Underlying Behavioural Differences Between Normal Egress and Emergency Evacuations. *Fire Safety Journal*, Vol. 85, 2016, pp. 1–9.
67. Tsai, J., E. Bowring, S. Marsella, and M. Tambe. Empirical Evaluation of Computational Fear Contagion Models in Crowd Dispersions. *Autonomous Agents and Multi-Agent Systems*, Vol. 27, No. 2, 2013, pp. 200–217.
68. Gu, Z., Z. Liu, N. Shiwakoti, and M. Yang. Video-based Analysis of School Students' Emergency Evacuation Behavior in Earthquakes. *International Journal of Disaster Risk Reduction*, Vol. 18, 2016, pp. 1–11.
69. Bernardini, G., E. Quagliarini, and M. D'Orazio. Towards Creating a Combined Database for Earthquake Pedestrians' Evacuation Models. *Safety Science*, Vol. 82, 2016, pp. 77–94.
70. Lawson, G. *Predicting Human Behaviour in Emergencies [Electronic Resource]*, 2011.
71. Haghani, M., and M. Sarvi. Crowd Behaviour and Motion: Empirical Methods. *Transportation Research Part B: Methodological*, Vol. 107, 2018, pp. 253–294.
72. Bosse, T., et al., A Multi-agent Model for Emotion Contagion Spirals Integrated within a Supporting Ambient Agent Model. In *Proceedings of the Principles of Practice in Multi-Agent Systems: 12th International Conference, PRIMA 2009*, Nagoya, Japan, 14–16 December 2009, (J.-J. Yang, Makoto Yokoo, Takayuki Ito, Zhi Jin, and Paul Scerri, eds.), Springer, Berlin, Heidelberg, 2009, pp. 48–67.

The Standing Committee on Emergency Evacuations (ABR30) peer-reviewed this paper (18-03898).

Article

TRR
JOURNAL OF THE TRANSPORTATION RESEARCH BOARD

Transportation Research Record
2018, Vol. 2672(1) 198–208

DOI: 10.1177/0361198118798736
journals.sagepub.com/home/trr
SAGE

Crowdsourcing Incident Information for Emergency Response using Open Data Sources in Smart Cities

Fan Zuo[1], Abdullah Kurkcu[2], Kaan Ozbay[3], and Jingqin Gao[1]

Abstract

Emergency events affect human security and safety as well as the integrity of the local infrastructure. Emergency response officials are required to make decisions using limited information and time. During emergency events, people post updates to social media networks, such as tweets, containing information about their status, help requests, incident reports, and other useful information. In this research project, the Latent Dirichlet Allocation (LDA) model is used to automatically classify incident-related tweets and incident types using Twitter data. Unlike the previous social media information models proposed in the related literature, the LDA is an unsupervised learning model which can be utilized directly without prior knowledge and preparation for data in order to save time during emergencies. Twitter data including messages and geolocation information during two recent events in New York City, the Chelsea explosion and Hurricane Sandy, are used as two case studies to test the accuracy of the LDA model for extracting incident-related tweets and labeling them by incident type. Results showed that the model could extract emergency events and classify them for both small and large-scale events, and the model's hyper-parameters can be shared in a similar language environment to save model training time. Furthermore, the list of keywords generated by the model can be used as prior knowledge for emergency event classification and training of supervised classification models such as support vector machine and recurrent neural network.

Emergency events such as hurricanes, floods, releases of hazardous or nuclear materials, building fires, and explosions dramatically affect the local infrastructure. In particular, decision makers are required to make immediate decisions based on the limited information available in sudden-onset crisis situations. Ozbay and Kachroo (*1*) was among the first studies to describe innovative approaches to predict incident management-related information such as online incident durations and delays using real-time data from sensors and incident response crews. More recently, Ozbay and Noyan (*2*) and Demiroluk and Ozbay (*3*) proposed novel methods for incident duration prediction using the traditional sensor data as well other data directly generated and made available by response personnel. However, systems and protocols similar to the ones cited above that are used for emergency/incident response today still lack accurate and abundant spatial and temporal resolution data, which are required to address emergency recovery needs promptly. On the other hand, open source social media data have the potential to be used as a source of valuable real-time information to supplement existing traditional data-collection systems.

Social media data have proven to be useful in emergency response (*4*). During emergency events, people post social media updates containing information about their statuses, help requests, incident reports, and other useful information. These data can be examined later to understand the spatial-temporal distribution of incidents. Researchers in social media analytics have built tools to extract information from users' updates. Natural language processing and real-time big data analysis are the new challenges in making that information available in a useful format to officials. If critical information shared on social media networks is immediately available to

[1]C2SMART Center, Department of Civil and Urban Engineering, Tandon School of Engineering, New York University, Brooklyn, NY
[2]C2SMART Center, Department of Civil & Urban Engineering, Tandon School of Engineering, Center for Urban Science + Progress (CUSP), New York University, Brooklyn, NY
[3]C2SMART Center (A Tier 1 USDOT UTC), Department of Civil and Urban Engineering & Center for Urban Science & Progress (CUSP), Tandon School of Engineering, New York University, Brooklyn, NY

Corresponding Author:
Address correspondence to Fan Zuo: fz380@nyu.edu

officials, emergency responders and relief organizations can provide more efficient service to victims of incidents.

Two recent events in New York City can be used as models of social media usage during emergency situations. During Hurricane Sandy, the second costliest hurricane in United States history (*5*), there were 69,010 geolocated tweets posted on Twitter. The hurricane's storm surge hit New York City on October 29, 2012, flooding streets, tunnels, and subway lines, and cutting power in and around the city. Another emergency example is the explosion that happened in Manhattan, New York City, on September 17, 2016. In this terrorist attack, an improvised explosive device was used, and the explosion injured 29 people.

The contributions of this study are threefold. First, the Latent Dirichlet Allocation (LDA) (*6*) model is utilized to generate topics from posted tweets automatically. Second, tweets are classified into groups by finding their representative topics. Finally, locally word-using features are extracted for future works. The main objective of this study is to propose a novel approach that can automatically extract incident-related tweets and classify incident types during emergencies. Although the LDA model can generate topics automatically, manual work is still required to build relationships between incidents and generated topics. Two different data sets mentioned above are used for case studies in this research. Hurricane Sandy data are used as the first case study, which contains two tasks. The first task involves finding tweets containing incident information during Hurricane Sandy. In the second task, only incident-related tweets are used to filter the sub-incident types of the posts by the proposed approach, and the model's accuracy in classifying tweets by incident type is evaluated. Chelsea explosion data are used as the second case study. The one task in the second case study is model accuracy, for both labeling incident-related tweets and event classification are evaluated against manually labeled data.

Background and Motivation

Intelligent transportation systems (ITS) over the last few decades have transformed capabilities to monitor and manage transportation systems in real time. Sensor networks, social media, and real-time web-based traffic information providers are key elements of such monitoring systems. With additional data coming from connected and autonomous vehicle technologies, the major challenge in the future will be to cooperatively integrate various levels of information with interoperable systems. These systems can be developed with existing data sources and extended with new resources as they emerge.

Alazawi et al. proposed an intelligent disaster management system based on cloud-enabled vehicular networks (*7, 8*). The proposed system automatically gathers and processes the data from multiple sources. Then, it constructs effective strategies and makes decisions based on real-time information. They focused on microscopic models to design and evaluate the system performance in two urban scenarios (*7*). The first scenario included traditional emergency response and management systems. The second scenario included the cloud-based management system. The results showed that with the proposed system, more people could evacuate from the city, and the use of transportation resources was improved.

Ashktorab et al. (*9*) developed "Tweedr" as a Twitter mining tool to extract useful information for disaster relief workers during disasters. They employed supervised LDA (sLDA) (*10*) for text classification, Bloom filter for clustering and merging, and conditional random fields (CRF) (*11*) for geo-info extraction. The disaster-related information was extracted from 1049 of the tweets generated during natural disasters, of which 793 were labeled as positive examples. Tweedr classifier has been tested on 12 different crises since 2006 with an *F*-score above 0.5 on predicting several categories such as missing persons and health infrastructures.

Kurkcu et al. (*4*) compared traditional incident data with Twitter data during Hurricane Sandy, and the results showed that Twitter could provide a more detailed geolocation-based picture of specific incidents, along with their intensity, duration, and impact on people. Furthermore, Twitter data provides some information that may not be easily obtained by traditional incident detection systems. A text classifier based on term frequency-inverse document frequency (TF-IDF) and naïve Bayesian methods was trained to process natural language in Twitter to extract incident-related information. One of the interesting key findings of the study was that the incident reporting was considerably faster on Twitter than 511NY, an official incident reporting website. Kurkcu et al. (*12*) introduced a method to collect and utilize geolocated tweets to report a quantitative assessment of human mobility. They used density-based spatial clustering of applications with noise (DBSCAN) (*13*) to identify users' trip origins. The study showed that the mobility of Twitter users follows a heavy-tailed Levy Fly curve (*14*). In other words, human movement contains many short trips and some low-frequency long journeys. Another study by Kurkcu et al. (*15*) used an extended virtual sensor (*16*) framework to provide an automated Twitter and travel time data-collection method as incidents occur.

For disaster response studies, Mandel et al. (*17*) clustered 65,000 Twitter messages during Hurricane Irene by users' gender and location. 408 pre-labeled messages were used to train a sentiment classifier based on the level of concern to find out the number of disaster-related tweets correlating with the temporal distribution of the event. The classifier was used to identify concerned or unconcerned Twitter messages by applying three different methods: logistic regression, naïve Bayes, and decision tree. The study found that analyzing hurricane-related tweets shows differences in sentiment with respect to users' gender or location.

Zhang et al. (*18*) analyzed "Twitter concentrations" to find out the relationship between public activities and unpredictable traffic surges. Unpredictable traffic surge aggravates congested conditions and worsens the level of service of road links, which may correlate with non-recurrent traffic patterns such as accidents, weather, signal malfunction, and festival parades. This study combined traffic data with Twitter data collected during 2014 in Northern Virginia. *K*-means algorithm was used to cluster traffic patterns to identify the social activities causing unpredictable traffic surges, and Akaike information criterion was employed to measure clustering quality. latent semantic analysis (LSA) was used to classify tweets topics based on event keywords. The results showed a local traffic surge could justify 77.4% of traffic-related Twitter concentrations from the test data set.

Chen et al. (*19*) mined Twitter data collected during 2014 from the Chicago area to identify incidents and dynamically manage traffic. The keyword list was extracted from official users' tweets by TF-IDF and another list generated by Fu et al. (*20*) was used to cluster Twitter data based on LSA. The *K*-means and DBSCAN algorithm were implemented to reduce the noise. Then, the labeled-LDA-based procedure addressed by Ritter et al. (*21*) was applied to extract the location information. Finally, an experiment was designed to prove that online Twitter data could improve the effectiveness of dynamic traffic management under incident conditions.

Most of the studies in the literature used Twitter for data-mining approaches to extract disaster-related tweets in large-scale areas only. This study aims to automatically extract tweets by topics in the selected area from raw Twitter data and cluster them by pre-defined event types, and then utilize classified Twitter data as a complementary source to supplement incident-reporting systems. Such data can provide additional information when existing reporting structures fail or are partially out of service (*4*). In addition, a well-trained LDA model can be continually utilized in a similar language environment without repeating the training process for similar tasks.

Smart Crowd-Sourced Emergency Response System (ERS)

Twitter-Based ERS Architecture

The fundamental process of the proposed system is described in (Figure 1). First, users need to decide the mission target geographical information, including attending location and area size, then identify keywords based on the requirements. Second, the system filters the Twitter stream in real time based on the decided time

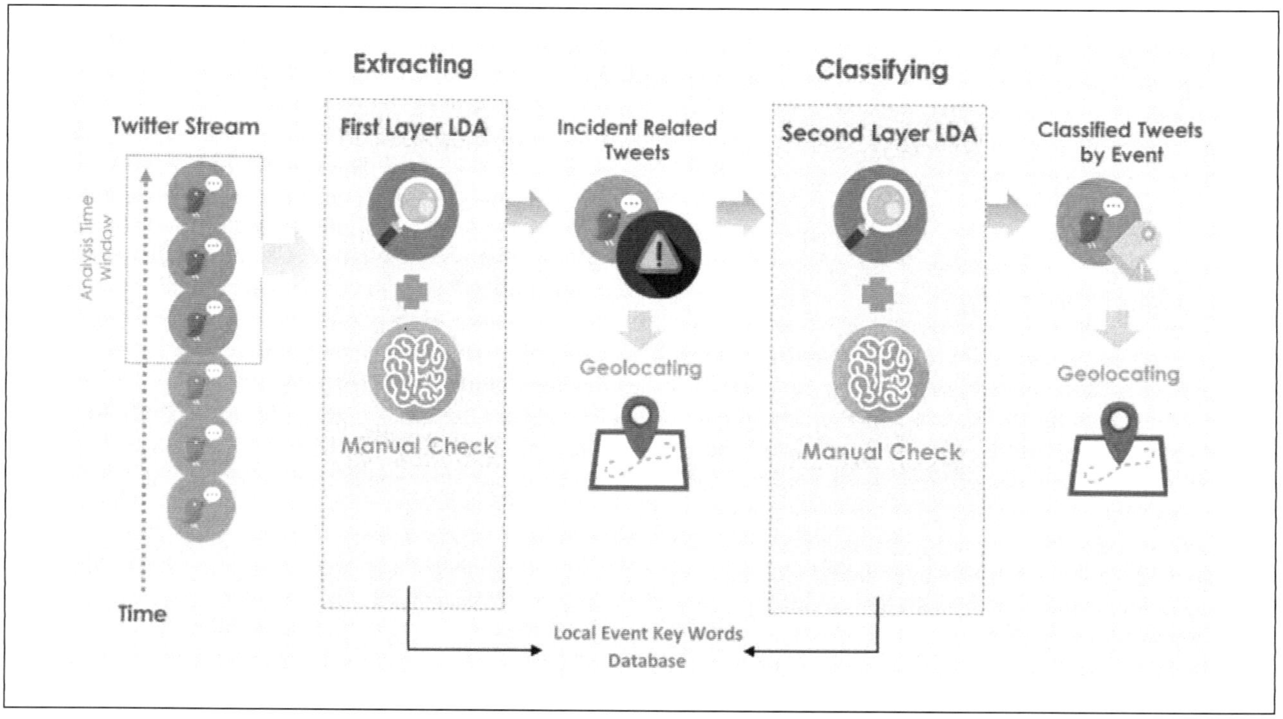

Figure 1. Proposed tweets-based emergency response system (ERS) architecture.

window and geolocation. Then, the system will run the first layer LDA model with given hyper-parameters on all the collected tweets and label them by generated topics. Users are required to manually check those topics to assign them to incident types. After that, the system will run the second LDA layer with another combination of hyper-parameters. Users can determine incident/event type of each topic. Finally, the system will deliver each tweet's relevant incident and corresponding locations. The hyper-parameters and keyword lists will be saved in a local database for further use.

Relationship between Tweets-Based ERS and Transportation

The relationship between an ERS using social media data sources and tranportation planning and engineering can be discussed on three levels.

The first level is the incident response. With the ERS presented in the paper, it is possible to report transportation-related incidents, such as road accidents, gas shortage, downed trees, or flooding, extracted from the open data sources. The transportation department can estimate the urgency and relevance levels and then send messages to warn people about these unavailable road facilities and public transit. Another utilization scenario is reporting of accidents. Traffic accidents during emergency events are critical as they may block the road, causing congestion and even secondary accidents. Previous research (*4*) pointed out that social media data can detect and then report accidents faster than official reporting channels used by the department of transportation or police department. Thus, accident information can be collected in a timelier manner, which could in turn increase safety and provide faster response times during emergency events.

Second is the recovery and resilience level. Extreme weather events have a significant impact on the transportation infrastructure, including road and subway networks. It is essential to decide the order of recovery and resilience for different types of infrastructure. For example, the power supply system should have priority over the transit system, and the transit system covering high-density population areas should have priority over roadways in low-density population areas. This study explores the spatiotemporal patterns of social media data sets, extracts the most frequent disaster locations and social sentiments of events. Such information can help transportation planners make the right decisions to identify recovery and resilience priorities. The logic behind this approach is as follows. If more people are discussing a particular transportation infrastructure failure, then this infrastructure may have more importance in terms of recovery than the others. Thus, merging different open data sources will be useful, especially for building incident-prediction models and generating incident maps under emergency conditions.

Third is the transportation planning level. Modern transportation models are built on data. It is safe to assume that the more data that exist, the better models will be developed. The ERS can report disasters and collect data where traditional means cannot. For example, traffic managers may install emergency power supplies at high-volume and high-impact intersections that have the highest possibility of losing power, or traffic planners can set warning signs at road segments that have a higher rate of accidents under certain conditions.

Retrieving Incident-Related Keywords

Statistical language models estimate the probability distribution of a sequence of words. Classic language models such as Bag-of-Words and TF-IDF perform well in larger documents for classification, but they lack accuracy in small documents. These models also suffer under data sparsity, making it difficult to represent large-context and long-range relationships. TF-IDF and Bag-of-Words models are based on the naïve Bayes model, meaning that those models can only identify "differences" between documents. The LDA model is a dynamic probabilistic model with interpretable topics, meaning that each topic generated by the model ranks words using words' appearance probability distribution as reference. Moreover, the topic's details provide opportunities to identify the meaningful clusters for documents. The following section will introduce the LDA model, and notations listed below are used throughout this paper:

Notation	Description
Indices	
n	Index of words
k	Index of topics
d	Index of documents
e	Index of event types
In LDA	
α	Dirichlet prior on the per-document topic distributions
η	Dirichlet prior on the per-topic word distributions
θ_d	Topic distribution of document d
β_k	Word distribution of topic k
W_d	Word collection of document d
$W_{d,n}$	Word n in W_d
$Z_{d,n}$	Topic assignment for word $W_{d,n}$ from document d
K	Number of topics
V	Number of words in the vocabulary
D	Number of documents
N_d	Number of words in document d
N	Total number of words in all documents

Latent Dirichlet Allocation Model. The LDA model, first introduced by Blei et al. (*6*), is a generative probabilistic topic model structured as three-layer Bayesian network. The unsupervised feature of the model is another advantage, as it does not require any pre-labeling or pre-annotation.

The LDA model defines K topics initially. Each topic k follows a distribution β_k over words in the vocabulary—the set contains all the unique words in all the documents—and β_k picked from a Dirichlet distribution $Dirichlet_V(\eta)$. Based on those randomly created topics, each document d built by a collection of words W_d is determined by a distribution θ_d over K topics from another Dirichlet distribution $Dirichlet_K(\alpha)$, which describes the topic assignment for each word in W_d, and then ranks each word $W_{d,n}$ based on θ_d.

For the n th word in document d, the LDA model samples a specific topic distribution $Z_{d,n} \in [1, K]$ from multinomial distribution $Multinomial_K(\theta_d)$, and the word distribution $W_{d,n}$ for each topic is sampled from multinomial distribution $Multinomial_V\left(\beta_{Z_{d,n}}\right)$. The process of the LDA model is showed in pseudocode below:

Topic Layer
for all topics $k \in [1, K]$ **do**
  sample mixture components $\beta_k \sim Dirichlet_V(\eta)$
Document Layer
for all documents $d \in [1, D]$ **do**
  sample mixture proportion $\theta_d \sim Dirichlet_K(\alpha)$
  Word Layer
  for each word $n \in [1, N]$ in document d **do**
    sample topic index $Z_{d,n} \sim Multinomial_K(\theta_d)$
    sample term for word $W_{d,n} \sim Multinomial_V\left(\beta_{Z_{d,n}}\right)$

$$p\left(\vec{\beta}, \vec{\theta}, \vec{Z}, \vec{W} | \vec{\alpha}, \vec{\eta}\right) = \left(\prod_{k=1}^{K} p\left(\beta_k | \vec{\eta}\right)\right) \left(\prod_{d=1}^{D} p\left(\theta_d | \vec{\alpha}\right) \prod_{n=1}^{N} p(Z_{d,n}|) p\left(W_{d,n} | \vec{\beta}_k,\right)\right) \quad (1)$$

$$\ell\left(\vec{\alpha}, \vec{\eta}\right) = \log p\left(D | \vec{\alpha}, \vec{\eta}\right) = \log \prod_{d=1}^{D} p\left(w_d | \vec{\alpha}, \vec{\eta}\right) = \sum_{d=1}^{D} \log p\left(w_d | \vec{\alpha}, \vec{\eta}\right) \quad (2)$$

Equation 1 shows the joint distribution of all variables. Equation 2 shows the objective function of model training. The model training objective is to find out the $\vec{\alpha}^*$ and $\vec{\eta}^*$ to maximize $\ell\left(\vec{\alpha}, \vec{\eta}\right)$. The approaches of the LDA model's inference include the use of variational expectation-maximization (VEM) (*6*) and Gibbs sampling (*22*). The LDA model utilizes the log likelihood value to measure training performance.

Topic Variation for Classification of Incident-Related Tweets. The posterior document–topic distribution θ_d will be generated after the LDA model is trained on Twitter data, which can be used to identify the topic-related event type. In each generated topic, words with top 10 highest posterior probability are selected as keywords, and their meanings are evaluated by experienced experts, and then the event type of each topic can be decided. Sometimes a topic will be related to multiple events; then the principal relevant event needs to be identified. Jensen Shannon divergence was used to measure the difference between topic/event keywords and topic/topic keywords, and the lower divergence score means higher level of similarity:

$$JSD(P||Q) = \frac{1}{2} D_{KL}(P||M) + \frac{1}{2} D_{KL}(Q||M) \quad (3)$$

where $M = \frac{1}{2}(P + Q)$. $D_{KL}(P||Q)$ is the Kullback–Leibler divergence (*23*). For discrete word distributions for different topics P and Q, the Kullback–Leibler divergence from P to Q is given by:

$$D_{KL}(P||Q) = \sum_x P(x) \log \frac{P(x)}{Q(x)} \quad (4)$$

Data Description

Twitter Data

Two different data sets were used in this study. The first Twitter data set contains user messages that were posted before and after Hurricane Sandy. They were collected by Zubiaga and Ji (*24*) between October 25, 2012 and November 5, 2012 . There are more than 14 million tweets posted by 5 million different users. Some 69,010 tweets that have geolocation information and fall within the study area are selected for monitoring hurricane-related incidents. The second Twitter data set contains tweets posted by users in Manhattan before and after the Chelsea explosion (between September 14, 2016 and September 19, 2016). Tweets are separated by the time that explosion occurred at 08:30 PM on September 17, 2016. Some 103,323 tweets before and 61,090 tweets after the explosion that have geolocation information were collected through Twitter API (*25*).

Case Studies

Data Pre-Processing and Model Inference

The raw Twitter data contain a large number of symbols, digits, and meaningless words, which are removed for statistical documents analysis. After that, each tweet is broken into a word list by any delimiting character such as space and hyphen (tokenizing). A standard "stop list" provided by Natural Language Toolkit (NLTK) (*26*) is utilized to remove the most common words like "a, the, of, to," and words that appeared less than four times in all documents are removed.

In this study, Gibbs sampling is applied to process LDA inference. Python package "lda" (*27*) providing a fast-collapsed Gibbs sampling is used to implement the algorithm. As an unsupervised learning model, the LDA has only three hyper-parameters that need to be optimized:

- **k**: The number of topics. This is the most important parameter directly affecting the training result. This will be further discussed in the case study section;
- **η**: The Dirichlet parameter over distribution $\beta_k \sim \text{Dirichlet}_V(\eta)$, the lower η means the lower number of potential topics per document, $\eta = 0.01$ was set as default;
- **α**: The Dirichlet parameter over distribution $\theta_d \sim \text{Dirichlet}_k(a)$, the lower α means the lower number of potential words per topic, $\alpha = 0.01$ was set as default;

After processing the LDA model, two types of posterior distributions will be obtained: the posterior topic distribution θ_d of each document d and posterior word distribution β_k of each topic k. For each document (tweet), the highest posterior probability $\theta_{d,k}$ topic is set as the document's topic.

Case Study 1: Hurricane Sandy

In the first case study, the LDA model is examined by two main tasks: (1) It is trained with raw Twitter data to extract the total number of incident-related tweets; (2) It is trained with only incident-related tweets generated by the LDA model to discover the accuracy of identifying tweets by incident type.

Task 1: Extract the Tweets that Contain Incident Information. The purpose of the first task is to evaluate the performance of the LDA model in extracting incident-related tweets. The number of topics (K) is sensitive and it has a direct effect on the result. Very few topics may result in aggregated topics containing multiple incidents, whereas too many topics may diffuse the information with the same words shared across many topics. As the number of topics increase, incident-related tweets found by the model should converge around the actual number of incident-related tweets. The first task is designed to test and evaluate different numbers of topics between $K \in [5, 100]$ and analyze the output to check the accuracy.

For this task, 258 manually labeled incident-related tweets are selected out of 3,131. The data set was randomly separated into 2,087 for training and 1,044 for testing. Regular stop words, some common words that often appeared in posts about the hurricane, and some location-related tweets, including "hurricane, sandy, nyc, hurricanesandy, th, st, ave, nj, ny, staten, island, new, york, city, jersey, like, looks" were removed.

Figure 2 shows the performance of the LDA model on the testing data for the first task. As expected, the number of incident-related tweets found by the model is converging to 8% of the total number of tweets as K increases, where the actual ratio is 7.6%. The precision, recall, accuracy, and F1 values are used to measure the performance of the model. When K is greater than 30, the accuracies are higher than 80% for all the measures. An interesting finding is that the recall value decreases as the number of topics increases, meaning a higher number of topics would filter relevant tweets out. Over six types of incidents are retrieved during the labeling process executed manually. Therefore, it can be concluded that there might be more than one topic describing the same incident type after the analysis.

Task 2: Discover the Accuracy of Tweet Classification by Content. For the second task, 1,880 tweets containing the most significant incident contents are chosen from geotagged Hurricane Sandy tweets. These tweets are manually labeled and grouped into eight significant incident types with details shown below: Gasoline (25.11%), Flooding (20.90%), Tree (18.35%), Transit (12.55%), Traffic (9.36%), Wind (6.38%), Communication (4.04%), Debris (3.30%).

In this case, the objective is to discover the LDA model's ability to identify incident-related tweets by type. By utilizing incident-related keywords with highest posterior probability in each topic, it was possible to manually label each topic into incident types. The number of tweets under each conjectured type was used to compare with the grand truth data, and accuracy, precision, recall and F1 value were used to measure the performance of the classification.

The 1,880 tweets were separated into 1,290 in the training and 590 in the testing set. Based on the knowledge gained from the first task, model training started with $K = 5$ and four types of incidents identified. Then $K = (6, 7, 8, 9, 10, 11)$ were tested till the topics could cover all the types. After that, $K = (14, 20, 30, 50)$ were examined to find the feature when the K value went high. From Figure 3, when $K = 20$, the output has the highest accuracy. The $K = 20$ scenario was used as an example to describe the conjecture process in the following part.

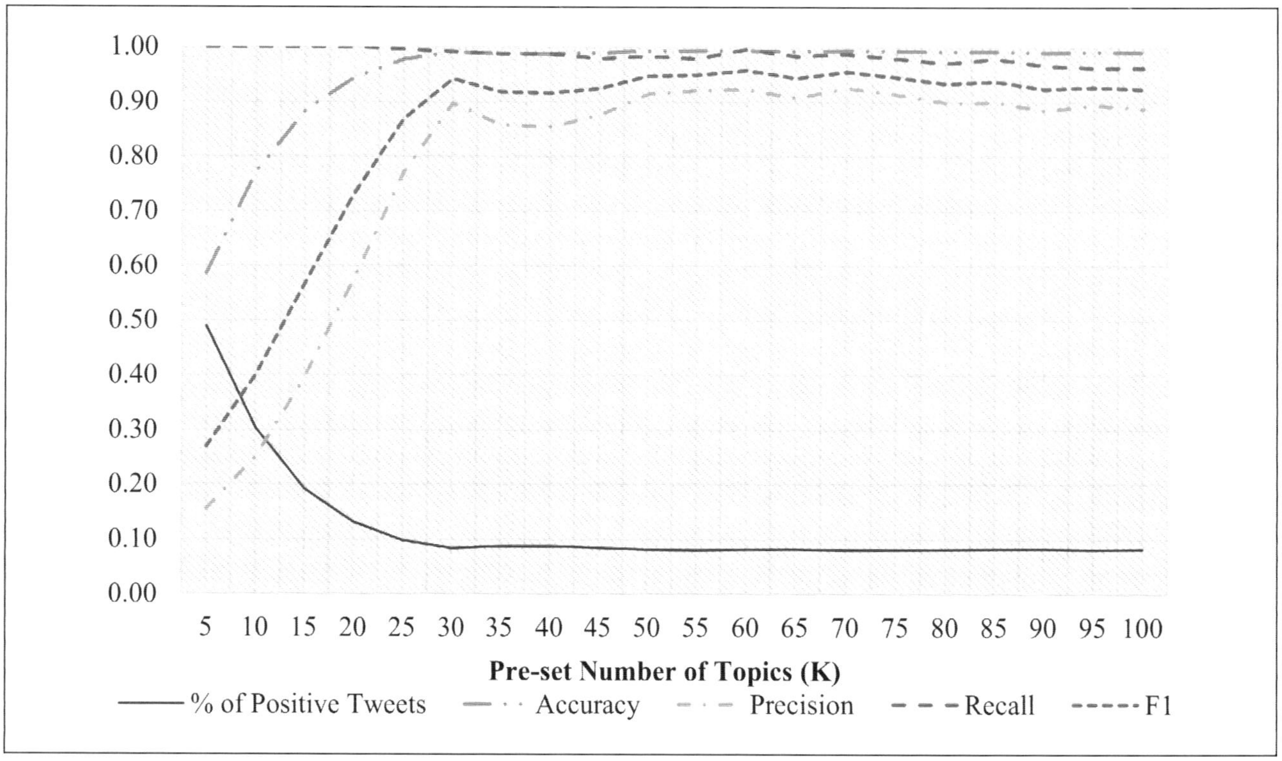

Figure 2. Event extraction performance of the LDA model for Hurricane Sandy.

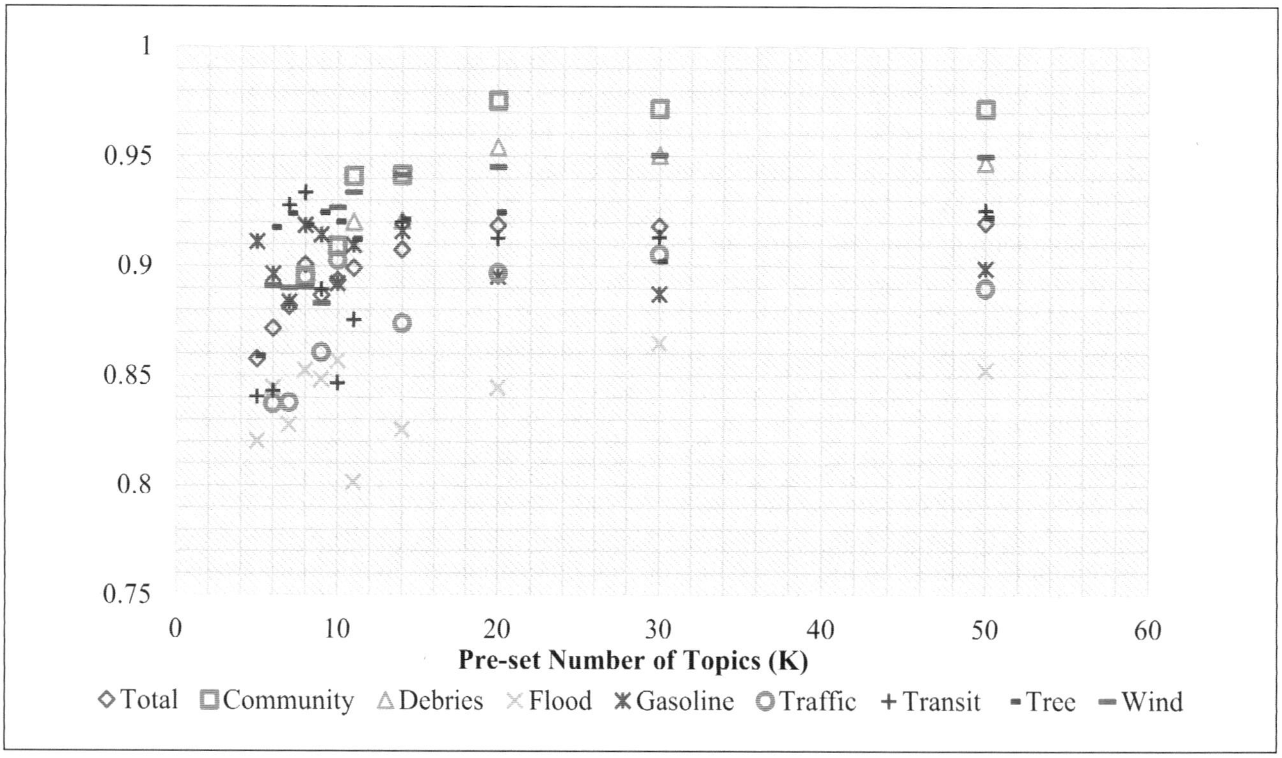

Figure 3. The accuracy distribution of LDA model for events classification training.

Table 1. The LDA Model Generated Topics With $K = 20$

Generated topic and top 10 keywords	Conjectured type
subway mta train service station bus transit pm trains buses	Transit
east fdr flooded drive beach ocean river street lake parkway	Flood
bus manhattan brooklyn line shuttle buses get mta bridge subway	Transit
winds wind mph rain trees flooding gusts safe strong tree	Wind
brooklyn traffic trees newyork storm due avenue rd bridge mta	Traffic
gas get getting trying im know driver traffic bad downtown	Gasoline
streets closed flooded manhattan first red starbucks hook time lot	Flood
tunnel brooklyn battery water rt point sea flooding flooded park	Flood
internet cable cell back service phone tv people working knocked	Community
beach debris streets street water post ocean avenue boardwalk sand	Debris
gas store boat even route lines street number pic water	Gasoline
gas line station cars waiting long wait hess hour mile	Gasoline
park tree trees uprooted brooklyn astoria central post aftermath downed	Tree
river water east street flood hudson flooded park already side	Flood
gas line lines long area stations way get waiting turnpike	Gasoline
gas lines line station long cars people donuts stations get	Gasoline
tree trees house one fell road car took fallen night	Tree
grand Central terminal bus others mta found minutes west whole	Transit
water flooded high dumbo tide flood atlantic street sad completely	Flood
bridge traffic post williamsburg work manhattan queensboro commute queens morning	Traffic

Table 2. Performances of Classification by the LDA Model with $K = 20$

Incident type	Percentage	Accuracy	Precision	Recall	F1
Performances of classification training by the LDA model with $K = 20$					
Community	4.04%	0.975531	0.720588	0.644737	0.680556
Debris	3.30%	0.954255	0.363636	0.516129	0.426667
Flood	20.90%	0.830851	0.587822	0.638677	0.612195
Gasoline	25.11%	0.895213	0.793177	0.788136	0.790648
Traffic	9.36%	0.896809	0.434783	0.340909	0.382166
Transit	12.55%	0.912766	0.627660	0.750000	0.683398
Tree	18.35%	0.924468	0.883019	0.678261	0.767213
Wind	6.38%	0.945213	0.559441	0.666667	0.608365
Total	100%	0.916888	0.621266	0.627939	0.618901
Performances of classification testing by the LDA model with $K = 20$					
Communication	3.39%	0.967797	0.529412	0.450000	0.486486
Debris[a]	3.56%	0.000000	0.000000	0.000000	0.000000
Flood	20.34%	0.791525	0.487603	0.491667	0.489627
Gas	24.24%	0.761017	0.505952	0.594406	0.546624
Traffic	9.49%	0.896610	0.456140	0.464286	0.460177
Transit	12.88%	0.854237	0.443182	0.513158	0.475610
Tree	19.49%	0.833898	0.613333	0.400000	0.484211
Wind	6.61%	0.923729	0.442308	0.589744	0.505495
Total	100.00%	0.861259	0.496847	0.500466	0.492604

[a]Event "Debris" was not detected.

Twenty topics would be generated and conjectured as a type of incident by the word's posterior probability. Table 1 shows the topic and top keywords list and conjectured incident.

As each tweet was labeled by the most correlated topic, they were also labeled as a conjectured incident by the conjectured type of topic. From the counting numbers, accuracy, precision, recall, and F1 value were calculated to measure the classification performance shown in Table 2. The bottom half of Table 2 shows the classification performance of the trained LDA model with $K = 20$ on the testing data. As a statistical model, topics extracted are restricted by the data size. In this case, the topic "Debris" was not detected.

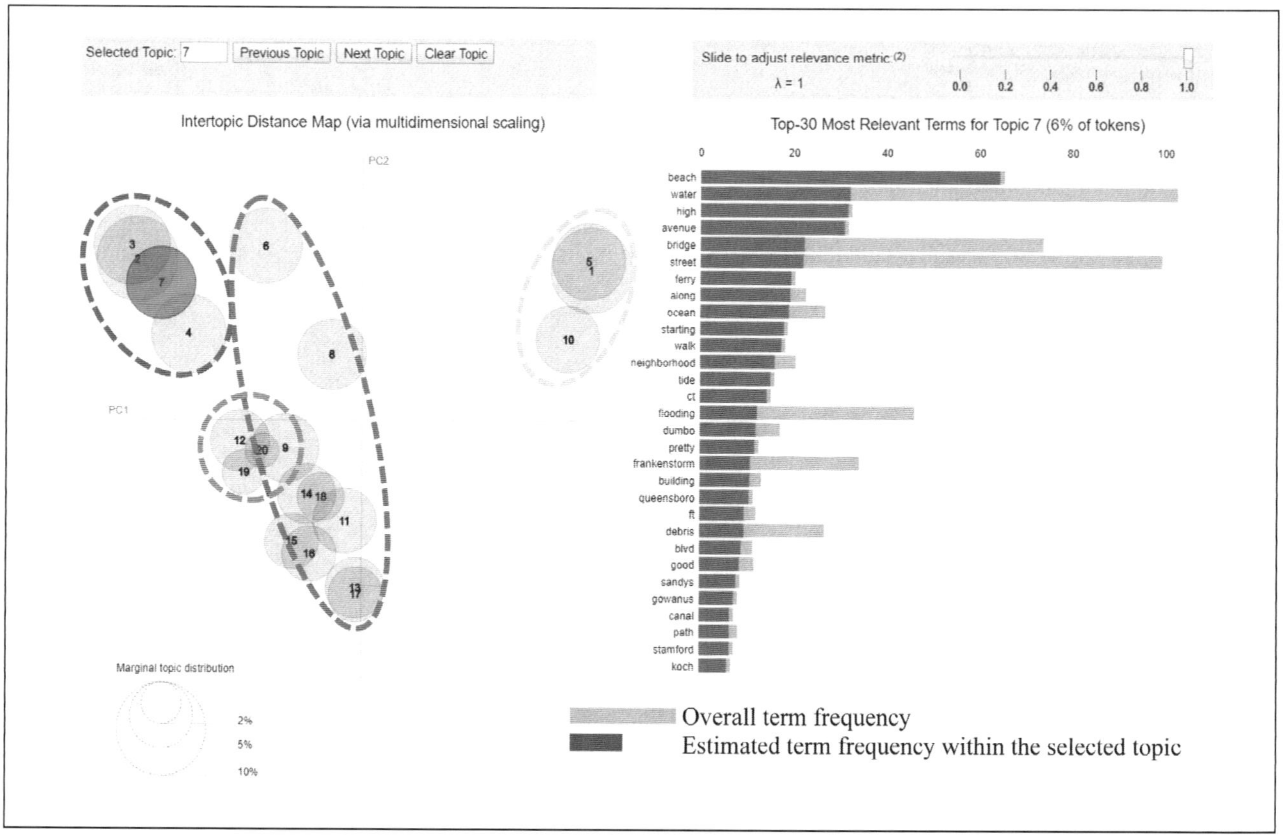

Figure 4. Inter-topic distance map via multidimensional scaling and manual clusters. Each circle presents one topic and the radius of circle presents the percentage of tweets under each topic. Topics are concentrated by similar context; for example, topics in the blue dashed circle are related to flood, the yellow dashed circle clusters gasoline shortage-related topics, the green dashed circle is about wind and downed trees, and the red dashed circle clusters traffic and transit topics.

Multiple topics may refer to the same incident. A measure of dissimilarity between sets of topics is required as a reference for deciding which topics should be combined or split. Moreover, after reducing the dimension of each tweet's topic distribution, the 2-D or 3-D space distribution is a good reference to analyze relations between topics. A python-based topic model visualization package *pyLDAvis* (*28*) was used in this case study. The Jensen Shannon divergence in Equation 3 was applied to determine the distance between topics. Figure 4 shows the 2-D relationship between each topic based on the training result under $K = 20$. The clusters' distribution also provided some interrelationships, such as the relationship between downed trees and transit service suspension.

Case Study 2: Chelsea Explosion (September 17, 2016)

In this case study, the LDA model is trained to prove the ability to extract explosion-related event topics and discover the number of tweets related to the target event. The design of the experiment is to test the number of topics $K \in [1, 80]$ and examine the correlation of output topics to the explosion event.

After this filtering process, there were 61,089 tweets after the explosion, the vocabulary of $V = 10213$ unique words with total document size $N = 321478$ words. The data set was randomly separated into 40,306 for training and 20,783 for testing. For comparison, 103,323 tweets before the explosion are also analyzed.

Table 3 illustrates the results of the experiments. The first column is the total number of topics. The second column presents the top 10 keywords of all the explosion-related topics. The third column shows the ratio of tweets under the selected topic to the total number of tweets. Keyword "Chelsea" and "explosion" starts appearing from $K = 6$, but those topics also contain top keywords that are not related to the explosion event until $K \geqslant 15$. After $K > 30$, the percentage of tweets labeled by event-related topics fluctuates around 2%. This result is similar to the first case scenario. By the threshold of $K = 30$, the number of Chelsea explosion-related tweets in the study data set are determined to be around

Table 3. The Ratio and Top 10 Keywords of Chelsea Explosion-Related Topics

K	Topic selected from generated topics correlate with Chelsea explosion (Presented by top 10 keywords)	% of total tweets
6	chelsea game day one today get night giants explosion go	27.81%
8	chelsea get shit know go really fuck im time explosion	18.27%
10	chelsea one explosion time know never us trump live get	13.67%
13	chelsea trump explosion safe news one bomb dvd cases know	8.34%
15	chelsea explosion safe one news everyone bomb last stay today	7.75%
20	chelsea explosion safe stay bomb everyone news night morning manhattan	5.48%
25	chelsea explosion safe stay bomb manhattan news united bombing police	4.08%
30	chelsea explosion safe stay everyone news bomb cases manhattan nypd	2.53%
	chelsea everyone safe stay know hope live going explosion real	1.12%
40	chelsea explosion bomb news bombing police safe manhattan alert nypd	2.14%
50	chelsea explosion safe bomb stay news manhattan everyone bombing police	2.03%
60	chelsea explosion safe stay manhattan everyone hope ok bomb away	1.74%
	chelsea bombing alert police nj rahami suspect ahmad khan act	1.27%
80	chelsea news explosion police alert bombing nypd bomb rahami suspect	1.11%
	chelsea safe stay everyone explosion hope manhattan away tonight heard	0.95%

2%*61089 = 1222. The highly correlated keywords include: "Chelsea," "explosion," "safe," "everyone," "stay," "bomb," "police," "alert," "Manhattan," "NYPD," and "suspect."

The generated topics before and after the Chelsea explosion are also compared. As expected, the explosion-related topics that appeared in the "After-explosion" data set do not show up in the "Before-explosion" data.

Conclusion and Discussion

This paper presented the use of a new model to automatically extract and classify incident-related tweets and incident types by using the LDA topic model (*6*) on Twitter data. Two different New York City-based datasets collected during Hurricane Sandy (October 29, 2012) and the Chelsea explosion (September 17, 2016) are used to evaluate the accuracy of the proposed model. The LDA topic model does not require any prior knowledge to identify topics as it is an unsupervised learning model. The generated topics are completely calculated from the statistical structure of the collected data. Using the posterior document–topic and topic–word distributions, the context of each topic and the variation of topics can be determined.

The results showed that the LDA topic model could extract incident-related tweets with high accuracy in both case studies when the number of topics is higher than 30. It can also generate topics that can explain some distinct incident types. There may be more than one topic that is related to the same incident type. Therefore, a manual grouping of the topics is required. However, this task is relatively easy and can be executed by checking the top 10 keywords in the topic list. With the presented system, high volumes of real-time social media data can be analyzed to find incident-related tweets, incident types, and the location of specific incidents. This information can be provided via smartphones and tablets for emergency response personnel on site, which will dramatically improve the emergency response in crisis situations.

The disadvantage of LDA model is that it requires many documents to provide better performance. In our experiments, 5% of a total number of tweets is used as the threshold. Those event types with lower than 5% covered tweets are difficult to identify. The well-trained supervised classification model can solve this problem. Machine learning techniques such as recurrent neural network can be used in future works. For training those models, high-quality keyword lists are one of the most important materials. Thus, this study may be the beginning of future research based on the keywords list generated by the LDA model.

Acknowledgments

The research work presented in this paper was partially supported by C2SMART Tier 1 University Transportation Center at New York University. Part of this material presented in this paper is based upon work supported by the National Science Foundation under Grant No. 1541164. The Twitter data used in this study is from a past research project by Arkaitz Zubiaga and Heng Ji (*24*).

Author Contributions

The authors confirm contribution to the paper as follows: study conception and design: Fan Zuo, Abdullah Kurkcu, Kaan Ozbay, Jingqin Gao; data collection: Abdullah Kurkcu, Fan Zuo; analysis and interpretation of results: Fan Zuo, Abdullah Kurkcu; draft manuscript preparation: Fan Zuo, Abdullah Kurkcu, Kaan Ozbay, Jingqin Gao. All authors reviewed the results and approved the final version of the manuscript.

References

1. Özbay, K., and P. Kachroo. *Incident Management for Intelligent Transportation Systems (ITS)*. Artech House, Boston, Massachusetts, 1999.
2. Ozbay, K., and N. Noyan. Estimation of Incident Clearance Times Using Bayesian Networks Approach. *Accident Analysis & Prevention*, Vol. 38, No. 3, 2006, pp. 542–555.
3. Demiroluk, S., and K. Ozbay. Adaptive learning in bayesian networks for incident duration prediction. *Transportation Research Record: Journal of the Transportation Research Board*, 2014. 2460: 77–85.
4. Kurkcu, A., F. Zuo, J. Gao, E. F. Morgul, and K. Ozbay. Crowdsourcing Incident Information for Disaster Response Using Twitter. Presented at 96th Annual Meeting of the Transportation Research Board, Washington, D.C., 2017.
5. Blake, E. S., T. B. Kimberlain, R. J. Berg, P. C. John, and J. L. Beven, II. *United States National Oceanic and Atmospheric Administration's National Weather Service*. Tropical Cyclone Report: Hurricane Sandy (AL182012). 22−29 October 2012, Natl Hurricane Cent, Miami, FL, 2013.
6. Blei, D. M., A. Y. Ng, and M. I. Jordan. Latent Dirichlet Allocation. *Journal of MACHINE LEARNING RESEARCH*, Vol. 3, 2003, pp. 993–1022.
7. Alazawi, Z., O. Alani, M. B. Abdljabar, S. Altowaijri, and R. Mehmood. A Smart Disaster Management System for Future Cities. *Proc., 2014 ACM International Workshop on Wireless and Mobile Technologies for Smart Cities*, ACM, 2014. pp. 1–10.
8. Alazawi, Z., S. Altowaijri, R. Mehmood, and M. B. Abdljabar. Intelligent Disaster Management System Based on Cloud-enabled Vehicular Networks. *2011 11th International Conference on ITS Telecommunications*, IEEE, St. Petersburg, FL, 2011, pp. 361–368.
9. Ashktorab, Z., C. Brown, M. Nandi, and A. Culotta. Tweedr: Mining Twitter to Inform Disaster Response. *Proc. of 11th ISCRAM Conference*, University Park, PA, 2014, pp. 354–358.
10. Mcauliffe, J. D., and D. M. Blei. Supervised Topic Models. *Advances in Neural Information Processing Systems*, 2008, pp. 121–128.
11. Lafferty, J., A. McCallum, and F. C. Pereira. *Conditional Random Fields: Probabilistic Models for Segmenting and Labeling Sequence Data*. 2001.
12. Kurkcu, A., K. Ozbay, and E. F. Morgul. Evaluating the Usability of Geo-Located Twitter as a Tool for Human Activity and Mobility Patterns: A Case Study for New York City. Presented at 95th Annual Meeting of the Transportation Research Board, Washington, D.C., 2016.
13. Ester, M., H.-P. Kriegel, J. Sander, and X. Xu. *A Density-Based Algorithm for Discovering Clusters in Large Spatial Databases with Noise*. No. 96, 1996. pp. 226–231.
14. Mandelbrot, B. B. *The Fractal Geometry of Nature*. WH Freeman, San Francisco, CA, 1982, Revised ed., Fractals, 1977.
15. Kurkcu, A., E. F. Morgul, and K. Ozbay. Extended Implementation Method for Virtual Sensors: Web-Based Real-Time Transportation Data Collection and Analysis for Incident Management. *Transportation Research Record: Journal of the Transportation Research Board*, 2015. 2528: 27–37.
16. Morgul, E., H. Yang, A. Kurkcu, K. Ozbay, B. Bartin, C. Kamga, and R. Salloum. Virtual Sensors: Web-Based Real-Time Data Collection Methodology for Transportation Operation Performance Analysis. *Transportation Research Record: Journal of the Transportation Research Board*, 2014. 2442: 106–116.
17. Mandel, B., A. Culotta, J. Boulahanis, D. Stark, B. Lewis, and J. Rodrigue. A demographic Analysis of Online Sentiment During Hurricane Irene. *Proc., Second Workshop on Language in Social Media*, Association for Computational Linguistics, 2012, pp. 27–36.
18. Zhang, Z., M. Ni, Q. He, J. Gao, J. Gou, and X. Li. Exploratory Study on Correlation Between Twitter Concentration and Traffic Surges. *Transportation Research Record: Journal of the Transportation Research Board*, 2016. 2553: 90–98.
19. Chen, Y., A. Mittal, and H. S. Mahmassani. Twitter or Chatter? Involving Social Media Data Analysis in Traffic Incident Management. Presented at 96th Annual Meeting of the Transportation Research Board, Washington, D.C., 2017.
20. Fu, K., R. Nune, and J. X. Tao. Social Media Data Analysis for Traffic Incident Detection and Management. Presented at 94th Annual Meeting of the Transportation Research Board, Washington, D.C., 2015.
21. Ritter, A., S. Clark, and O. Etzioni. Named entity recognition in tweets: an experimental study. *Proc., Conference on Empirical Methods in Natural Language Processing*, Association for Computational Linguistics, 2011, pp. 1524–1534.
22. Griffiths, T. L., and M. Steyvers. Finding Scientific Topics. *Proc., National Academy of Sciences*, Vol. 101, Suppl 1, 2004, pp. 5228–5235.
23. Kullback, S., and R. A. Leibler. On Information and Sufficiency. *The Annals of Mathematical Statistics*, Vol. 22, No. 1, 1951, pp. 79–86.
24. Zubiaga, A., and H. Ji. Tweet, But Verify: Epistemic Study of Information Verification on Twitter. *Social Network Analysis and Mining*, Vol. 4, No. 1, 2014, p. 163.
25. Makice, K. *Twitter API: Up and Running: Learn How to Build Applications with the Twitter API*. O'Reilly Media, Inc., Sebastopol, CA, 2009.
26. Bird, S., E. Klein, and E. Loper. *Natural Language Processing with Python: Analyzing Text with the Natural Language Toolkit*. O'Reilly Media, Inc., Beijing, China, 2009.
27. Latent Dirichlet Allocation. *Latent Dirichlet Allocation: Topic Modeling with Latent Dirichlet Allocation v1.0.3*. http://pythonhosted.org/lda/index.html.
28. Sievert, C., and K. E. Shirley. LDAvis: A Method for Visualizing and Interpreting Topics. *Proc., Workshop on Interactive Language Learning, Visualization, and Interfaces*, 2014, pp. 63–70.

The Standing Committee on Transportation Economics (ABE20) peer-reviewed this paper (18-06211).

The views and opinions expressed in this article are those of the authors and do not necessarily reflect the official policy or position of any agency of the U.S. government and other funding agencies.

Sage Publications USA
sagepub.com

ISBN 978-1-5108-7821-1
9 781510 878211

CURRAN ASSOCIATES INC.
proceedings.com

Socioeconomic, Sustainability, Health, and Human Factors 2018

Transportation Research Record: Journal of the Transportation Research Board Volume 2672, Issue 3

SAGE Publishing

Socioeconomic, Sustainability, Health, and Human Factors 2018

Transportation Research Record: Journal of the Transportation Research Board Volume 2672, Issue 3

SAGE Publishing